Mineralogical Crystallography

Mineralogical Crystallography

Editor

Vladislav V. Gurzhiy

MDPI • Basel • Beijing • Wuhan • Barcelona • Belgrade • Manchester • Tokyo • Cluj • Tianjin

Editor
Vladislav V. Gurzhiy
Saint Petersburg State University
Russian Federation

Editorial Office
MDPI
St. Alban-Anlage 66
4052 Basel, Switzerland

This is a reprint of articles from the Special Issue published online in the open access journal *Crystals* (ISSN 2073-4352) (available at: https://www.mdpi.com/journal/crystals/special_issues/mineralogical_crystallography).

For citation purposes, cite each article independently as indicated on the article page online and as indicated below:

LastName, A.A.; LastName, B.B.; LastName, C.C. Article Title. *Journal Name* **Year**, *Article Number*, Page Range.

ISBN 978-3-03936-974-4 (Hbk)
ISBN 978-3-03936-975-1 (PDF)

© 2020 by the authors. Articles in this book are Open Access and distributed under the Creative Commons Attribution (CC BY) license, which allows users to download, copy and build upon published articles, as long as the author and publisher are properly credited, which ensures maximum dissemination and a wider impact of our publications.

The book as a whole is distributed by MDPI under the terms and conditions of the Creative Commons license CC BY-NC-ND.

Contents

About the Editor vii

Vladislav V. Gurzhiy
Mineralogical Crystallography
Reprinted from: *Crystals* 2020, *10*, 805, doi:10.3390/cryst10090805 1

Vladislav V. Gurzhiy, Ivan V. Kuporev, Vadim M. Kovrugin, Mikhail N. Murashko,
Anatoly V. Kasatkin and Jakub Plášil
Crystal Chemistry and Structural Complexity of Natural and Synthetic Uranyl Selenites
Reprinted from: *Crystals* 2019, *9*, 639, doi:10.3390/cryst9120639 5

Olga S. Tyumentseva, Ilya V. Kornyakov, Sergey N. Britvin, Andrey A. Zolotarev and
Vladislav V. Gurzhiy
Crystallographic Insights into Uranyl Sulfate Minerals Formation: Synthesis and Crystal
Structures of Three Novel Cesium Uranyl Sulfates
Reprinted from: *Crystals* 2019, *9*, 660, doi:10.3390/cryst9120660 33

Igor V. Pekov, Natalia V. Zubkova, Ilya I. Chaikovskiy, Elena P. Chirkova,
Dmitry I. Belakovskiy, Vasiliy O. Yapaskurt, Yana V. Bychkova, Inna Lykova,
Sergey N. Britvin and Dmitry Yu. Pushcharovsky
Krasnoshteinite, $Al_8[B_2O_4(OH)_2](OH)_{16}Cl_4 \cdot 7H_2O$, a New Microporous Mineral with a Novel
Type of Borate Polyanion
Reprinted from: *Crystals* 2020, *10*, 301, doi:10.3390/cryst10040301 47

Sergey N. Britvin, Maria G. Krzhizhanovskaya, Vladimir N. Bocharov and
Edita V. Obolonskaya
Crystal Chemistry of Stanfieldite, $Ca_7M_2Mg_9(PO_4)_{12}$ (M = Ca, Mg, Fe^{2+}), a Structural Base of
$Ca_3Mg_3(PO_4)_4$ Phosphors
Reprinted from: *Crystals* 2020, *10*, 464, doi:10.3390/cryst10060464 61

Paola Comodi, Azzurra Zucchini, Tonci Balić-Žunić, Michael Hanfland and Ines Collings
The High Pressure Behavior of Galenobismutite, $PbBi_2S_4$: A Synchrotron Single Crystal X-ray
Diffraction Study
Reprinted from: *Crystals* 2019, *9*, 210, doi:10.3390/cryst9040210 75

Yunfan Miao, Youwei Pang, Yu Ye, Joseph R. Smyth, Junfeng Zhang, Dan Liu, Xiang Wang
and Xi Zhu
Crystal Structures and High-Temperature Vibrational Spectra for Synthetic Boron and
Aluminum Doped Hydrous Coesite
Reprinted from: *Crystals* 2019, *9*, 642, doi:10.3390/cryst9120642 93

Alina R. Izatulina, Anton M. Nikolaev, Mariya A. Kuz'mina, Olga V. Frank-Kamenetskaya
and Vladimir V. Malyshev
Bacterial Effect on the Crystallization of Mineral Phases in a Solution Simulating Human Urine
Reprinted from: *Crystals* 2019, *9*, 259, doi:10.3390/cryst9050259 111

Aleksei V. Rusakov, Mariya A. Kuzmina, Alina R. Izatulina and Olga V. Frank-Kamenetskaya
Synthesis and Characterization of $(Ca,Sr)[C_2O_4] \cdot nH_2O$ Solid Solutions: Variations of Phase
Composition, Crystal Morphologies and in Ionic Substitutions
Reprinted from: *Crystals* 2019, *9*, 654, doi:10.3390/cryst9120654 123

Alejandro De la Rosa-Tilapa, Agustín Maceda and Teresa Terrazas
Characterization of Biominerals in Cacteae Species by FTIR
Reprinted from: *Crystals* **2020**, *10*, 432, doi:10.3390/cryst10060432 . . . . **135**

Donata Konopacka-Łyskawa
Synthesis Methods and Favorable Conditions for Spherical Vaterite Precipitation: A Review
Reprinted from: *Crystals* **2019**, *9*, 223, doi:10.3390/cryst9040223 . . . . **147**

Min Tang and Yi-Liang Li
A Complex Assemblage of Crystal Habits of Pyrite in the Volcanic Hot Springs from Kamchatka, Russia: Implications for the Mineral Signature of Life on Mars
Reprinted from: *Crystals* **2020**, *10*, 535, doi:10.3390/cryst10060535 . . . . **163**

About the Editor

Vladislav V. Gurzhiy is an associate professor at the Crystallography Department of the Institute of Earth Sciences, Saint Petersburg State University, and holds the position of Chairman of the Institute of Earth Sciences' Scientific Committee. Dr. Gurzhiy graduated from St. Petersburg State University in 2007 and completed his PhD in 2009, with a dissertation entitled "Crystal chemistry of uranyl selenates with organic and inorganic cations". His main research interests are related to the crystal chemistry of minerals and their synthetic analogs, in particular, compounds bearing uranium and transuranium elements, and biominerals. Dr. Gurzhiy is a full member of the Russian Mineralogical Society and the Saint-Petersburg Society of Naturalists. He has published over 150 peer-reviewed journal articles and several book chapters. Dr. Gurzhiy has been honored with several awards, including the Yu.T. Struchkov Prize and Academia Europaea Prize, for research works in the fields of X-ray crystallography and Earth sciences.

Editorial

Mineralogical Crystallography

Vladislav V. Gurzhiy

Department of Crystallography, Institute of Earth Sciences, St. Petersburg State University, University Emb. 7/9, St. Petersburg 199034, Russia; vladislav.gurzhiy@spbu.ru or vladgeo17@mail.ru

Received: 4 September 2020; Accepted: 9 September 2020; Published: 11 September 2020

Keywords: mineral; crystallography; crystal chemistry; X-ray diffraction; crystal structure; crystal growth; mineral evolution

---

Crystallography remains, for mineralogy, one of the main sources of information on natural crystalline substances. A description of mineral species shape is carried out according to the principles of geometric crystallography; the crystal structure of minerals is determined using X-ray crystallography techniques, and physical crystallography approaches allow one to evaluate various properties of minerals, etc. However, the reverse comparison should not be forgotten as well: the crystallography science, in its current form, was born in the course of mineralogical research, long before preparative chemistry received such extensive development. It is worth noting that, even today, investigations of the crystallographic characteristics of minerals regularly open up new horizons in materials science, because the possibilities of nature (fascinating chemical diversity; great variation of thermodynamic parameters; and, of course, almost endless processing time) are still not available for reproduction in any of the world's laboratories. This Special Issue is devoted to mineralogical crystallography, the oldest branch of crystallographic science, and combines important surveys covering such topics as: discovery of new mineral species; crystal chemistry of minerals and their synthetic analogs; behavior of minerals at non-ambient conditions; biomineralogy; and crystal growth techniques.

We hope that the current set of reviews and articles will arouse genuine interest among readers and, perhaps, push them to their own successful research in the field of mineralogical crystallography.

1. Crystal Chemistry of Minerals and Their Synthetic Analogs

Gurzhiy et al. [1] reviewed the crystal chemistry of the family of natural and synthetic uranyl selenite compounds, paying special attention to the pathways of synthesis and topological analysis of the known crystal structures. Crystal structures of two minerals were refined. The H atoms positions belonging to the interstitial H_2O molecules in the structure of demesmaekerite, $Pb_2Cu_5[(UO_2)_2(SeO_3)_6(OH)_6](H_2O)_2$, were assigned. The refinement of the guilleminite crystal structure allowed the determination of an additional site arranged within the void of the interlayer space and occupied by an H_2O molecule, which suggests the new formula of guilleminite to be written as $Ba[(UO_2)_3(SeO_3)_2O_2](H_2O)_4$. This paper could be regarded as the first review on the mineralogy and crystal chemistry of the named group of compounds.

Tyumentseva et al. [2] studied the alteration of the uranyl oxide hydroxy-hydrate mineral schoepite $[(UO_2)_8O_2(OH)_{12}](H_2O)_{12}$ at mild hydrothermal conditions in the presence of sulfate oxyanions, which resulted in the crystallization of three novel compounds.

Comparison of the isotypic natural and synthetic uranyl-bearing compounds [1,2] suggests that formation of all uranyl selenite and of the majority of uranyl sulfate minerals requires heating, which most likely can be attributed to the radioactive decay. The temperature range could be assumed from the manner of the interpolyhedral linkage.

2. Discovery of New Mineral Species

Pekov et al. [3] discovered the new hydrous aluminum chloroborate mineral krasnoshteinite ($Al_8[B_2O_4(OH)_2](OH)_{16}Cl_4 \cdot 7H_2O$), with a zeolite-like microporous structure and a three-dimensional system of wide channels containing Cl^- anions and weakly bonded H_2O molecules. The crystal structure of krasnoshteinite is also remarkable due to the presence of a novel insular borate polyanion $[B_2O_4(OH)_2]^{4-}$.

Britvin et al. [4] reported on the crystal structure of natural Ca-Mg-phosphate stanfieldite, $Ca_7M_2Mg_7(PO_4)_{12}$ (M = Ca, Mg, Fe^{2+}), derived from the pallasite meteorite Brahin for the first time. The authors reviewed the existing analytical data and showed that there is no evidence that the phosphor base with the formula $Ca_3Mg_3(PO_4)_4$ exists.

3. Behavior of Minerals at Non-Ambient Conditions

Comodi et al. [5] studied the transformation of the crystal structure of galenobismutite, $PbBi_2S_4$, under pressure up to 20.9 GPa. The structure undergoes reversible and completely elastic transitions. The size and the shape of Bi- and Pb-centered polyhedra suggest that the high-pressure structure of galenobismutite can host Na and Al in the lower mantle, which are incompatible with the periclase or perovskite crystal structures.

Hydrous coesite crystals, a high-pressure SiO_2 polymorph, were synthesized with various B^{3+} and Al^{3+} contents and in situ high-temperature Raman and FTIR spectra were collected at ambient pressure by Miao et al. [6]. Crystals were observed to be stable up to 1500 K. Al substitution significantly reduces the H^+ concentration in coesite, so the mechanism is controlled by oxygen vacancies, while the B incorporation may prefer the electrostatically coupled substitution ($Si^{4+} = B^{3+} + H^+$).

4. Biomineralogy

Izatulina et al. [7] studied the effect of bacteria that are present in human urine on the crystallization of oxalate and phosphate mineral phases, the most common constituents of renal stones. It was shown that the inflammatory process will contribute to the decrease in oxalate supersaturation in urine due to calcium oxalate crystallization, while the change in urine pH and the products of bacterial metabolism will be of major importance in the case of phosphate mineralization.

Rusakov et al. [8] reported on the mechanisms of Sr-to-Ca substitution in the structures of calcium oxalate minerals that were found in lichen thalli on Sr-bearing apatite rock. It was shown that the incorporation of Sr ions is less preferable than Ca into the structures of whewellite and weddellite, and substitution rates are slightly higher for weddellite than for whewellite, which is most likely caused by the denser manner of the interpolyhedral linkage in the latter structure.

Five Cacteae species were studied using various experimental techniques to characterize the biomineral composition within their different tissues by De la Rosa-Tilapa et al. [9]. Calcium carbonates and silicate phases were detected along with common calcium oxalates.

5. Crystal Growth Techniques

Konopacka-Łyskawa [10] reviewed the state of the art of the vaterite crystallization techniques. Vaterite is known to be the least thermodynamically stable anhydrous calcium carbonate polymorph, very rarely found in nature. However, synthetic vaterite has large potential in pharmacology and manufacturing. Well-known classical and new methods used for vaterite precipitation were discussed with particular attention to the parameters affecting the formation of spherical particles.

Tang and Yi-Liang [11] revealed that specific geochemical microenvironments and the bacterial activities in the long-lived volcanic hot springs from Kamchatka result in the development and preservation of the complex pyrite crystal habits. Application of similar techniques to other systems may help in the identification of biogenic iron sulfides in sediments on Earth and other planets.

Funding: This research received no external funding.

Acknowledgments: As the Guest Editor, I would like to acknowledge all the authors for their valuable contribution to this Special Issue, which is expressed in fascinating and inspiring papers.

Conflicts of Interest: The author declares no conflict of interest.

References

1. Gurzhiy, V.; Kuporev, I.; Kovrugin, V.; Murashko, M.; Kasatkin, A.; Plášil, J. Crystal chemistry and structural complexity of natural and synthetic uranyl selenites. *Crystals* **2019**, *9*, 639. [CrossRef]
2. Tyumentseva, O.; Kornyakov, I.; Britvin, S.; Zolotarev, A.; Gurzhiy, V. Crystallographic insights into uranyl sulfate minerals formation: Synthesis and crystal structures of three novel cesium uranyl sulfates. *Crystals* **2019**, *9*, 660. [CrossRef]
3. Pekov, I.; Zubkova, N.; Chaikovskiy, I.; Chirkova, E.; Belakovskiy, D.; Yapaskurt, V.; Bychkova, Y.; Lykova, I.; Britvin, S.; Pushcharovsky, D. Krasnoshteinite, $Al_8[B_2O_4(OH)_2](OH)_{16}Cl_4 \cdot 7H_2O$, a new microporous mineral with a novel type of Borate Polyanion. *Crystals* **2020**, *10*, 301. [CrossRef]
4. Britvin, S.; Krzhizhanovskaya, M.; Bocharov, V.; Obolonskaya, E. Crystal chemistry of Stanfieldite, $Ca_7M_2Mg_9(PO_4)_{12}$ (M = Ca, Mg, Fe^{2+}), a Structural Base of $Ca_3Mg_3(PO_4)_4$ Phosphors. *Crystals* **2020**, *10*, 464. [CrossRef]
5. Comodi, P.; Zucchini, A.; Balić-Žunić, T.; Hanfland, M.; Collings, I. The high pressure behavior of galenobismutite, $PbBi_2S_4$: A synchrotron single crystal X-ray diffraction study. *Crystals* **2019**, *9*, 210. [CrossRef]
6. Miao, Y.; Pang, Y.; Ye, Y.; Smyth, J.; Zhang, J.; Liu, D.; Wang, X.; Zhu, X. Crystal structures and high-temperature vibrational spectra for synthetic boron and aluminum doped hydrous coesite. *Crystals* **2019**, *9*, 642. [CrossRef]
7. Izatulina, A.; Nikolaev, A.; Kuz'mina, M.; Frank-Kamenetskaya, O.; Malyshev, V. Bacterial effect on the crystallization of mineral phases in a solution simulating human urine. *Crystals* **2019**, *9*, 259. [CrossRef]
8. Rusakov, A.; Kuzmina, M.; Izatulina, A.; Frank-Kamenetskaya, O. Synthesis and characterization of $(Ca,Sr)[C_2O_4] \cdot nH_2O$ solid solutions: Variations of phase composition, crystal morphologies and in ionic substitutions. *Crystals* **2019**, *9*, 654. [CrossRef]
9. De la Rosa-Tilapa, A.; Maceda, A.; Terrazas, T. Characterization of biominerals in *Cacteae* species by FTIR. *Crystals* **2020**, *10*, 432. [CrossRef]
10. Konopacka-Łyskawa, D. Synthesis methods and favorable conditions for spherical vaterite precipitation: A review. *Crystals* **2019**, *9*, 223. [CrossRef]
11. Tang, M.; Li, Y. A complex assemblage of crystal habits of pyrite in the volcanic hot springs from Kamchatka, Russia: Implications for the mineral signature of life on Mars. *Crystals* **2020**, *10*, 535. [CrossRef]

© 2020 by the author. Licensee MDPI, Basel, Switzerland. This article is an open access article distributed under the terms and conditions of the Creative Commons Attribution (CC BY) license (http://creativecommons.org/licenses/by/4.0/).

Review

Crystal Chemistry and Structural Complexity of Natural and Synthetic Uranyl Selenites

Vladislav V. Gurzhiy [1,*], Ivan V. Kuporev [1], Vadim M. Kovrugin [1], Mikhail N. Murashko [1], Anatoly V. Kasatkin [2] and Jakub Plášil [3]

[1] Institute of Earth Sciences, St. Petersburg State University, University Emb. 7/9, St. Petersburg 199034, Russian; st054910@student.spbu.ru (I.V.K.); kovrugin_vm@hotmail.com (V.M.K.); mzmurashko@gmail.com (M.N.M.)
[2] Fersman Mineralogical Museum of the Russian Academy of Sciences, Leninskiy pr. 18, 2, Moscow 119071, Russian; kasatkin@inbox.ru
[3] Institute of Physics, The Academy of Sciences of the Czech Republic, v.v.i., Na Slovance 2, 18221 Praha 8, Czech Republic; plasil@fzu.cz
* Correspondence: vladislav.gurzhiy@spbu.ru or vladgeo17@mail.ru

Received: 10 November 2019; Accepted: 28 November 2019; Published: 30 November 2019

Abstract: Comparison of the natural and synthetic phases allows an overview to be made and even an understanding of the crystal growth processes and mechanisms of the particular crystal structure formation. Thus, in this work, we review the crystal chemistry of the family of uranyl selenite compounds, paying special attention to the pathways of synthesis and topological analysis of the known crystal structures. Comparison of the isotypic natural and synthetic uranyl-bearing compounds suggests that uranyl selenite mineral formation requires heating, which most likely can be attributed to the radioactive decay. Structural complexity studies revealed that the majority of synthetic compounds have the topological symmetry of uranyl selenite building blocks equal to the structural symmetry, which means that the highest symmetry of uranyl complexes is preserved regardless of the interstitial filling of the structures. Whereas the real symmetry of U-Se complexes in the structures of minerals is lower than their topological symmetry, which means that interstitial cations and H_2O molecules significantly affect the structural architecture of natural compounds. At the same time, structural complexity parameters for the whole structure are usually higher for the minerals than those for the synthetic compounds of a similar or close organization, which probably indicates the preferred existence of such natural-born architectures. In addition, the reexamination of the crystal structures of two uranyl selenite minerals guilleminite and demesmaekerite is reported. As a result of the single crystal X-ray diffraction analysis of demesmaekerite, $Pb_2Cu_5[(UO_2)_2(SeO_3)_6(OH)_6](H_2O)_2$, the H atoms positions belonging to the interstitial H_2O molecules were assigned. The refinement of the guilleminite crystal structure allowed the determination of an additional site arranged within the void of the interlayer space and occupied by an H_2O molecule, which suggests the formula of guilleminite to be written as $Ba[(UO_2)_3(SeO_3)_2O_2](H_2O)_4$ instead of $Ba[(UO_2)_3(SeO_3)_2O_2](H_2O)_3$.

Keywords: uranyl; selenite; selenate; crystal structure; topology; structural complexity; demesmaekerite; guillemenite; haynesite

1. Introduction

All natural compounds of U(VI) and selenium are selenites. Uranyl selenites can be justifiably attributed to rare mineral species. Nowadays, there are only seven uranyl selenite mineral species approved by the International Mineralogical Association as of 20 October 2019 (for comparison, there are >40 uranyl sulfates and ~50 uranyl phosphates): Guilleminite, $Ba[(UO_2)_3(SeO_3)_2O_2](H_2O)_3$ [1], demesmaekerite, $Pb_2Cu_5[(UO_2)_2(SeO_3)_6(OH)_6](H_2O)_2$ [2], marthozite, $Cu[(UO_2)_3(SeO_3)_2O_2](H_2O)_8$ [3], derriksite, $Cu_4[(UO_2)(SeO_3)_2](OH)_6$ [4], haynesite, $[(UO_2)_3(SeO_3)_2(OH)_2](H_2O)_5$ [5], piretite, $Ca(UO_2)_3(SeO_3)_2(OH)_4 \cdot 4H_2O$ [6], and larisaite, $Na(H_3O)[(UO_2)_3(SeO_3)_2O_2](H_2O)_4$ [7]. Their occurrence is limited to just a few localities. First, these are Musonoi and Shinkolobwe mines in DR Congo [6], two of the minerals were only found in the Repete mine (San Juan County, Utah, USA) [5], and a few more occurrences in Europe could be mentioned (small uranium deposit Zálesí in the Czech Republic, Liauzun in France, and La Creusaz U prospect in Switzerland) [8]. Nevertheless, apart from mineralogy, uranyl selenites are of great interest from the geochemical and radiochemical points of view. It is known that fission products contain 53 g per ton [9] of long-lived ^{79}Se isotope with a half-life of 1.1×10^6 years [10] after three years of nuclear fuel irradiation in the reactor. Thus, an understanding of the processes of mineral formation in nature and their synthetic analogs in laboratories can help in the processing of nuclear wastes. Crystal chemical and structural investigations are key points in such a material's scientific studies due to the essential knowledge of how the variation in the chemical composition and growth conditions affects the crystal structure formation.

Herein, we review the topological diversity and growth conditions of natural and synthetic uranyl selenites. Crystal structures of two uranyl selenite minerals guilleminite and demesmaekerite were refined. The structural complexity approach was implemented to determine the preference of a particular topological type, taking into account existing geometrical isomers.

2. Materials and Methods

2.1. Occurrence

The samples of minerals studied in this work were taken from the Fersman Mineralogical Museum, Museum of Natural History in Luxembourg and private collections of authors of the current paper (V.V.G., A.V.K.). Guilleminite: from the Museum (69465 and 82312), from V.V.G. (6111). Demesmaekerite: from J.P. Haynesite: from the Fersman Museum (88922 and 94267), from V.V.G. (5767), from A.V.K. (247X). The samples of guilleminite and demesmaekerite originate from the Musonoi, DR Congo. The samples of haynesite originate from the Repete mine, Utah, UT, USA.

2.2. Single-Crystal X-Ray Diffraction Study

A single crystal of guilleminite (0.08 × 0.04 × 0.01 mm^3) was selected under binoculars, encased in viscous cryoprotectant, and mounted on cryo-loop. Diffraction data were collected using a Bruker Kappa Duo diffractometer (Bruker AXS, Madison, WI, USA) equipped with a CCD (charge-coupled device) Apex II detector operated with monochromated microfocused MoKα radiation (λ[MoKα] = 0.71073 Å) at 45 kV and 0.6 mA. Diffraction data were collected at 100 K with frame widths of 0.5° in ω and φ, and an exposure of 70 s per frame. Diffraction data were integrated, and background, Lorentz, and polarization correction were applied. An empirical absorption correction based on spherical harmonics implemented in the SCALE3 ABSPACK algorithm was applied in the CrysAlisPro program [11]. The unit-cell parameters were refined using the least-squares techniques. The crystal structure of guilleminite was solved by a dual-space algorithm and refined using the SHELX programs [12,13] incorporated in the OLEX2 program package [14]. The final model includes coordinates and anisotropic displacement parameters for all non-H atoms. The H atoms of H$_2$O molecules were localized from difference Fourier maps and were included in the refinement, with U_{iso}(H) set to 1.5U_{eq}(O) and O–H restrained to 0.95 Å.

A dark-olive green prismatic crystal of demesmaekerite (0.034 × 0.032 × 0.022 mm^3) was mounted on a glass fiber, and diffraction intensities were measured at room temperature with a Rigaku SuperNova (Oxford, UK) single-crystal diffractometer. The diffraction experiment was done using MoKα radiation from a micro-focus X-ray source collimated and monochromatized by mirror-optics and the detection of the reflected X-rays was done by an Atlas S2 CCD detector. X-ray diffraction data were collected at room-temperature with frame widths of 1.0° in ω and an exposure of 80 s per frame. Diffraction data were integrated, and background, Lorentz, and polarization correction were applied. An empirical absorption correction based on spherical harmonics implemented in the SCALE3 ABSPACK algorithm was applied in the CrysAlisPro program [11]. The structure was solved by the charge-flipping algorithm [12] and refined using the Jana2006 program [15]. The final refinement cycles were undertaken considering all atoms (except of hydrogen) refined with anisotropic atomic displacement parameters. The H atoms of H_2O molecules were localized from the difference Fourier maps and were subsequently refined with $U_{iso}(H)$ set to $1.2*U_{eq}$ of the donor O atom and O–H softly restrained to 0.95 Å.

Supplementary crystallographic data were deposited in the Inorganic Crystal Structure Database (ICSD) and can be obtained by quoting the depository numbers CSD 1963864 and 1964420 for guilleminite and demesmaekerite, respectively, at https://www.ccdc.cam.ac.uk/structures/ (see Supplementary Materials).

2.3. Coordination of U and Se

The crystal structures of all the natural and synthetic compounds described herein are based on the chained or layered substructural units built by the linkage of U- and Se-centered coordination polyhedra. U(VI) atoms form approximately linear UO_2^{2+} uranyl ions (Ur) with two short $U^{6+}\equiv O^{2-}$ bonds. Ur cation is coordinated in the equatorial plane by other four to six oxygen atoms, to form a tetra-, penta-, or hexagonal bipyramid, respectively, as a coordination polyhedron of U^{6+} atoms. The selenite group has a configuration of a trigonal pyramid with its apical vertex occupied by the Se^{4+} cation possessing a stereochemically active lone-electron pair. In the crystal structures of number of synthetic uranyl selenium compounds, there are also Se(VI) species that form $[SeO_4]^{2-}$ tetrahedra.

2.4. Graphical Representation and Anion Topologies

For topological analysis, the theory of graphical (nodal) representation of crystal structures [16] and the anion topology method [17] were used along with the classification suggested in [18]. Anion topologies were used to describe the layered complexes having edge-sharing polymerization of uranyl coordination polyhedra. For the rest of the structures, graphical representation was used. Each graph has a special index ccD–U:Se–#, where cc means "cation-centered", D indicates dimensionality (1—chains; and 2—sheets), U:Se ratio, # is the registration number of the unit. Each anion topology is indicated by a, so called, ring symbol, $p_1{}^{r_1}p_2{}^{r_2}\ldots$, where p is the sum of vertices in a topological cycle, and r is the number of the respective cycles in the reduced section of the layer.

Three-connected selenate tetrahedra, sharing three of its corners with adjacent uranyl bipyramids, and 2- or 3-connected selenite pyramids, can possess the fourth non-shared corner or lone electron pair, respectively, oriented either *up*, *down*, or disordered relative to the plane of the chain, layer, or, in particular, to the equatorial plane of the uranyl bipyramid. Such ambiguity gives rise to geometric isomerism with various orientations of the Se-centered polyhedra. To distinguish the isomers, their orientation matrices were assigned using symbols **u** (*up*), **d** (*down*), **m** (orientation *up-down* topologically equivalent), or □ (white vertex, Se-centered polyhedron, is missing in the graph).

2.5. Complexity Calculations

In order to characterize and quantify the impact of each substructural units on the formation of a particular architecture, the structural complexity approach recently developed by S.V. Krivovichev [19–23], which allows comparison of the structures in terms of their information content, was used.

The complexity of the crystal structure was estimated as a Shannon information content per atom (I_G) and per unit cell ($I_{G,total}$) using the following equations:

$$I_G = -\sum_{i=1}^{k} p_i \log_2 p_i \qquad \text{(bits/atom)}, \qquad (1)$$

$$I_{G,total} = -v\,I_G = -v\sum_{i=1}^{k} p_i \log_2 p_i \quad \text{(bits/cell)}, \tag{2}$$

where k is the number of different crystallographic orbits (independent sites) in the structure and p_i is the random choice probability for an atom from the i-th crystallographic orbit, that is:

$$p_i = m_i/v, \tag{3}$$

where m_i is a multiplicity of a crystallographic orbit (i.e., the number of atoms of a specific Wyckoff site in the reduced unit cell), and v is the total number of atoms in the reduced unit cell.

The reliable correlation of structural complexity parameters is possible only for compounds with the same or very close chemical composition (e.g., polymorphs), whereas changes in the hydration state, nature of interstitial complexes, and size and shape of organic molecules could significantly affect the overall complexity behavior. In this light, within the current crystal chemical review, structural complexity parameters of various building blocks (uranyl selenite units, interstitial structure, H-bonding system) were calculated to analyze their contributions to the complexity of the whole structure. This approach suggested by S.V. Krivovichev [24] and recently successfully implemented in [25,26] allows the factors that influence the symmetry preservation or reduction of uranyl selenite units to be revealed, and it shows which of the multiple blocks plays the most important role in a particular structure formation.

3. Results

3.1. Uranyl Selenite Minerals

Guilleminite, $Ba[(UO_2)_3(SeO_3)_2O_2](H_2O)_3$ [1,27], and demesmaekerite, $Pb_2Cu_5[(UO_2)_2(SeO_3)_6(OH)_6](H_2O)_2$ [2,28], were the first uranyl selenites found in nature (Table 1). These minerals occur in the lower part of the oxidized zone of the copper-cobalt deposit of Musonoi (Katanga, DR Congo). The first mineral was named after the general director of the Union Minière du Haut-Katanga (UMHK), co-founder of the International Mineralogical Association, French chemist and mineralogist, Jean-Claude Guillemin. Guilleminite crystallizes in the orthorhombic $Pmn2_1$ space group and forms small tabular crystals and canary yellow crusts. It occurs in association with malachite, uranophane-α, wulfenite, etc. The second mineral was named in honor of the director of the geological department of the UMHK, Belgian geologist Gaston Demesmaeker. Demesmaekerite crystallizes in the triclinic P-1 space group in the form of lamellar and elongated crystals of bottle-green to dark olive-green color in association with malachite, uranophane-α, chalcomenite, and other uranyl-selenites: Namely marthozite and derriksite as well as guilleminite.

Marthozite, $Cu[(UO_2)_3(SeO_3)_2O_2](H_2O)_8$ [3,29], was also found in the Musonoi mine within a few years after, and named to honor Aimé Marthoz, former director of the UMHK. Marthozite crystallizes in the orthorhombic $Pbn2_1$ space group, in the form of well-faceted green crystals, in association with the other selenites, including guilleminite and demesmaekerite, as well as kasolite, cuprosklodowskite, malachite, chalcomenite, and sengierite. Mineral is isotypic with guilleminite.

A few years later, derriksite, $Cu_4[(UO_2)(SeO_3)_2](OH)_6$ [4,30], was found at the same deposit in Congo, and named after Jean-Marie François Joseph Derriks, a Belgian geologist and administrator of the UMHK. Derriksite crystallizes in the orthorhombic $Pn2_1m$ space group, as sub-green up to bottle-green-colored crystals, elongated at [001] or incrustations and fine-crystalline crusts on digenite and the mineral is associated with marthozite, demesmaekerite, kasolite, malachite, etc.

Table 1. Crystallographic characteristics of natural uranyl selenites.

No.	Formula/Mineral Name	Topology	Sp. Gr.	a, Å/α, °	b, Å/β, °	c, Å/γ, °	Reference
	Chains						
1	$Cu_4[(UO_2)(SeO_3)_2](OH)_6$ derriksite	cc1-1:2–1	$Pn2_1m$	5.570(2)/90	19.088(8)/90	5.965(2)/90	[2]
2	$Pb_2Cu_5[(UO_2)_2(SeO_3)_6(OH)_6](H_2O)_2$ demesmaekerite	cc1-1:3–2	P-1	11.9663(9)/89.891(8)	10.0615(14)/100.341(11)	5.6318(8)/91.339(9)	This work, [4]
	Layers with edge-linkage						
3	$Cu[(UO_2)_3(SeO_3)_2O_2](H_2O)_8$ marthozite		$Pbn2_1$	6.9879(4)/90	16.454(1)/90	17.223(1)/90	[17]
4	$Ba[(UO_2)_3(SeO_3)_2O_2](H_2O)_4$ guilleminite	$6^15^24^23^2$	$Pmn2_1$	16.762(1)/90	7.2522(5)/90	7.0629(4)/90	This work, [18]
5	$Na(H_3O)[(UO_2)_3(SeO_3)_2O_2](H_2O)_4$ larisaite		$P11m$	6.9806(9)/90	7.646(1)/90	17.249(2)/90.039(4)	[19]
6	$[(UO_2)_3(SeO_3)_2(OH)_2](H_2O)_5$ haynesite		$Pnc2$ or $Pncm$	6.935/90	8.025/90	17.430/90	[21]
7	$Ca[(UO_2)_3(SeO_3)_2(OH)_4](H_2O)_4$ piretite		$Pmn2_1$ or $Pmmm$	7.010(3)/90	17.135(7)/90	17.606(4)/90	[22]

Next, natural uranyl selenite was discovered in 20 years across the Atlantic, in the Repete mine (Utah, USA). Haynesite, $[(UO_2)_3(SeO_3)_2(OH)_2](H_2O)_5$ [5,31,32] is named after the American geologist Patrick Eugene Haynes. Haynesite is orthorhombic, occurs as amber-yellow tablets, transparent to translucent, elongated at [001], and as acicular prismatic rosettes up to 3 mm in diameter, and is associated with andersonite, boltwoodite, gypsum, and calcite as crusts on mudstones and sandstones.

Piretite, $Ca(UO_2)_3(SeO_3)_2(OH)_4 \cdot 4H_2O$ [6], calcium uranyl selenite from Shinkolobwe mine (Katanga, DR Congo) is named after the Belgian crystallographer Paul Piret. Piretite is orthorhombic, it crystallizes as lemon-yellow elongated tablets, irregular in outline and up to 3 mm, flattened on (001), or as needle-prismatic crystals up to 5 mm. It occurs in association with a masuyite-like uranyl-lead oxide as crusts on uraninite. It should be noted that crystal structures of haynesite and piretite have still not been determined.

The last to date, uranyl selenite mineral, larisaite, $Na(H_3O)[(UO_2)_3(SeO_3)_2O_2](H_2O)_4$ [7], was found in the Repete mine (Utah, UT, USA) and named in honor of Larisa Nikolaevna Belova, a Russian mineralogist and crystallographer who made a significant contribution to the knowledge on uranium minerals. Larisaite occurs as canary-yellow lamellar crystals up to 1 mm long, and as radial aggregates up to 2 mm across; most crystals are fissured and ribbed. The mineral is a supergene product associated with calcite, quartz, gypsum, montmorillonite, wölsendorfite, andersonite, haynesite, and uranophane–α in sedimentary rocks.

3.2. Synthetic Uranyl Compounds with Selenite Ions

The first synthetic and the simplest uranyl selenite, $[(UO_2)(SeO_3)]$, was obtained in 1978 [33] (and its neptunyl analog has been recently reported [34] as well). Further, the research undertaken by V. E. Mistryukov and Yu. N. Mikhailov from the Kurnakov Institute of General and Inorganic Chemistry RAS (Russian Federation), and by V.N. Serezhkin and L.B. Serezhkina from the Samara State University (Russian Federation) should be mentioned, who studied uranyl selenites with electroneutral ligands and the first Na-bearing synthetic uranyl selenite compounds. Nearly half of the synthetic compounds described within this review were synthesized and characterized by T.E. Albrecht-Schmitt and co-workers (Table 2). The significant impact of their works on the development of uranyl selenites' structural chemistry should be especially noted.

Synthetic compounds, whose structures are based on inorganic units with the linkage of *Ur* to selenite oxyanions (Table 2), could be divided into two groups: Pure inorganic and organically templated phases.

Table 2. Crystallographic characteristics of synthetic uranyl selenites and selenite-selenates.

No.	Formula	Topology	Sp. Gr.	a, Å/α, °	b, Å/β, °	c, Å/γ, °	Reference
	Chains						
8	[(UO$_2$)(HSeO$_3$)$_2$(H$_2$O)]	cc1-1:2-1	A2/a	6.354(1)/90	12.578(2)/82.35(1)	9.972(2)/90	[35]
9	[(UO$_2$)(HSeO$_3$)$_2$](H$_2$O)		C2/c	9.924(5)/90	12.546(5)/98.090(5)	6.324(5)/90	[36]
10	Ca[(UO$_2$)(SeO$_3$)$_2$]	cc1-1:2-14	P–1	5.5502(6)/104.055(2)	6.6415(7)/93.342(2)	11.013(1)/110.589(2)	[37]
11	Sr[(UO$_2$)(SeO$_3$)$_2$]		P–1	5.6722(4)/104.698(1)	6.7627(5)/93.708(1)	11.2622(8)/109.489(1)	[38]
12	Sr[(UO$_2$)(SeO$_3$)$_2$](H$_2$O)$_2$	cc1-1:2-15	P–1	7.0545(5)/106.995(1)	7.4656(5)/108.028(1)	10.0484(6)/98.875(1)	[37]
13	Na$_3$[H$_3$O][(UO$_2$)(SeO$_3$)$_2$]$_2$(H$_2$O)		P–1	9.543(6)/66.69(2)	9.602(7)/84.10(2)	11.742(8)/63.69(1)	[39]
	Layers with corner-linkage						
14	[NH$_4$]$_2$[(UO$_2$)(SeO$_3$)$_2$](H$_2$O)$_{0.5}$		P2$_1$/c	7.193(5)/90	10.368(5)/91.470(5)	13.823(5)/90	[36]
15	[NH$_4$][(UO$_2$)(SeO$_3$)(HSeO$_3$)]		P2$_1$/n	8.348(2)/90	10.326(2)/97.06(2)	9.929(2)/90	[40]
16	K[(UO$_2$)(HSeO$_3$)(SeO$_3$)]		P2$_1$/n	8.4164(4)/90	10.1435(5)/97.556(1)	9.6913(5)/90	[41]
17	Rb[(UO$_2$)(HSeO$_3$)(SeO$_3$)]		P2$_1$/n	8.4167(5)/90	10.2581(6)/96.825(1)	9.8542(5)/90	[41]
18	Cs[(UO$_2$)(HSeO$_3$)(SeO$_3$)]	cc2-1:2-4	P2$_1$/c	13.8529(7)/90	10.6153(6)/101.094(1)	12.5921(7)/90	[41,42]
19	Cs[((U,Np)O$_2$)(HSeO$_3$)(SeO$_3$)]		P2$_1$/n	8.4966(2)/90	10.3910(3)/93.693(1)	10.2087(3)/90	[42]
20	Tl[(UO$_2$)(HSeO$_3$)(SeO$_3$)]		P2$_1$/n	8.364(3)/90	10.346(4)/97.269(8)	9.834(4)/90	[41]
21	Cs[(UO$_2$)(SeO$_3$)(HSeO$_3$)](H$_2$O$_3$		P2$_1$/n	8.673(2)/90	10.452(3)/105.147(4)	13.235(4)/90	[43]
22	Na[(UO$_2$)(SeO$_3$)(HSeO$_3$)](H$_2$O)$_4$		P2$_1$/n	8.8032(5)/90	10.4610(7)/105.054(2)	13.1312(7)/90	[44]
23	[H$_3$O][(UO$_2$)(SeO$_4$)(HSeO$_3$)]		P2$_1$/n	8.668(2)/90	10.655(2)/97.88(2)	9.846(2)/90	[45]
24	Ag$_2$[(UO$_2$)(SeO$_3$)$_2$]	cc2-1:2-5	P2$_1$/n	5.8555(6)/90	6.5051(7)/96.796(2)	21.164(2)/90	[41]
	Layers with edge-linkage						
25	Pb[(UO$_2$)(SeO$_3$)$_2$]	cc2-1:2-19	Pmc2$_1$	11.9911(7)/90	5.7814(3)/90	111.2525(6)/90	[41]
26	Ba[(UO$_2$)(SeO$_3$)$_2$]	cc2-1:2-21	P2$_1$/c	7.3067(6)/90	8.1239(7)/100.375(2)	13.651(1)/90	[37]
27	[(UO$_2$)(SeO$_3$)]	6^{1}3^2	P2$_1$/m	5.408(2)/90	9.278(11)/93.45(10)	4.254(1)/90	[33]
28	Sr[(UO$_2$)$_3$(SeO$_3$)$_2$O$_2$](H$_2$O)$_4$	6^{1}5^{2}4^{2}3^2	C2/m	17.014(2)/90	7.0637(7)/100.544(2)	7.1084(7)/90	[38]
29	Li$_2$[(UO$_2$)$_3$(SeO$_3$)$_2$O$_2$](H$_2$O)$_6$		P2$_1$/c	7.5213(9)/90	7.0071(8)/98.834(2)	17.328(2)/90	[46]
30	Cs$_2$[(UO$_2$)$_4$(SeO$_3$)$_5$](H$_2$O)$_2$	6^{1}5^{4}6^{2}3^5	P2$_1$/n	10.913(3)/90	12.427(3)/90.393(3)	18.448(4)/90	[46]
31	Cs$_2$[(UO$_2$)$_7$(SeO$_4$)$_2$(SeO$_3$)$_2$(OH)$_4$O$_2$](H$_2$O)$_5$	6^{1}5^{4}6^{3}6	P2$_1$/m	9.1381(3)/90	15.0098(5)/91.171(1)	15.1732(5)/90	[46]
32	UO$_2$Se$_2$O$_5$	8^{1}5^{2}3^8	P–1	9.405(2)/93.01(3)	11.574(3)/93.66(3)	6.698(2)/109.69(1)	[47]

Table 2. *Cont.*

No.	Formula	Topology	Sp. Gr.	a, Å/α, °	b, Å/β, °	c, Å/γ, °	Reference
	Organically templated						
33	[C$_4$H$_{12}$N][(UO$_2$)(SeO$_3$)(NO$_3$)]	cc1-1:2-12	C2/m	21.888(3)/90	6.950(1)/97.618(3)	8.350(1)/90	[48]
34	[C$_6$H$_{14}$N$_2$]$_{0.5}$[(UO$_2$)(HSeO$_3$)(SeO$_3$)](H$_2$O)$_{0.5}$(CH$_3$CO$_2$H)$_{0.5}$		Pnma	13.086(1)/90	17.555(1)/90	10.5984(7)/90	[49]
35	[C$_4$H$_{12}$N$_2$]$_{0.5}$[(UO$_2$)(HSeO$_3$)(SeO$_3$)]		P2$_1$/c	10.9378(5)/90	8.6903(4)/90.3040(8)	9.9913(5)/90	[49]
36	[(C$_2$H$_8$N$_2$)H$_2$][(UO$_2$)(SeO$_3$)(HSeO$_3$)](NO$_3$)(H$_2$O)$_{0.5}$	cc2-1:2-4	Pbca	13.170(3)/90	11.055(2)/90	18.009(4)/90	[50]
37	[C$_5$H$_{14}$N][(UO$_2$)(SeO$_4$)(HSeO$_3$)]		P2$_1$/n	11.553(2)/90	10.645(2)/108.05(2)	12.138(2)/90	[51]
38	[C$_2$H$_8$N][(UO$_2$)(SeO$_4$)(HSeO$_3$)]		P2$_1$/n	8.475(3)/90	12.264(2)/95.23(3)	10.404(3)/90	[52]
39	[C$_5$H$_6$N][(UO$_2$)(SeO$_4$)(HSeO$_3$)]		P2$_1$/n	8.993(3)/90	13.399(5)/108.230(4)	10.640(4)/90	[53]
40	[C$_9$H$_{24}$N$_2$][(UO$_2$)(SeO$_4$)(HSeO$_3$)](NO$_3$)		P–1	10.748(1)/109.960(1)	13.885(1)/103.212(2)	14.636(1)/90.409(1)	[54]
41	[C$_2$H$_8$N][(H$_5$O$_2$)(H$_2$O)][(UO$_2$)$_2$(SeO$_4$)$_3$(H$_2$SeO$_3$)](H$_2$O)	cc2-1:2-14	P2$_1$/n	14.798(1)/90	10.024(1)/111.628(1)	16.418(1)/90	[55]
42	[C$_4$H$_{15}$N$_3$][H$_3$O]$_{0.5}$[(UO$_2$)$_2$(SeO$_4$)$_{2.93}$(SeO$_3$)$_{0.07}$(H$_2$O)](NO$_3$)$_{0.5}$	cc2-2:3-4	P2$_1$/c	11.1679(4)/90	10.9040(4)/98.019(1)	17.991(1)/90	[56]
43	[C$_5$H$_{14}$N]$_4$[(UO$_2$)$_3$(SeO$_4$)$_4$(HSeO$_3$)(H$_2$O)](H$_2$SeO$_3$)(HSeO$_4$)	cc2-3:5-3	P–1	11.707(1)/73.90(1)	14.817(1)/76.22(1)	16.977(2)/89.36(1)	[57]
44	[C$_2$H$_8$N]$_3$(C$_2$H$_7$N)[(UO$_2$)$_{2/3}$(SeO$_4$)$_4$(HSeO$_3$)(H$_2$O)]		Pnma	11.659(1)/90	14.956(2)/90	22.194(2)/90	[56]
45	[C$_2$H$_8$N]$_2$[H$_3$O][(UO$_2$)$_{2/3}$(SeO$_4$)$_4$(HSeO$_3$)(H$_2$O)](H$_2$SeO$_3$)$_{0.2}$		P2$_1$/m	8.3116(4)/90	18.636(4)/97.582(1)	11.5623(5)/90	[56]
46	[C$_8$H$_{15}$N$_2$]$_2$[(UO$_2$)$_4$(SeO$_3$)$_5$]	6^{1}5^{3}4^{6}3^5	Pnma	18.860(2)/90	18.010(2)/90	11.140(1)/90	[58]

Most of the inorganic uranyl selenites were obtained during low or medium temperature hydrothermal experiments in the temperature range of 100 to 220 °C using Teflon-lined steel autoclaves. Various reagents were used as the source of uranium (U(VI) oxide, uranyl hydroxide, uranyl nitrate hexahydrate, uranyl acetate dihydrate), whereas selenous acid (H_2SeO_3) was the only source of Se(IV). To be precise, either acid itself or SeO_2 were used in the reactions, but Se(IV) dioxide reacts with water to form selenous acid. H_2SeO_3 is a very weak acid and it hardly dissociates at room temperature, which explains the required heating for the reaction. Several compounds obtained in different ways should be mentioned separately. Compounds **22** [44] and **23** [45] were obtained during evaporation at room temperature. The first compound was obtained from the reaction of $UO_2(NO_3)_2 \cdot 6H_2O$ with selenic acid (H_2SeO_4) in aqueous medium for 1 year, which could be explained by the reduction of Se(VI) to Se(IV) in the solution during the experiment. Moreover, as it was recently shown, the hydronium ions usually enter the structure at the very latest crystallization stages, when there are no more other cations in the solution [52,59,60]. The Na-bearing compound was obtained in the presence of sodium oxalate, which probably could be regarded as a catalyst of the uranyl selenite crystallization process. Another five compounds, **11** [38], **21** [43], **27** [33], **28** [38], and **32** [47], were obtained in the gas–solid or hydrothermal reactions using sealed tubes. In the case of the last three compounds, the temperature reached over 425 °C.

The majority of the organically templated compounds are actually uranyl selenites-selenates. The selenite anions that are arranged in the structures of such compounds are in minor amounts with respect to the selenate groups. Such a tendency comes from the experimental conditions, in which the source of Se was the selenic acid. Selenic acid is less stable in environmental conditions than selenous acid, and it reduces to the latter during storage. Initially, pure H_2SeO_4 reagent after a few months of storage contains significant amount of $[SeO_3]^{2-}$ and $[HSeO_3]^-$ ions, which participate in the structure formation along with $[SeO_4]^{2-}$ groups. There are only five structures of organically templated uranyl selenites known without $[SeO_4]^{2-}$ oxyanions (Table 2), three of which (**34** [49], **35** [49], and **46** [58]) were obtained during mild hydrothermal experiments (130–150 °C) when the source of Se was again selenic acid. Here, the temperature and amine molecules or ionic liquids [58] acted as a reduction agent for H_2SeO_4, since it is known that the selenic acid is easily reduced to $H_2Se^{4+}O_3$ and oxygen upon heating above 160 °C [61]. The other two organically templated uranyl selenites (**33** [48] and **36** [50]) were obtained during evaporation at room temperature from the aqueous solution of $UO_2(NO_3)_2 \cdot 6H_2O$, SeO_2, and respective amine. Since SeO_2 transforms to weak selenous acid in water, the low dissociation ability of the latter [62] and the presence of $[NO_3]^-$ groups in the structures of both compounds explains the long crystallization process of 1 to 2 months. It is likely that dissociation of uranyl nitrate and the presence of amine finally helped to create an environment sufficient for the selenous acid dissociation, and thus to start the crystallization of uranyl selenites. Nitrate groups, in these cases, act as additional oxyanions involved in structure formation with a shortage of $[SeO_3]^{2-}$ groups.

3.3. Topological Analysis

The vast majority of the uranyl selenite crystal structures are based on the layered complexes of various topologies (Tables 1 and 2), and only nine compounds have chain-based crystal structures. However, among those nine compounds, two are uranyl selenite minerals.

Crystal structures of derriksite and another two synthetic compounds (Tables 1 and 2, Figure 1a–c), which are actually the same but were refined in different space groups, are based on the 1D units of the cc1-1:2-1 topological type (graph is an infinite chain of four-membered vertex-sharing rings). The graph corresponds to the type of chains, which were observed in the kröhnkite [63]. This topology is one of the most common and simplest chain topologies among U(VI)-bearing compounds with the $[TO_m]^{n-}$ groups (m = 3,4; T = S, Se, P, As). It was observed in the structures of uranyl-sulfate minerals as svornostite, $K_2Mg[(UO_2)(SO_4)_2]_2(H_2O)_8$ [64], rietveldite, $Fe(UO_2)(SO_4)_2(H_2O)_5$ [65], and their Mg-bearing synthetic analogues $Mg[(UO_2)(TO_4)_2(H_2O)](H_2O)_4$ (T = S, Se) [66]. Although the topology of chains is the same, their structures are remarkably different, representing two different isomers.

In the case of derriksite (Figure 1c), U^{6+} atoms present in the tetragonal bipyramidal coordination, where all four equatorial O atoms are shared with the $[SeO_3]^{2-}$ groups, and each selenite group in turn has only two O atoms shared with two neighbors' *Ur*. Uranyl selenite chains in the structure of derriksite are directed along [001] and the equatorial planes of uranyl bipyramids are arranged parallel to the (101). In between the chains, Cu-centered tri-octahedral layers are observed as being arranged parallel to (010), in which each Cu atom has four OH⁻ groups shared with the neighbor Cu atoms and two more vertices in the *trans*-orientation are the third vertices of selenite pyramids, non-shared with U-centered bipyramids. Selenite groups are arranged in such a way that lone electron pairs from one side of the U-Se chain are directed in one way, and from the other side, in the opposite direction (*up* or *down*), relative to the equatorial planes of uranyl bipyramids. Thus, the sequence of orientation symbols could be written as (**u**)(**d**). The latter has been termed an *orientation matrix*. In the structures of synthetic $[(UO_2)(HSeO_3)_2(H_2O)]$ [35,36] compounds, U^{6+} atoms are arranged in the center of pentagonal bipyramids, in which four equatorial O atoms are shared with the $[HSeO_3]^-$ groups and the fifth vertex is occupied by the H_2O molecule. Hydrogen selenite groups also have two O atoms shared with two neighboring *Ur* and the third vertex is attributed to the OH⁻ group. The linkage of chains into the 3D structure is carried out by the means of H-bonding between the neighbor chains only. The arrangement of lone electron pairs relative to the equatorial planes of uranyl bipyramids is staggered on both sides of the chain, so the orientation matrix for the current geometrical isomer is (**ud**)(**du**).

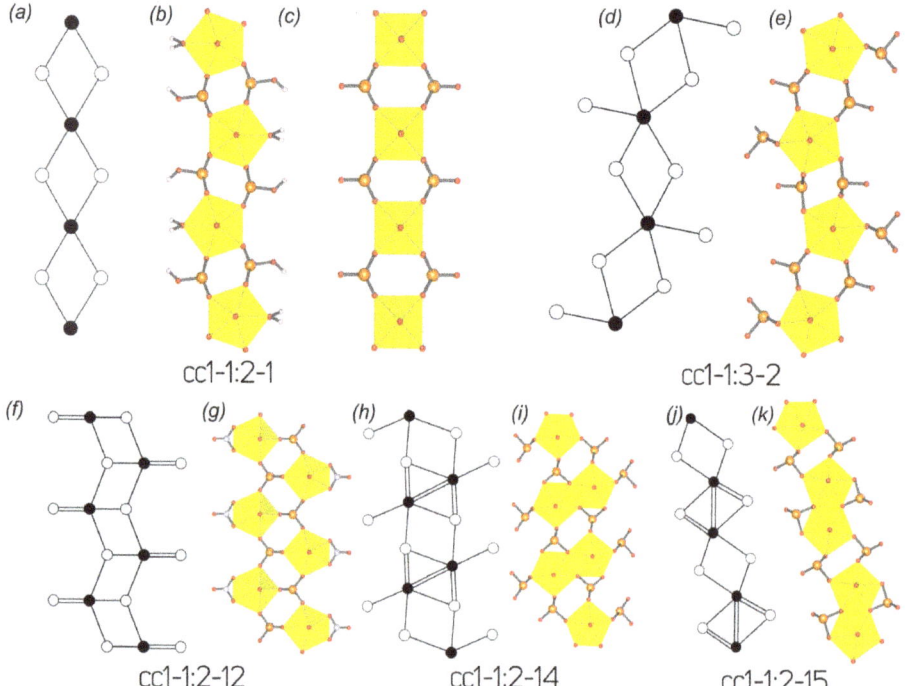

Figure 1. (**a**–**k**) 1D complexes in the crystal structures of natural and synthetic uranyl selenites (**a**–**k**: see text for details). Legend: U-bearing coordination polyhedra = yellow; Se atoms = orange; O atoms = red; H atoms = white; N atoms = light blue; black nodes = U atoms, white nodes = Se atoms. $Se^{IV}O_3$ trigonal pyramids and NO_3 groups are shown in a ball-and-stick mode.

The crystal structure of demesmaekerite is based on the chains of the *cc*1–1:3–2 topology (Figure 1d,e), which is very similar to the previous type. The graph is a vertex-sharing infinite chain of four-membered rings with additional one-connected selenite group to each Ur. This topology is quite rare and has been observed in the structures of two synthetic uranyl chromates $Na_4[(UO_2)(CrO_4)_3]$ [67] and $K_5[(UO_2)(CrO_4)_3](NO_3)(H_2O)_3$ [68], two uranyl molybdates $Na_3Tl_5[(UO_2)(MoO_4)_3]_2(H_2O)_3$ and $Na_{13}Tl_3[(UO_2)(MoO_4)_3]_4(H_2O)_5$ [69], and one uranyl selenate $(C_2H_8N)_3[(UO_2)(SeO_4)_2(HSeO_4)]$ [52]. U^{6+} atoms are arranged in the centers of pentagonal bipyramids, so that four equatorial vertices of which are shared with two-connected selenite groups (as in previous type), and the fifth vertex that was occupied by H_2O molecule, now is replaced by another one-connected $[SeO_3]^{2-}$ pyramid. Uranyl selenite chains are passing along the (101), and stacked one above the other, forming blocks parallel to (010). These blocks are separated by the sheets of edge-shared Cu- and Pb-centered coordination polyhedra. There are three types of Cu^{2+}-centered octahedra in the structure of demesmaekerite, $[CuO_4(OH)_2]^{8-}$, $[CuO_3(OH)_3]^{7-}$, and $[CuO_2(OH)_3(H_2O)]^{6-}$, and the single type of ninefold $[Pb^{2+}O_6(OH)_3]^{13-}$ complexes. Lone electron pairs of one- and two-connected selenite groups from one side of the U-Se chain are oriented in the same direction, while on the other side the direction is the opposite, thus the orientation matrix could be written as (**u**)(**d**).

The crystal structure of the organically templated compound 33 [48] is based on the uranyl selenite nitrate 1D complexes that belong to the cc1–1:2–12 topological type (Figure 1f,g). This topology has been observed in the structures of several uranyl and neptunyl sulfates and selenates, for example, see [70–73], and represents an infinite chain of edge-shared four-memebered cycles, in which each uranyl polyhedron has three equatorial vertices shared with three selenite groups while the left pair of O atoms is edge-shared with the $[NO_3]^-$ group. Being three-connected to the neighbor Ur, $[SeO_3]^{2-}$ pyramids have a lone electron pair oriented either *up* or *down* relative to the equatorial planes of uranyl bipyramids in the (**ud**)$_\infty$ sequence.

The crystal structures of Ca- [37] and Sr-bearing [38] isotypic uranyl selenites are based on 1D complexes of the cc1–1:2–14 topological type (Figure 1h,i), which are built by the dimers of edge-sharing uranyl pentagonal bipyramids that are interlinked by the pair of edge- and vertex-sharing selenite groups with another one-connected selenite group decorating the fifth non-shared equatorial vertices of U polyhedra from both sides of such a double-wide chain. It should be noted that $[SeO_3]^{2-}$ pyramids, which are involved in the linkage of U dimers, have lone electron pairs oriented *up* from one side of the chain, and *down* from the other side, thus illustrating the (**ud**)$_\infty$ sequence. This type of chains occurs in the structures of two uranyl minerals: Parsonite, $Pb_2[(UO_2)(PO_4)_2$, [74] and hallmondite, $Pb_2[(UO_2)(AsO_4)_2](H_2O)_n$, [75].

The crystal structures of two more Sr- [37] and Na-hydronium-bearing [39] compounds are based on the uranyl selenite chains with an edge-sharing motif, similar to the previous one. Chains belong to the cc1–1:2–15 topological type (Figure 1j,k), and are built by the dimers of edge-sharing uranyl pentagonal bipyramids, which, in contrary to the aforementioned topology, are interlinked by a pair of only vertex-sharing selenite groups, while edge-sharing selenite pyramids in this case decorate both sides of the chain. Both compounds represent two different geometrical isomers, assuming the orientation of lone electron pairs. Thus, Sr uranyl selenite possesses the same (**ud**)$_\infty$ sequence, as in a previous case, while the Na-bearing compound has a (**u**)$_\infty$ sequence. This type of topology has been observed in several synthetic uranyl chromates, phosphates, and arsenates, as well as in lakebogaite, $CaNa(Fe^{3+})_2[H(UO_2)_2(PO_4)_4(OH)_2](H_2O)_8$ [76].

The crystal structures of 17 synthetic uranyl selenites are based on the layers, which belong to the cc2–1:2–4 topological type (Figure 2a,b), the most common among the uranyl selenite compounds and among the layered uranyl compounds, generally. The topology consists of dense four-membered cycles and large hollow eight-membered rings. It is worth noting, that almost all sheets of this topology contain protonated $[HSeO_3]^-$ groups with the H-bonds arranged inside the eight-membered cycles. Although the topology of the sheets remains the same, their real architecture is quite diverse, which occurs due to various blocks involved in the structure formation. Thus, the structures of these compounds are formed via combination of the $[UO_7]^{8-}$, $[HSeO_3]^-$, $[SeO_3]^{2-}$, and $[SeO_4]^{2-}$ coordination polyhedra through common oxygen atoms. Uranyl pentagonal bipyramids share all of five equatorial O atoms with the selenite or selenate groups, while Se-bearing oxyanions act as two- or three-connected units. Such a diversity of building blocks opens up the possibility of a large number of geometric isomers' existence. Within the uranyl selenite and selenite-selenate compounds of *cc*2–1:2–4 topology, three isomers are distinguished: Layers, containing only selenite groups; those, having selenite and hydrogen selenite groups; and those with hydrogen selenite groups and selenate tetrahedra. However, what is the most interesting, is that all three isomers have a similar orientation of lone electron pairs and fourth non-shared vertices (for tetrahedra), which is described by the very simple (**ud**) matrix. Only except for the compound **36**, which has the (**ud**)(**du**) matrix.

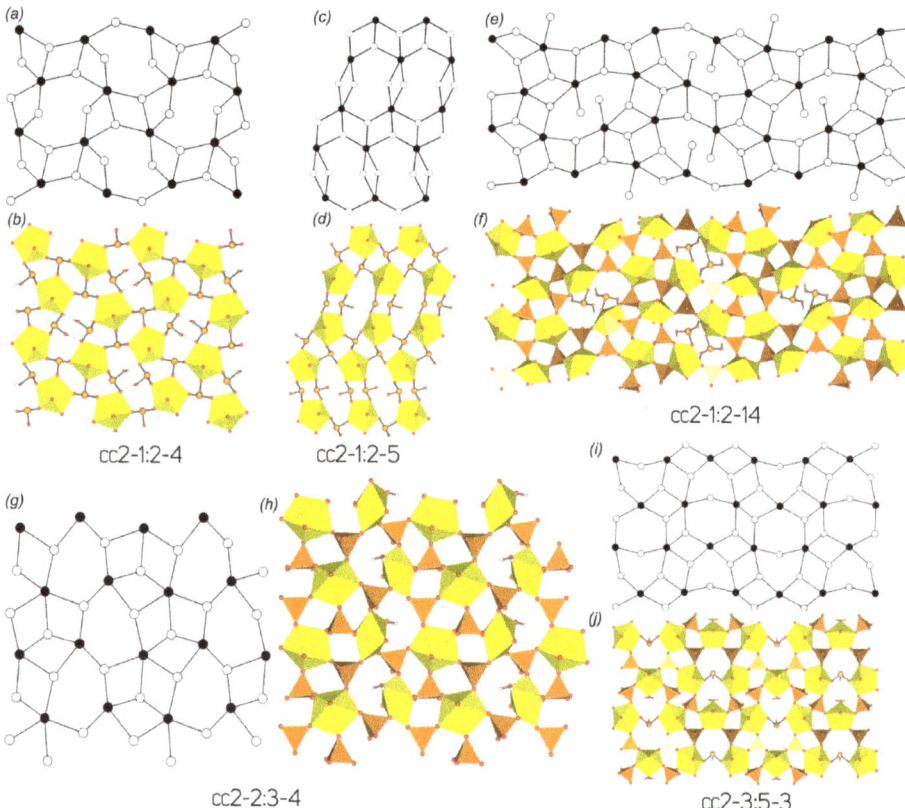

Figure 2. (a–j) 2D complexes based on corner-sharing linkage in the crystal structures of synthetic uranyl selenites and selenite-selenates (a–j: see text for details). Legend: see Figure 1; $Se^{VI}O_4$ groups = orange tetrahedra.

The crystal structures of Ag-bearing uranyl selenite [41] is based on the layered complex of *cc*2–1:2–5 topological type (Figure 2c,d). This type of topology has been observed in the structures of several synthetic uranyl and neptunyl molybdates as $Na_2(UO_2)(MoO_4)_2$ [77] and $K_3NpO_2(MoO_4)_2$ [78]. Topological types *cc*2–1:2–4 and *cc*2–1:2–5 have nearly identical chemical composition and looks quite similar. Those graphs are built from the similar four- and eight-membered rings, and even have the same connectivity of black and white vertices (U and Se polyhedra, respectively), but the topologies are different due to differences in coordination sequence [18]. Such chemically identical, but topologically different structural units are called topological or structural isomers. It should be noted that the *cc*2–1:2–4 topology is much more representative among the inorganic oxysalt compounds than cc2–1:2–5. If the lone electron pair of the selenite pyramid would be equated to the fourth non-shared vertex of the selenate tetrahedron, the current isomer can be described by the (**uddu**)(**dduu**) matrix.

The crystal structure of **41** [55] is based on the 2D complexes, possessing unprecedented topology for both the structural chemistry of uranium and the chemistry of inorganic oxysalts in general, of the *cc*2–1:2–14 type (Figure 2e,f). U atoms are arranged in the centers of pentagonal bipyramids. Each $[SeO_4]^{2-}$ group is three-connected, coordinating three uranyl ions, whereas protonated selenite groups coordinate one uranyl ion each. The topology is remarkable due to the presence of one-connected branches inside eight-membered cycles, which are actually selenous acid groups.

The crystal structure of **25** [41] is based on the layered complexes of *cc*2–1:2–19 topological type (Figure 3a,b), which is a derivative of the autunite topology [18], where each uranyl pentagonal bipyramid has only one edge shared with the selenite group. The graph of the layer consists of eight-membered rings only. The current isomer can be described by the (**uudd**)(**uddu**)(**dduu**)(**duud**) matrix.

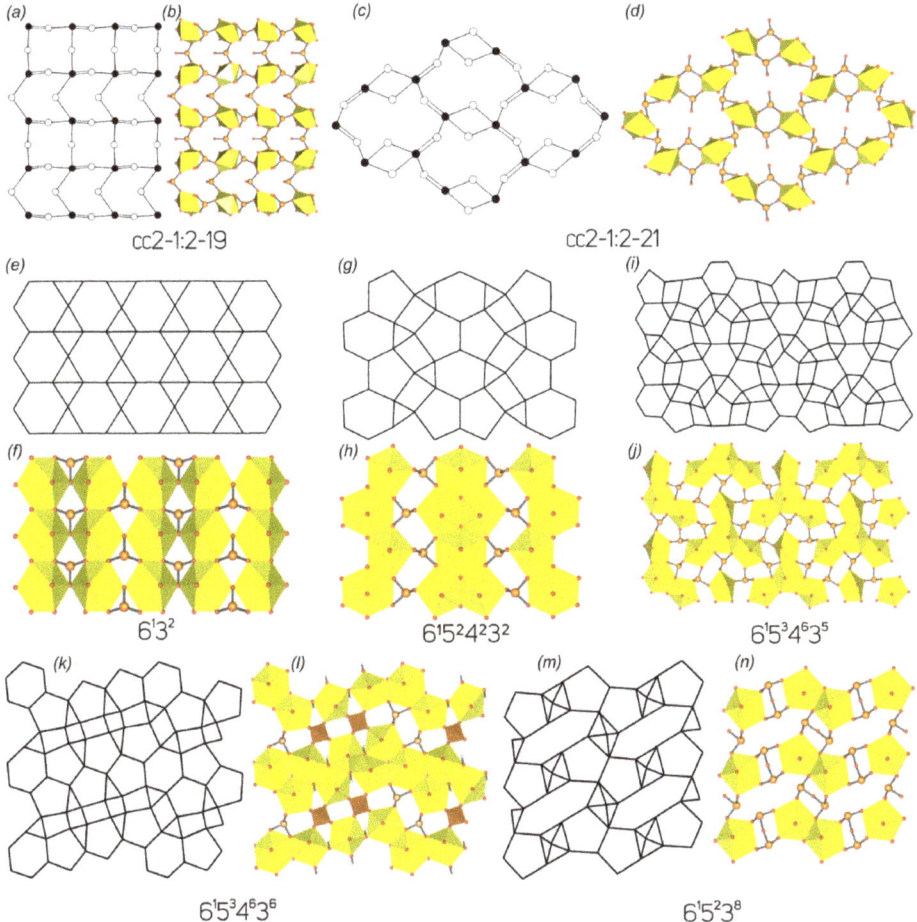

Figure 3. (a–n) 2D complexes based on edge-sharing linkage in the crystal structures of natural and synthetic uranyl selenites and selenite-selenates (a–n: see text for details). Legend: see Figures 1 and 2.

Compound **26** [37] is the only known uranyl selenite, which crystal structure is based on the layered complexes of *cc*2–1:2–21 topological type (Figure 3c,d). The graph of the U-bearing sheet consists of dense 4-membered and large 12-membered rings. Double links between the black and white vertices in a graph indicate sharing of an edge between uranyl coordination polyhedra and the selenite pyramid. Despite the fact that [SeO$_3$]$^{2-}$ groups are two-connected, edge-sharing coordination generates a possibility for an orientational isomerism of the lone electron pair arrangement. Current isomer can be described by the (**ud**) matrix. It is of interest that interlayer Ba^{2+} cations are actually arranged within the layer, inside the 12-membered rings.

Next compound, **42** [56], got into the review with a large tolerance. There are three nonequivalent positions of Se in the structure, only one of which was occupied by both Se(VI) and Se(IV), and the amount of the latter is very small (~0.07 per formula unit). The current topology of the *cc*2–2:3–4 type (Figure 2g,h) is one of the most common among synthetic uranyl sulfates, chromates, and selenates (>30 structures are known), but it has not been observed for any compound with a higher content of selenite ions than here.

The crystal structures of three organically templated compounds, **43** [57], **44**, and **45** [56], are based upon the layers with U:Se = 3:5 formed as a result of condensation of the [UO$_2$]$^{2+}$, [UO$_2$(H$_2$O)]$^{2+}$, [Se^{VI}O$_4$]$^{2-}$, and [HSe^{IV}O$_3$]$^-$ coordination polyhedra by sharing common oxygen atoms. The corresponding graph of *cc*2–3:5–3 topology is built by four- and six-membered rings (Figure 2i,j). This topology of inorganic complexes is typical for uranyl selenite-selenates but has also been observed in some pure uranyl selenates, for instance, Rb$_4$[(UO$_2$)$_3$(SeO$_4$)$_5$(H$_2$O)] [79]. The presence of two-connected selenite trigonal pyramids and three-connected selenate tetrahedra gives rise to geometric isomerism. Thus, the orientation matrices can be written as (**ududud**)(**ud◻du◻**) for the first and second, and (**duuudd**)(**ud◻du◻**) for the third compound, respectively.

The simplest uranyl selenite, at least from the chemical point of view, [(UO$_2$)(SeO$_3$)] [33], has a layered structure (Figure 3e,f). According to Lussier et al. [80], the anionic topology of the layer of this compound belongs to the topology consisting of triangles and hexagons. The topology of the layer in this compound is the same as in mineral rutherfordine, [(UO$_2$)(CO$_3$)] [81,82], which is why it is called a rutherfordine anion topology. This topology consists of parallel chains of edge-sharing hexagons divided by dimers of edge-sharing triangles. Each of the hexagons is occupied by U*r*, and one triangle per dimer is occupied by the [SeO$_3$]$^{2-}$ group. The other half of the triangles is vacant. Electroneutral sheets are linked together by van der Waals interactions only. It should be noted that recently, an isotypic neptunyl compound has been reported [34].

One of the most remarkable topological types within the uranyl selenite family of compounds is the phosphuranylite topology (Figure 3g,h): The crystal structures of marthozite, guilleminite, and larisaite are based on such layers, while haynesite and piretite (although their structures are still unknown) are supposed to have topologically the same architecture due to the similarity of their unit-cell parameters. Except for minerals, two more Li- and Sr-bearing synthetic uranyl selenites have structures based on the 2D units belonging to the phosphuranilite anion topology. The phosphuranilite topology contains two types of alternating infinite chains: Edge-sharing dimers of pentagons that are further linked by edge-sharing hexagons, and zig-zag chains of edge-sharing triangles and squares [80,83]. The topology can be described by the 6^{1}5^{2}4^{2}3^2 ring symbol with pentagons and hexagons occupied by U*r*, triangles are occupied by selenite anions, while squares stay vacant. In the crystal structures of natural and synthetic compounds, additional mono-, divalent cations, and H$_2$O molecules are arranged in between the layers forming covalent and H-bonding systems to build the 3D structure. In the structure of marthozite, there are Cu^{2+} cations arranged in between the layers and octahedrally coordinated by two O atoms of uranyl ions from the above and underlying layers and four O atoms of H$_2$O molecules from the interlayer space. There are also four 'zeolite'-like H$_2$O molecules arranged in the interlayer space, which are not covalently bonded to cations and held in the structure by H-bonds only. Na$^+$ and K$^+$ sites in the structure of larisaite are characterized by partial occupancies, as well as H$_2$O molecules and hydronium cations, which are statistically distributed over six sites within the interlayer space. Thus, there are also two types of H$_2$O molecules, those which coordinate alkali cations and 'zeolite'-like, as in the structure of marthozite. Na$^+$ and K$^+$ cations in the crystal structure of larisaite alternately occupy neighbor cavities in the interlayer space, while in the structure of guilleminite, those cavities are equivalent and occupied only by Ba^{2+} cations. It is of interest that according to previous works [1,27], only two sites of H_2O molecules coordinating Ba^{2+} cations have been determined in the structure of guilleminite, leaving rather a large cavity to be vacant. Our single crystal XRD studies at low temperatures allowed us to determine the third site arranged within the void and occupied by the H_2O molecule, which suggests a change to the formula of guilleminite to $Ba[(UO_2)_3(SeO_3)_2O_2](H_2O)_4$. Such ambiguity allows reference to the variable character of H_2O molecules' amount within these structures, which could depend on the chemical composition and conditions, and the temperature and humidity storage of samples. Another interesting feature is that the structures of natural and synthetic compounds belong to different geometrical isomers. The **(ud)(du)** isomer was determined in the structures of Li- and Sr-bearing synthetic uranyl selenites, while **(ud)(ud)** isomer was observed in the crystal structures of all three minerals. It should be noted that implementation of the **(ud)(du)** isomer results in formation of stepped layers, in which each subsequent chain of edge-sharing uranyl polyhedra is located above the level of the previous chain, whereas the **(ud)(ud)** isomer results in the formation of zig-zag uranyl selenite layers, in which the chains of edge-sharing uranyl polyhedra are alternately located above or below the mean plane of the layer (Figure 4).

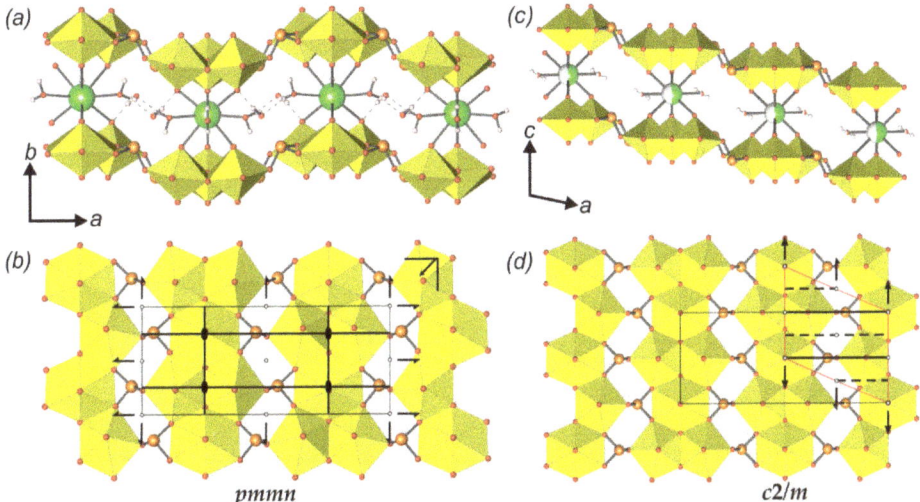

Figure 4. (a–d) The crystal structure projections along the layers, uranyl selenite layers, symmetry elements, and the respective layer symmetry groups for guilleminite (a,b) and $Sr[(UO_2)_3(SeO_3)_2O_2](H_2O)_4$ (c,d). Legend: see Figure 1.

Another topology that consists of hexagons, pentagons, squares, and triangles can be described by the $6^15^34^63^5$ ring symbol (Figure 3i,j), and is quite rare. There are only three compounds known, whose structures are based on the layers of this type. Two of them are Cs-bearing [46] and organically templated [58] uranyl selenites, and the third one is a very exotic $Cs_2[(UO_2)_4(Co(H_2O)_2)_2(HPO_4)(PO_4)]$ uranyl phosphate compound [84]. Layers are formed by the specific heptamers, and the uranyl hexagonal bipyramid is in the center, sharing each even equatorial edge with three uranyl pentagonal bipyramids, while the odd edges are shared with $[SeO_3]^{2-}$ groups. The linkage of these heptamers occurs via the third non-shared vertex of the selenite group and by the two additional selenite groups of each pentagonal bipyramid, which share all three O atoms with three neighbor heptamers. Thus, all pentagons and hexagons in the anion topology are occupied by the uranyl ions, triangles, and by the selenite groups while squares are vacant. It should be noted, that the arrangement of the lone electron pair in the structures of both uranyl selenites is different. In the structure of Cs-bearing uranyl selenite, the orientation of the lone electron pairs around the core of uranyl bipyramids is uneven and can be described by the (**uuudduuuudd**) matrix, while that in the structure of organically templated uranyl selenite is uniform (**uududuuddudd**), but it does not result in any visible differences in the distortion or undulations between the layers.

The crystal structure of the Cs-bearing uranyl selenite-selenate **31** [46] phase is based on the layers of a highly remarkable anion topology with the $6^15^64^63^6$ ring symbol (Figure 3k,l), which could be assumed as the modular structure, composed of blocks from both the phosphuranylite and zippeite anion topologies. The latter, for instance, is one of the most common topologies among the natural uranyl sulfates [25]. The zippeite fragment of the topology includes selenate tetrahedra, and the phosphuranylite fragment contains selenite groups.

The crystal structure of the only uranyl diselenite compound [47] is based on the sheets of miscellaneous anion topology of the $8^15^23^8$ type (Figure 3m,n), consisting of octagons, pentagons, and triangles. The layered complex is built by the dimers of edge-sharing uranyl pentagonal bipyramids, which are arranged similarly as in the structures of such minerals as deliensite, $Fe[(UO_2)_2(SO_4)_2(OH)_2](H_2O)_7$ [85] or plášilite, $Na(UO_2)(SO_4)(OH)(H_2O)_2$ [86], but the linkage character is remarkably different. Instead of isolated groups, uranyl dimers are interlinked length- and side-ways through the vertex-sharing diselenite groups; besides, lone electron pairs within the $[Se_2O_5]^{2-}$ oxyanions are co-directed. In those diselenite groups, which are arranged along the extension of the uranyl dimers, lone electron pairs are oriented towards one side relative to the plane of the sheet, and in those groups, arranged side-ways, the direction of the lone electron pair is the opposite. The crystal structure of **32** is anhydrous and free of additional ions, thus electroneutral layered complexes are linked into the 3D structure by the means of electrostatic interactions involving lone electron pairs only.

3.4. Structural and Topological Complexity

Calculation was performed in several stages and the main results are summarized in Table 3 and Figures 5 and 6. First, the topological complexity (**Tl**), according to the maximal rod (for chains) or layer symmetry group, was calculated, since these are the basic structural units. Second, the structural complexity (**Sl**) of the units was analyzed taking into account its real symmetry. The next contribution to information comes from the stacking (**LS**) of chained and layered complexes (if more than one layer or chain is in the unit cell). The fourth contribution to the total structural complexity is given by the interstitial structure (**IS**). The last portion of information comes from the interstitial H bonding system (**H**). It should be noted that the H atoms related to the U-bearing chains and layers were considered as a part of those complexes but not within the contribution of the H-bonding system. Complexity parameters for the whole structures were calculated using *ToposPro* package [87].

Table 3. Structural and topological complexity parameters for the uranyl selenite and selenite-selenate compounds.

No.	Formula	Topology	Complexity Parameters of the Crystal Structure				Structural Complexity of the U-Se Unit				Topological Complexity of the U-Se Unit			
			Sp. Gr.	v	I_G	$I_{G,total}$	Layer or Rod Sym. Gr.	v	I_G	$I_{G,total}$	Layer or Rod Sym. Gr.	v	I_G	$I_{G,total}$
	Chains													
1	$Cu_4[(UO_2)(SeO_3)_2]$ $(OH)_6$/**derriksite**	cc1-1:2-1	$Pn2_1m$	54	4.236	228.764	t_cm11	11	3.096	34.054	$t_a2/m11$	11	2.187	24.054
8	$[(UO_2)(HSeO_3)_2(H_2O)]$		$A2/a$	32	3.125	100.000	$t_a12/a1$	32	3.125	100.000	$t_a12/a1$	32	3.125	100.000
9	$[(UO_2)(HSeO_3)_2](H_2O)$		$C2/c$											
2	$Pb_2Cu_5[(UO_2)_2(SeO_3)_6(OH)_6]$ $(H_2O)_2$/**demesmaekerite**	cc1-1:3-2	$P-1$	55	4.800	263.975	$t-1$	30	3.907	117.207	$t_a2_1/m11$	30	3.374	101.207
33	$[C_4H_{12}N][(UO_2)(SeO_3)(NO_3)]$	cc1-1:2-12	$C2/m$	56	4.236	237.212	$t_a2_1/m11$	22	3.096	68.108	$t_a2_1/m11$	22	3.096	68.108
10	$Ca[(UO_2)(SeO_3)_2]$	cc1-1:2-14	$P-1$	24	3.585	86.039	$t-1$	22	3.459	76.108	$t-1$	22	3.459	76.108
11	$Sr[(UO_2)(SeO_3)_2]$													
12	$Sr[(UO_2)(SeO_3)_2](H_2O)_2$	cc1-1:2-15	$P-1$	36	4.170	150.117	$t-1$	22	3.459	76.108	$t-1$	22	3.459	76.108
13	$Na_3[H_3O][(UO_2)(SeO_3)_2]_2(H_2O)$			64	5.000	320.000								
	Layers with corner-linkage													
14	$[NH_4]_2[(UO_2)(SeO_3)_2](H_2O)_{0.5}$		$P2_1/c$	94	4.576	430.131	$p2_1/b$	44	3.459	152.196		44	3.459	152.196
15	$[NH_4][(UO_2)(SeO_3)(HSeO_3)]$		$P2_1/n$	68	4.087	277.947	$p2_1/b$	48	3.585	172.080				
16	$K[(UO_2)(HSeO_3)(SeO_3)]$		$P2_1/n$	52	3.700	192.423	$p2_1/b$	48	3.585	172.080				
17	$Rb[(UO_2)(HSeO_3)(SeO_3)]$		$P2_1/n$	52	3.700	192.423	$p2_1/b$	48	3.585	172.080				
18	$Cs[(UO_2)(HSeO_3)(SeO_3)]$		$P2_1/n$	104	4.700	488.846	$p2_1$	48	4.585	172.080				
19	$Cs[((U,Np)O_2)(HSeO_3)(SeO_3)]$		$P2_1/n$	52	3.700	192.423	$p2_1/b$	48	3.585	172.080	$p2_1/b$	48	3.585	172.080
20	$Tl[(UO_2)(HSeO_3)(SeO_3)]$		$P2_1/n$	52	3.700	192.423	$p2_1/b$	48	3.585	172.080				
21	$Cs[(UO_2)(SeO_3)(HSeO_3)](H_2O)_3$	cc2-1:2-4	$P2_1/n$	88	4.459	392.430	$p2_1/b$	48	3.585	172.080				
22	$Na[(UO_2)(SeO_3)(HSeO_3)](H_2O)_4$		$P2_1/n$	100	4.644	464.386	$p2_1/b$	48	3.585	172.080				
34	$[C_6H_{14}N_2]_{0.5}[(UO_2)$ $(HSeO_3)(SeO_3)](H_2O)_{0.5}(CH_3CO_2H)_{0.5}$		$Pnma$	228	4.991	1137.899	$p2_1/b$	48	3.585	172.080				
35	$[C_4H_{12}N_2]_{0.5}[(UO_2)(HSeO_3)(SeO_3)]$		$P2_1/c$	84	4.392	368.955	$p2_1/b$	48	3.585	172.080				
36	$[(C_2H_8N_2)H_2][(UO_2)(SeO_3)(HSeO_3)]$ $(NO_3)(H_2O)_{0.5}$		$Pbca$	264	5.044	1331.720	$p2_1/b$	48	3.585	172.080				

Table 3. Cont.

No.	Formula	Topology	Complexity Parameters of the Crystal Structure				Structural Complexity of the U-Se Unit				Topological Complexity of the U-Se Unit			
			Sp. Gr.	v	I_G	$I_{G,total}$	Layer or Rod Sym. Gr.	v	I_G	$I_{G,total}$	Layer or Rod Sym. Gr.	v	I_G	$I_{G,total}$
23	$[H_3O][(UO_2)(SeO_4)(HSeO_3)]$		$P2_1/n$	68	4.087	277.947								
37	$[C_5H_{14}N][(UO_2)(SeO_4)(HSeO_3)]$		$P2_1/n$	132	5.044	665.860								
38	$[C_2H_8N][(UO_2)(SeO_4)(HSeO_3)]$		$P2_1/n$	96	4.585	440.156	$p2_1/b$	52	3.700	192.423		52	3.700	192.423
39	$[C_5H_6N][(UO_2)(SeO_4)(HSeO_3)]$		$P2_1/n$	100	4.644	464.386								
40	$[C_9H_{24}N_2][(UO_2)(SeO_4)(HSeO_3)](NO_3)$		$P-1$	208	6.700	1393.691	$p-1$	52	4.700	244.423				
24	$Ag_2[(UO_2)(SeO_3)_2]$	cc2-1:2-5	$P2_1/n$	52	3.700	192.423	$p2_1/b$	44	3.459	152.215	$p2_1/b$	44	3.459	152.215
41	$[C_2H_8N][(H_5O_2)(H_2O)][(UO_2)_2(SeO_4)_3(H_2SeO_3)](H_2O)$	cc2-1:2-14	$P2_1/n$	204	5.672	1157.175	$p2_1/b$	104	4.755	513.528	$p2_1/b$	104	4.755	513.528
42	$[C_4H_{15}N_3][H_3O]_{0.5}[(UO_2)_2(SeO_4)_{2.93}(SeO_3)_{0.07}(H_2O)](NO_3)_{0.15}$	cc2-2:3-4	$P2_1/c$	212	5.728	1214.319	$p2_1$	48	4.585	220.078	$p2_1$	48	4.585	220.078
43	$[C_5H_{14}N]_4[(UO_2)_3(SeO_4)_4(HSeO_3)(H_2O)][(H_2SeO_3)(HSeO_4)]$	cc2-3:5-3	$P-1$	258	7.011	1808.897	$p-1$	74	5.209	385.500	$p2_1/m$	74	4.399	325.500
44	$[C_2H_8N]_3(C_2H_7N)[(UO_2)_3(SeO_4)_4(HSeO_3)(H_2O)]$		$Pnma$	364	5.629	2048.837	$p2_1/m$	74	4.399	325.500				
45	$[C_2H_8N]_2[H_3O][(UO_2)_3(SeO_4)_4(HSeO_3)(H_2O)](H_2SeO_3)_{0.2}$		$P2_1/m$	134	5.200	696.856								
Layers with Edge-Linkage														
25	$Pb[(UO_2)(SeO_3)_2]$	cc2-1:2-19	$Pmc2_1$	48	3.835	184.078	$p2_1ma$	44	3.641	160.215	$p2_1ma$	44	3.641	160.215
26	$Ba[(UO_2)(SeO_3)_2]$	cc2-1:2-21	$P2_1/c$	48	3.585	172.078	$p2_1/a$	44	3.459	152.215	$p2_1/a$	44	3.459	152.215
27	$[(UO_2)(SeO_3)]$	$6^1 3^2$	$P2_1/m$	14	2.236	31.303	$p2_1/m$	14	2.236	31.303	$p2_1/m$	14	2.236	31.303
3	$Cu[(UO_2)_3(SeO_3)_2O_2]$ $(H_2O)_8$/marthozite		$Pbn2_1$	176	5.459	960.860	pm	38	4.248	161.421	$pmmm$	38	3.195	121.421
4	$Ba[(UO_2)_3(SeO_3)_2O_2]$ $(H_2O)_4$/guilleminite	$6^1 5^4 4^2 3^2$	$Pmm2_1$	70	4.386	307.050	$p2_1mn$	38	3.511	133.421				
5	$Na(H_3O)[(UO_2)_3(SeO_3)_2O_2]$ $(H_2O)_4$/larisaite		$P11m$	73	5.395	393.857	pm	38	4.511	171.421				
28	$Sr[(UO_2)_3(SeO_3)_2O_2](H_2O)_4$		$C2/m$	28	3.450	96.606	$c2/m$	19	3.090	58.711	$c2/m$	19	3.090	58.711
29	$Li_2[(UO_2)_3(SeO_3)_2O_2](H_2O)_6$		$P2_1/c$	78	4.311	336.261	$p2_1/a$	38	3.301	125.421				

Table 3. Cont.

No.	Formula	Topology	Complexity Parameters of the Crystal Structure				Structural Complexity of the U-Se Unit				Topological Complexity of the U-Se Unit			
			Sp. Gr.	v	I_G	$I_{G,total}$	Layer or Rod Sym. Gr.	v	I_G	$I_{G,total}$	Layer or Rod Sym. Gr.	v	I_G	$I_{G,total}$
30	$Cs_2[(UO_2)_4(SeO_3)_5](H_2O)_2$	$6^15^34^63^5$	$P2_1/n$	160	5.322	851.508	pn	64	5.000	320.000	$p2_1mn$	64	4.250	272.000
46	$[C_8H_{15}N_2]_2[(UO_2)_4(SeO_3)_5]$		$Pnma$	328	5.455	1789.277	$p2_1mn$	64	4.250	272.000				
31	$Cs_2[(UO_2)_7(SeO_4)_2(SeO_3)_2(OH)_4O_2](H_2O)_5$	$6^15^44^63^6$	$P2_1/m$	132	5.226	689.860	$p-1$	49	4.635	227.121	$p-1$	49	4.635	227.121
32	$UO_2Se_2O_5$	$8^15^23^8$	$P-1$	40	4.322	172.877	$p1$	20	4.322	86.439	$p2$	20	3.422	68.439

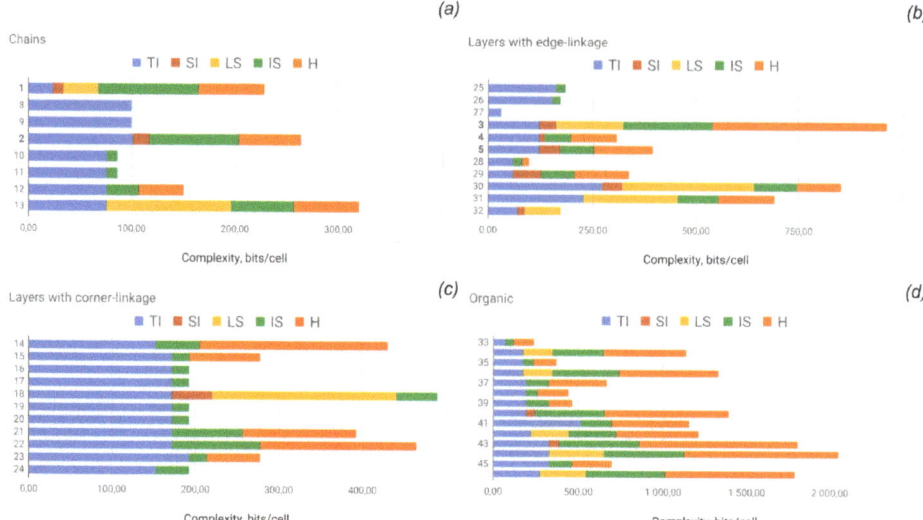

Figure 5. Ladder diagrams showing contributions of various factors to structural complexity in terms of bits per unit cell for the structures based on chains (**a**), layers with edge-linkage of polyhedra (**b**), layers with corner-linkage of polyhedra (**c**) and organically templated compounds (**d**). Legend: TI = topological information; CI = cluster information (valid for Prw and Nsb: See text for details); SI = structural information; LS = layer stacking; IS = interstitial structure; HB = hydrogen bonding. See Table 3 and text for details.

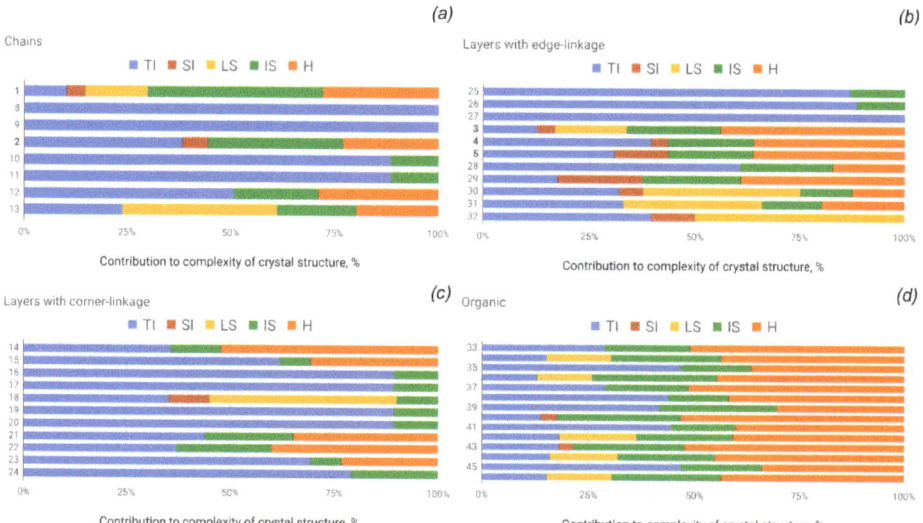

Figure 6. Ladder diagrams showing normalized contributions (in %) of various factors to structural complexity for the structures based on chains (**a**), layers with edge-linkage of polyhedra (**b**), layers with corner-linkage of polyhedra (**c**) and organically templated compounds (**d**). Legend: see Figure 5. See Table 3 and text for details.

4. Discussion

Structural features of natural uranyl selenites make one think about the conditions of their formation in nature. Analogies with synthetic compounds, which have a similar structure, allow some of the most probable pathways to be suggested. The formation of structural units with edge-sharing polyhedra in most cases indicates their hydrothermal origin, and the synthetic uranyl selenites **28** and **29**, whose structures are built upon the layers with a phosphuranylite topology (Figure 3g), are no exception. Both compounds were obtained from the aqueous medium at temperatures above 220 °C. In the case of compounds with structures based upon 1D units, the situation is somewhat more complicated. Topological type cc1–1:2–1, which is one of the most common among the U(VI)-bearing oxysalts, was repeatedly observed in the structures of compounds obtained at room temperature. However, synthetic uranyl selenite **9** was grown at slightly higher temperatures of 80 °C. Moreover, the presence of rather specific uranyl tetragonal bipyramids in the structure of derriksite refers to a family of isotypic uranyl phosphate [88], molybdate [89], and tellurite [90] compounds, which were obtained during hydrothermal (above 180 °C) or high temperature solid state (above 650 °C) syntheses. Analogously, the crystal structures of synthetic uranyl chromates [67,68] and molybdates [69], which are isotypic to that one of demesmaekerite, were obtained at hydrothermal conditions (above 120 °C) or solid state reactions (at 300 °C). Nevertheless, based on laboratory [91,92] and field observations, namely of the mineral association from Zálesí (Czech Republic) [8], it is clear that demesmaekerite and piretite (and several other unnamed or poorly identified U-Se phases) formed as a result of supergene alteration processes, which exclude hydrothermal activity. These observations are supported by the radioanalytical dating of demesmaekerite.

The crystal structure of derriksite is built on the 1D uranyl selenite complexes, whose symmetry is described by the $t_c m11$ rod symmetry group. However, its highest (topological) symmetry is described by the centrosymmetric $t_a 2/m11$ rod group (Figure 7a). Stacking of chains doubles the complexity contribution of the uranyl selenite block (68.107 bits/cell) into the whole structure, but is still less than the contribution of the Cu-O interstitial block (96.370 bits/cell) and nearly equal to the contribution of the interstitial H-bonding system (64.287 bits/cell; Figure 5 and 6). Alteration of uranyl tetragonal bipyramids by pentagonal ones with the additional H$_2$O molecule in the equatorial plane of Ur preserves the topology, but it doubles the size of the reduced segment of a chain and changes its maximal symmetry to the $t_a 2/m11$ rod group (Figure 7b). The absence of the interstitial substructure makes the topological complexity parameters be equal to those for the whole structure of **8** and **9**.

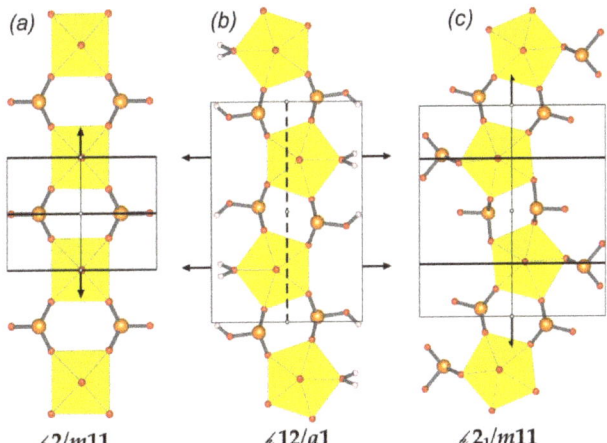

Figure 7. 1D uranyl selenite units and their highest rod symmetry groups for derriksite (**a**), [(UO$_2$)(HSeO$_3$)$_2$(H$_2$O)] (**b**) and demesmaekerite (**c**). Legend: see Figure 1.

The topological symmetry of the uranyl selenite chain in the structure of demesmaekerite is monoclinic $t_a2_1/m11$ and is higher than its real triclinic t-1 symmetry (Figure 7c). In this case, the uranyl selenite substructure (117.207 bits/cell) makes the largest contribution to the complexity of the whole structure. The interstitial complex contributes a slightly lower amount of information (85.926 bits/cell), and even less is accounted for in the H-bonding system (60.842 bits/cell; Figure 5 and 6).

The crystal structures of 17 uranyl selenites and selenite-selenates are built upon the layers of cc1–1:2–4 topological type and those are distributed almost equally between pure inorganic and organically templated compounds having various monovalent inorganic ions and protonated amine molecules of different shapes and sizes as an interstitial block. Moreover, this topology preserves changes in the chemical composition of uranyl-bearing layers, which involves the occurrence of uranyl selenites, selenite-hydrogen selenites, and hydrogen selenite-selenates. It is of interest that all isomers within this family of compounds, including chemical substitutions and two geometrical isomers (see Chapter 3.3.), have the highest symmetry of the layer described by the $p2_1/b$ layer group (Figure 8). Furthermore, the topological symmetry is preserved in the structures of almost all compounds, except for two of them (Table 3). All three aforementioned cases point to the fact that the current topological type is unusually resistant and one of the most preferable in the systems with the U:T ratio = 1:2. As for the complexity calculations, certainly, those will primarily depend on the number of orbits (atoms). Thus, the H-free uranyl selenite layer has the lowest amount of information (152.196 bits/cell), next in a row would be the uranyl selenite-hydrogen selenite complex (172.080 bits/cell), and finally those containing selenate oxyanions (192.423 bits/cell). Analogously, complexity parameters for the whole structure majorly depend on the size of the aliphatic part of organic molecules.

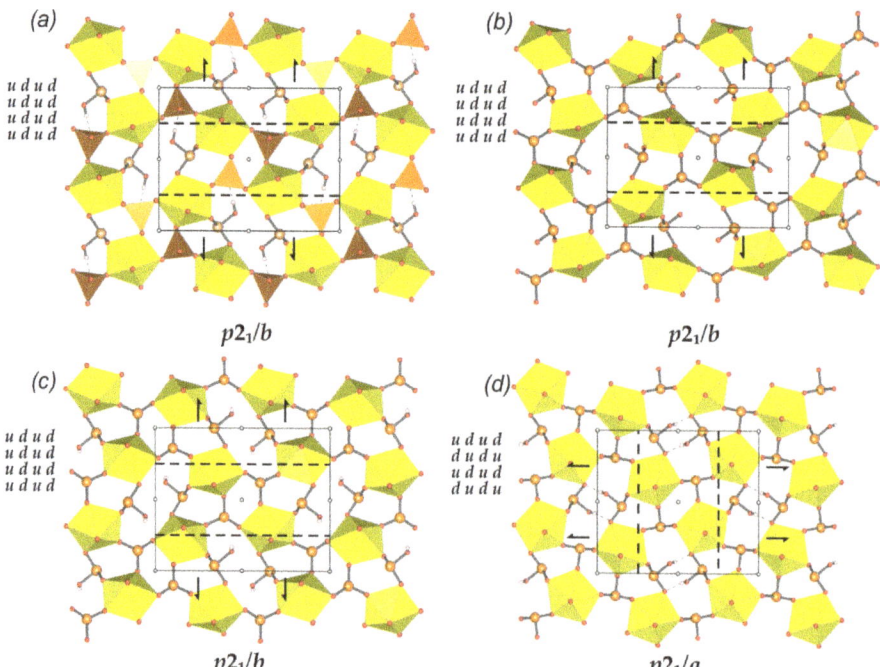

Figure 8. (**a**–**d**) Uranyl selenite layers, symmetry elements, and the respective layer symmetry groups for various isomers of the cc1–1:2–4 topological type (**a**–**d**: see text for details). Legend: see Figures 1 and 2.

The crystal structures of three uranyl selenite minerals and two synthetic compounds are based on dense layers with a phosphuranylite anion topology. It is of interest that natural and synthetic compounds are described by the different orientation matrices (Figure 4). The (**ud**)(**du**) orientation of the lone electron pairs in the structures of synthetic uranyl selenites **28** and **29** resulted in the formation of layers with the $c2/m$ topological symmetry (58.711 bits/cell), whereas the highest symmetry of those in natural compounds is described by the (**ud**)(**ud**) matrix and orthorhombic $pmmn$ layer symmetry group (121.421 bits/cell). It should be noted that only Sr-bearing synthetic compound **28** has the real symmetry of the layer equal to the topological one. In the cases of marthozite, guilleminite, larisaite, and Li-bearing synthetic compound, topological symmetry is significantly reduced by the interstitial cations and H_2O molecules (Table 3, Figures 4 and 5). The contribution of each of the components makes the marthozite the most complex inorganic uranyl selenite (960.860 bits/cell; Figures 5 and 6). It is of interest that the formation of a particular isomer causes the specific arrangement of the layers, and it appears that the (**ud**)(**ud**) isomer of the posphuranylite topology, which results in the formation of zig-zag layers, is more stable and most likely thermodynamically preferable among the others, since it has only been observed in the structures of natural layered uranyl selenites.

5. Conclusions

The refinement of the demesmaekerite crystal structure makes it possible to determine the H atoms positions belonging to the interstitial H_2O molecules. The refinement of the guilleminite crystal structure allows the determination of one additional site arranged within the void of the interlayer space and occupied by the H_2O molecule, which suggests the formula of guilleminite to be written as $Ba[(UO_2)_3(SeO_3)_2O_2](H_2O)_4$ instead of $Ba[(UO_2)_3(SeO_3)_2O_2](H_2O)_3$.

Our numerous attempts to determine the crystal structure of haynesite, $[(UO_2)_3(SeO_3)_2(OH)_2](H_2O)_5$, were unsuccessful. However, several assumptions could be made. First, the presence of $(OH)^-$ groups within such a dense layer is doubtful. At least, there is no other evidence for O atoms' protonation in the structures of all the other natural and synthetic uranyl selenites, whose structures are built upon layers of the same topology. Moreover, the structure of haynesite has to be electroneutral, and H_2O molecules from the interlayer space should be replaced by H_3O^+ cations, which points to the similarity and, probably, to the common genesis or even the closest relationships between the haynesite and larisaite, $Na(H_3O)[(UO_2)_3(SeO_3)_2O_2](H_2O)_4$.

Comparison of the isotypic natural and synthetic uranyl-bearing compounds suggests that uranyl selenite mineral formation requires heating, which most likely, keeping in mind their surface or near-surface occurrence conditions, can be attributed to the radioactive decay.

Structural complexity studies revealed an interesting tendency in that the majority of synthetic compounds have the topological symmetry of uranyl selenite building blocks equal to the structural symmetry, which means that the highest symmetry of uranyl complexes is preserved regardless of the interstitial filling of the structures. Whereas the real symmetry of chained and layered complexes in the structures of uranyl selenite minerals is lower than their topological symmetry, which means that interstitial cations and H_2O molecules significantly affect the structural architecture of natural compounds. At the same time, structural complexity parameters for the whole structure are usually higher for the minerals than that for synthetic compounds of a similar or close organization, which probably indicates the preferred existence of such natural-born architectures.

Supplementary Materials: The following are available online at http://www.mdpi.com/2073-4352/9/12/639/s1: Cif files for guilleminite and demesmaekerite.

Author Contributions: Conceptualization, V.V.G. and J.P.; Methodology, V.V.G., V.M.K. and I.V.K.; Investigation, V.V.G., I.V.K., V.M.K., M.N.M., A.V.K. and J.P.; Writing-Original Draft Preparation, V.V.G., I.V.K., V.M.K., M.N.M., A.V.K. and J.P.; Writing-Review & Editing, V.V.G. and J.P.; Visualization, V.V.G. and I.V.K.

Funding: This research was funded by the Russian Science Foundation (grant 18-17-00018 to V.V.G. and I.V.K.) and through the project of the Ministry of Education, Youth and Sports National sustainability program I of the Czech Republic (project No. LO1603 to J.P.).

Acknowledgments: The XRD measurements of guilleminite have been performed at the X-ray Diffraction Centre of the St. Petersburg State University. We are grateful to reviewers for useful comments.

Conflicts of Interest: The authors declare no conflict of interest.

References

1. Pierrot, R.; Toussaint, J.; Verbeek, T. La guilleminite, une nouvelle espèce minérale. *B. Soc. Fr. Minéral. Cr.* **1965**, *88*, 132–135. [CrossRef]
2. Cesbron, F.; Bachet, B.; Oosterbosch, R. La demesmaekerite, sélénite hydraté d'uranium, cuivre et plomb. *Bulletin B. Soc. Fr. Minéral. Cr.* **1965**, *88*, 422–425. [CrossRef]
3. Cesbron, F.; Oosterbosch, R.; Pierrot, R. Une nouvelle espèce minérale: La marthozite. Uranyl-sélénite de cuivre hydraté. *B. Soc. Fr. Minéral. Cr.* **1969**, *92*, 278–283. [CrossRef]
4. Cesbron, F.; Pierrot, R.; Verbeek, T. La derriksite, $Cu_4(UO_2)(SeO_3)_2(OH)_6 \cdot H_2O$, une nouvelle espèce minérale. *B. Soc. Fr. Minéral. Cr.* **1971**, *94*, 534–537.
5. Deliens, M.; Piret, P. La haynesite, sélénite hydraté d'uranyle, nouvelle espèce minérale de la Mine Repete, Comté de San Juan, Utah. *Can. Mineral.* **1991**, *29*, 561–564.
6. Vochten, R.; Blaton, N.; Peeters, O.; Deliens, M. Piretite, $Ca(UO_2)_3(SeO_3)_2(OH)_4 \cdot 4H_2O$, a new calcium uranyl selenite from Shinkolobwe, Shaba, Zaire. *Can. Mineral.* **1996**, *34*, 1317–1322.
7. Chukanov, N.V.; Pushcharovsky, D.Y.; Pasero, M.; Merlino, S.; Barinova, A.V.; Möckel, S.; Pekov, I.V.; Zadov, A.E.; Dubinchuk, V.T. Larisaite, $Na(H_3O)(UO_2)_3(SeO_3)_2O_2 \cdot 4H_2O$, a new uranyl selenite mineral from Repete mine, San Juan County, Utah, U.S.A. *Eur. J. Mineral.* **2004**, *16*, 367–374. [CrossRef]
8. Sejkora, J.; Škoda, R.; Pauliš, P. Selenium mineralization of the uranium deposit Zálesí, Rychlebské Hory Mts., Czech Republic. *Mineral. Pol.* **2006**, *28*, 196–198.
9. Gelfort, E. Nutzung der spaltprodtikte nach aufarbeitung ausgedienter brennelemente. *Atomwirtsch. Atomtech.* **1985**, *30*, 32–36.
10. Chen, F.; Burns, P.C.; Ewing, R.C. ^{79}Se: Geochemical and crystallo-chemical retardation mechanisms. *J. Nucl. Mater.* **1999**, *275*, 81–88. [CrossRef]
11. *CrysAlisPro Software System*; Version 1.171.38.46; Rigaku Oxford Diffraction: Oxford, UK, 2015.
12. Sheldrick, G.M. SHELXT—Integrated space-group and crystal structure determination. *Acta Crystallogr.* **2015**, *A71*, 3–8. [CrossRef] [PubMed]
13. Sheldrick, G.M. Crystal structure refinement with SHELXL. *Acta Crystallogr.* **2015**, *C71*, 3–8.
14. Dolomanov, O.V.; Bourhis, L.J.; Gildea, R.J.; Howard, J.A.K.; Puschmann, H. OLEX2: A complete structure solution, refinement and analysis program. *J. Appl. Cryst.* **2009**, *42*, 339–341. [CrossRef]
15. Petříček, V.; Dušek, M.; Palatinus, L. Crystallographic computing system JANA2006: General features. *Z. Kristallogr.* **2014**, *229*, 345–352. [CrossRef]
16. Krivovichev, S.V. Combinatorial topology of salts of inorganic oxoacids: Zero-, one- and two-dimensional units with corner-sharing between coordination polyhedra. *Crystallogr. Rev.* **2004**, *10*, 185–232. [CrossRef]
17. Burns, P.C.; Miller, M.L.; Ewing, R.C. U^{6+} minerals and inorganic phases: A comparison and hierarchy of structures. *Can. Mineral.* **1996**, *34*, 845–880.
18. Krivovichev, S.V. *Structural Crystallography of Inorganic Oxysalts*; Oxford University Press: Oxford, UK, 2008; 303p.
19. Krivovichev, S.V. Topological complexity of crystal structures: Quantitative approach. *Acta Crystallogr.* **2012**, *A68*, 393–398. [CrossRef]
20. Krivovichev, S.V. Structural complexity of minerals: Information storage and processing in the mineral world. *Mineral. Mag.* **2013**, *77*, 275–326. [CrossRef]
21. Krivovichev, S.V. Which inorganic structures are the most complex? *Angew. Chem. Int. Ed.* **2014**, *53*, 654–661. [CrossRef]
22. Krivovichev, S.V. Structural complexity of minerals and mineral parageneses: Information and its evolution in the mineral world. In *Highlights in Mineralogical Crystallography*; Danisi, R., Armbruster, T., Eds.; Walter de Gruyter GmbH: Berlin, Germany; Boston, MA, USA, 2015; pp. 31–73.
23. Krivovichev, S.V. Structural complexity and configurational entropy of crystalline solids. *Acta Crystallogr.* **2016**, *B72*, 274–276.
24. Krivovichev, S.V. Ladders of information: What contributes to the structural complexity in inorganic crystals. *Z. Kristallogr.* **2018**, *233*, 155–161. [CrossRef]

25. Gurzhiy, V.V.; Plasil, J. Structural complexity of natural uranyl sulfates. *Acta Crystallogr.* **2019**, *B75*, 39–48. [CrossRef]
26. Krivovichev, V.G.; Krivovichev, S.V.; Charykova, M.V. Selenium minerals: Structural and chemical diversity and complexity. *Minerals* **2019**, *9*, 455. [CrossRef]
27. Cooper, M.A.; Hawthorne, F.C. The crystal structure of guilleminite, a hydrated Ba–U–Se sheet structure. *Can. Mineral.* **1995**, *33*, 1103–1109.
28. Ginderow, D.; Cesbron, F. Structure de la demesmaekerite, $Pb_2Cu_5(SeO_3)_6(UO_2)_2(OH)_6 \cdot 2H_2O$. *Acta Crystallogr.* **1983**, *C39*, 824–827.
29. Cooper, M.A.; Hawthorne, F.C. Structure topology and hydrogen bonding in marthozite, $Cu^{2+}[(UO_2)_3(SeO_3)_2O_2](H_2O)_8$, a comparison with guilleminite, $Ba[(UO_2)_3(SeO_3)_2O_2](H_2O)_3$. *Can. Mineral.* **2001**, *39*, 797–807. [CrossRef]
30. Ginderow, D.; Cesbron, F. Structure da la derriksite, $Cu_4(UO_2)(SeO_3)_2(OH)_6$. *Acta Crystallogr.* **1983**, *C39*, 1605–1607.
31. Cejka, J.; Sejkora, J.; Deliens, M. To the infrared spectrum of haynesite, a hydrated uranyl selenite, and its comparison with other uranyl selenites. *Neues Jahbuch Mineral. Monatschefte* **1999**, *6*, 241–252.
32. Frost, R.L.; Weier, M.L.; Reddy, B.J.; Cejka, J. A Raman spectroscopic study of the uranyl selenite mineral haysenite. *J. Raman Spectrosc.* **2006**, *37*, 816–821. [CrossRef]
33. Loopstra, B.O.; Brandenburg, N.P. Uranyl selenite and uranyl tellurite. *Acta Crystallogr.* **1978**, *B34*, 1335–1337. [CrossRef]
34. Diefenbach, K.; Lin, J.; Cross, J.N.; Dalal, N.S.; Shatruk, M.; Albrecht-Schmitt, T.E. Expansion of the rich structures and magnetic properties of neptunium selenites: Soft ferromagnetism in $Np(SeO_3)_2$. *Inorg. Chem.* **2014**, *53*, 7154–7159. [CrossRef]
35. Mistryukov, V.E.; Mikhailov, Y.N. Structural features of the selenite group in uranyl complexes with neutral ligands. *Koordinats. Khim.* **1983**, *9*, 97–102.
36. Koskenlinna, M.; Mutikainen, I.; Leskela, T.; Leskela, M. Low-temperature crystal structures and thermal decomposition of uranyl hydrogen selenite monohydrate, $[(UO_2)(HSeO_3)_2](H_2O)$ and diammonium uranyl selenite hemihydrate, $[NH_4]_2[(UO_2)(SeO_3)_2](H_2O)_{0.5}$. *Acta Chem. Scand.* **1997**, *51*, 264–269. [CrossRef]
37. Almond, P.M.; Peper, S.; Bakker, E.; Albrecht-Schmitt, T.E. Variable dimensionality and new uranium oxide topologies in the alkaline-earth metal uranyl selenites $AE[(UO_2)(SeO_3)_2]$ (AE = Ca, Ba) and $Sr[(UO_2)(SeO_3)_2] \cdot 2H_2O$. *J. Solid State Chem.* **2002**, *168*, 358–366. [CrossRef]
38. Almond, P.M.; Albrecht-Schmitt, T.E. Hydrothermal synthesis and crystal chemistry of the new strontium uranyl selenites, $Sr[(UO_2)_3(SeO_3)_2O_2]\cdot 4H_2O$ and $Sr[UO_2(SeO_3)_2]$. *Am. Mineral.* **2004**, *89*, 976–980. [CrossRef]
39. Serezhkina, L.B.; Vologzhanina, A.V.; Marukhnov, A.V.; Pushkin, D.V.; Serezhkin, V.N. Synthesis and crystal structure of $Na_3(H_3O)[UO_2(SeO_3)_2]_2 \cdot H_2O$. *Crystallogr. Rep.* **2009**, *54*, 852–857. [CrossRef]
40. Koskenlinna, M.; Valkonen, J. Ammonium uranyl hydrogenselenite selenite. *Acta Crystallogr.* **1996**, *52*, 1857–1859. [CrossRef]
41. Almond, P.; Albrecht-Schmitt, T.E. Hydrothermal syntheses, structures, and properties of the new uranyl selenites $Ag_2(UO_2)(SeO_3)_2$, $M[(UO_2)(HSeO_3)(SeO_3)]$ (M = K, Rb, Cs, Tl), and $Pb(UO_2)(SeO_3)_2$. *Inorg. Chem.* **2002**, *41*, 1177–1183. [CrossRef]
42. Meredith, N.A.; Polinski, M.J.; Lin, J.; Simonetti, A.; Albrecht-Schmitt, T.E. Incorporation of Neptunium(VI) into a uranyl selenite. *Inorg. Chem.* **2012**, *51*, 10480–10482. [CrossRef]
43. Burns, W.L.; Ibers, J.A. Syntheses and structures of three f-element selenite/hydroselenite compounds. *J. Solid State Chem.* **2009**, *182*, 1457–1461. [CrossRef]
44. Marukhnov, A.V.; Pushkin, D.V.; Peresypkina, E.V.; Virovets, A.V.; Serezhkina, L.B. Synthesis and structure of $Na[(UO_2)(SeO_3)(HSeO_3)](H_2O)_4$. *Rus. J. Inorg. Chem.* **2008**, *53*, 831–836. [CrossRef]
45. Krivovichev, S.V. Crystal chemistry of selenates with mineral-like structures: VII. The structure of $(H_3O)[(UO_2)(SeO_4)(SeO_2OH)]$ and some structural features of selenite-selenates. *Geol. Ore Depos.* **2009**, *51*, 663–667. [CrossRef]
46. Wylie, E.M.; Burns, P.C. Crystal structures of six new uranyl selenate and selenite compounds and their relationship with uranyl mineral structures. *Can. Mineral.* **2012**, *50*, 147–157. [CrossRef]
47. Trombe, J.C.; Gleizes, A.; Galy, J. Structure of a uranyl diselenite, $UO_2Se_2O_5$. *Acta Crystallogr.* **1985**, *C41*, 1571–1573. [CrossRef]

48. Liu, D.-S.; Huang, G.-S.; Luo, Q.-Y.; Xu, Y.-P.; Li, X.-F. Poly[tetramethylammonium [nitratouranyl-μ_3-selenito]]. *Acta Crystallogr.* **2006**, *E62*, 1584–1585.
49. Almond, P.M.; Albrecht-Schmitt, T.E. Do secondary and tertiary ammonium cations act as structure-directing agents in the formation of layered uranyl selenites? *Inorg. Chem.* **2003**, *42*, 5693–5698. [CrossRef]
50. Liu, D.S.; Kuang, H.M.; Chen, W.T.; Luo, Q.Y.; Sui, Y. Synthesis, structure, and photoluminescence properties of an organically-templated uranyl selenite. *Z. Anorg. Allg. Chem.* **2015**, *641*, 2009–2013. [CrossRef]
51. Krivovichev, S.V.; Tananaev, I.G.; Kahlenberg, V.; Myasoedov, B.F. Synthesis and crystal structure of the first uranyl Selenite(IV)-Selenate(VI) [$C_5H_{14}N$][$(UO_2)(SeO_4)(SeO_2OH)$]. *Dokl. Phys. Chem.* **2005**, *403*, 124–127. [CrossRef]
52. Gurzhiy, V.V.; Krivovichev, S.V.; Tananaev, I.G. Dehydration-driven evolution of topological complexity in ethylammonium uranyl selenates. *J. Solid State Chem.* **2017**, *247*, 105–112. [CrossRef]
53. Jouffret, L.J.; Wylie, E.M.; Burns, P.C. Influence of the organic species and Oxoanion in the synthesis of two uranyl sulfate hydrates, $(H_3O)_2[(UO_2)_2(SO_4)_3(H_2O)]\cdot 7H_2O$ and $(H_3O)_2[(UO_2)_2(SO_4)_3(H_2O)]\cdot 4H_2O$, and a uranyl Selenate-Selenite [C_5H_6N][$(UO_2)(SeO_4)(HSeO_3)$]. *Z. Anorg. Allg. Chem.* **2012**, *638*, 1796–1803. [CrossRef]
54. Gurzhiy, V.V.; Krivovichev, S.V.; Burns, P.C.; Tananaev, I.G.; Myasoedov, B.F. Supramolecular templates for the synthesis of new nanostructured uranyl compounds: Crystal structure of [$NH_3(CH_2)_9NH_3$][$(UO_2)(SeO_4)(SeO_2OH)$](NO_3). *Radiochemistry* **2010**, *52*, 1–6. [CrossRef]
55. Kovrugin, V.M.; Gurzhiy, V.V.; Krivovichev, S.V.; Tananaev, I.G.; Myasoedov, B.F. Unprecedented layer topology in the crystal structure of a new organically templated uranyl selenite-selenate. *Mendeleev Commun.* **2012**, *22*, 11–12. [CrossRef]
56. Gurzhiy, V.V.; Kovrugin, V.M.; Tyumentseva, O.S.; Mikhailenko, P.A.; Krivovichev, S.V.; Tananaev, I.G. Topologically and geometrically flexible structural units in seven new organically templated uranyl selenates and selenite–selenates. *J. Solid State Chem.* **2015**, *229*, 32–40. [CrossRef]
57. Krivovichev, S.V.; Tananaev, I.G.; Kahlenberg, V.; Myasoedov, B.F. Synthesis and crystal structure of a new uranyl selenite(IV)-selenate(VI), [$C_5H_{14}N$]$_4$[$(UO_2)_3(SeO_4)_4(HSeO_3)(H_2O)$]($H_2SeO_3$)($HSeO_4$). *Radiochemistry* **2006**, *48*, 217–222. [CrossRef]
58. Wylie, E.M.; Smith, P.A.; Peruski, K.M.; Smith, J.S.; Dustin, M.K.; Burns, P.C. Effects of ionic liquid media on the cation selectivity of uranyl structural units in five new compounds produced using the ionothermal technique. *CrystEngComm* **2014**, *16*, 7236–7243. [CrossRef]
59. Gurzhiy, V.V.; Tyumentseva, O.S.; Tyshchenko, D.V.; Krivovichev, S.V.; Tananaev, I.G. Crown-ether-templated uranyl selenates: Novel family of mixed organic-inorganic actinide compounds. *Mendeleev Commun.* **2016**, *26*, 309–311. [CrossRef]
60. Gurzhiy, V.V.; Tyumentseva, O.S.; Britvin, S.N.; Krivovichev, S.V.; Tananaev, I.G. Ring opening of azetidine cycle: First examples of 1-azetidinepropanamine molecules as a template in hybrid organic-inorganic compounds. *J. Mol. Struct.* **2018**, *1151*, 88–96. [CrossRef]
61. Kovrugin, V.M.; Colmont, M.; Siidra, O.I.; Gurzhiy, V.V.; Krivovichev, S.V.; Mentre, O. Pathways for synthesis of new selenium-containing oxo-compounds: Chemical vapor transport reactions, hydrothermal techniques and evaporation method. *J. Cryst. Growth* **2017**, *457*, 307–313. [CrossRef]
62. Kovrugin, V.M.; Colmont, M.; Terryn, C.; Colis, S.; Siidra, O.I.; Krivovichev, S.V.; Mentre, O. pH controlled pathway and systematic hydrothermal phase diagram for elaboration of synthetic lead nickel selenites. *Inorg. Chem.* **2015**, *54*, 2425–2434. [CrossRef]
63. Hawthorne, F.C.; Ferguson, R.B. Refinement of the crystal structure of kroehnkite. *Acta Crystallogr.* **1975**, *B31*, 1753–1755. [CrossRef]
64. Plášil, J.; Hloušek, J.; Kasatkin, A.V.; Novak, M.; Cejka, J.; Lapcak, L. Svornostite, $K_2Mg[(UO_2)(SO_4)_2]_2\cdot 8H_2O$, a new uranyl sulfate mineral from Jáchymov, Czech Republic. *J. Geosci.* **2015**, *60*, 113–121. [CrossRef]
65. Kampf, A.R.; Sejkora, J.; Witzke, T.; Plášil, J.; Čejka, J.; Nash, B.P.; Marty, J. Rietveldite, $Fe(UO_2)_2(SO_4)_2(H_2O)_5$, a new uranyl sulfate mineral from Giveaway-Simplot mine (Utah, USA), Willi Agatz mine (Saxony, Germany) and Jáchymov (Czech Republic). *J. Geosci.* **2017**, *62*, 107–120. [CrossRef]
66. Gurzhiy, V.V.; Tyumentseva, O.S.; Izatulina, A.R.; Krivovichev, S.V.; Tananaev, I.G. Chemically induced polytypic phase transitions in the $Mg[(UO_2)(TO_4)_2(H_2O)](H_2O)_4$ (T = S, Se) system. *Inorg. Chem.* **2019**, *58*, 14760–14768. [CrossRef] [PubMed]

67. Krivovichev, S.V.; Burns, P.C. First sodium uranyl chromate, $Na_4[(UO_2)(CrO_4)_3]$: Synthesis and crystal structure determination. *Z. Anorg. Allg. Chem.* **2003**, *629*, 1965–1968. [CrossRef]
68. Krivovichev, S.V.; Burns, P.C. Crystal chemistry of K uranyl chromates: Crystal structures of $K_8[(UO_2)(CrO_4)_4](NO_3)_2$, $K_5[(UO_2)(CrO_4)_3](NO_3)(H_2O)_2$, $K_4[(UO_2)_3(CrO_4)_5](H_2O)_8$ and $K_2[(UO_2)_2(CrO_4)_3(H_2O)_2](H_2O)_4$. *Z. Kristallogr.* **2003**, *218*, 725–732.
69. Krivovichev, S.V.; Burns, P.C. Crystal chemistry of uranyl molybdates. VIII. Crystal structures of $Na_3Tl_3[(UO_2)(MoO_4)_4]$, $Na_{13}Tl_3[(UO_2)(MoO_4)_3]_4(H_2O)_5$, $Na_3Tl_5[(UO_2)(MoO_4)_3]_2(H_2O)_3$ and $Na_2[(UO_2)(MoO_4)_2](H_2O)_4$. *Can. Mineral.* **2003**, *41*, 707–720. [CrossRef]
70. Grigor'ev, M.S.; Fedoseev, A.M.; Budantseva, N.A.; Yanovskii, A.I.; Struchkov, Y.T.; Krot, N.N. Synthesis, crystal and molecular structure of complex neptunium(V) sulfates $(Co(NH_3)_6)(NpO_2(SO_4)_2) \cdot 2H_2O$ and $(Co(NH_3)_6) H_8O_3(NpO_2(SO_4)_3)$. *Sov. Radiokhem.* **1999**, *33*, 54–60.
71. Norquist, A.J.; Doran, M.B.; Thomas, P.M.; O'Hare, D. Structural diversity in organically templated sulfates. *Dalton Trans.* **2003**, 1168–1175. [CrossRef]
72. Forbes, T.Z.; Burns, P.C. Structures and syntheses of four Np^{5+} sulfate chain structures: Divergence from U^{6+} crystal chemistry. *J. Solid State Chem.* **2005**, *178*, 3455–3462. [CrossRef]
73. Gurzhiy, V.V.; Tyumentseva, O.S.; Krivovichev, S.V.; Tananaev, I.G. Novel type of molecular connectivity in one-dimensional uranyl compounds: $[K@(18-crown-6)(H_2O)][(UO_2)(SeO_4)(NO_3)]$, a new potassium uranyl selenate with 18-crown-6 ether. *Inorg. Chem. Commun.* **2014**, *45*, 93–96. [CrossRef]
74. Burns, P.C. A new uranyl phosphate chain in the structure of parsonsite. *Am. Mineral.* **2000**, *85*, 801–805. [CrossRef]
75. Locock, A.J.; Burns, P.C.; Flynn, T.M. The role of water in the structures of synthetic hallimondite, $Pb_2[(UO_2)(AsO_4)]_2(H_2O)_n$ and synthetic parsonsite, $Pb_2[(UO_2)(PO_4)]_2(H_2O)_n$, $0 \leq n \leq 0.5$. *Am. Mineral.* **2005**, *90*, 240–246. [CrossRef]
76. Mills, S.J.; Birch, W.D.; Kolitsch, U.; Mumme, W.G.; Grey, I.E. Lakebogaite, $CaNaFe_2^{3+}H(UO_2)_2(PO_4)_4(OH)(H_2O)_8$, a new uranyl phosphate with a unique crystal structure from Victoria, Australia. *Am. Mineral.* **2008**, *93*, 691–697. [CrossRef]
77. Krivovichev, S.V.; Finch, R.; Burns, P.C. Crystal chemistry of uranyl molybdates. V. Topologically different uranyl molybdate sheets in structures of $Na_2[(UO_2)(MoO_4)_2]$ and $K_2[(UO_2)(MoO_4)_2](H_2O)$. *Can. Mineral.* **2002**, *40*, 193–200. [CrossRef]
78. Grigor'ev, M.S.; Charushnikova, I.A.; Fedoseev, A.M.; Budantseva, N.A.; Yanovskii, A.I.; Struchkov, Y.T. Crystal and molecular structure of neptunium(V) complex molybdate $K_3NpO_2(MoO_4)_2$. *Sov. Radiokhem.* **1992**, *34*, 7–12.
79. Krivovichev, S.V.; Kahlenberg, V. Structural diversity of sheets in Rb uranyl selenates: Synthesis and crystal structures of $Rb_2[(UO_2)_2(SeO_4)_2(H_2O)](H_2O)$, $Rb_2[(UO_2)_2(SeO_4)_3(H_2O)_2](H_2O)_4$, $Rb_4[(UO_2)_3(SeO_4)_5(H_2O)]$. *Z. Anorg. Allg. Chem.* **2005**, *631*, 739–744. [CrossRef]
80. Lussier, A.J.; Lopez, R.A.K.; Burns, P.C. A revised and expanded structure hierarchy of natural and synthetic hexavalent uranium compounds. *Can. Mineral.* **2016**, *54*, 177–283. [CrossRef]
81. Christ, C.L.; Clark, J.R.; Evans, H.T., Jr. Crystal structure of rutherfordine, UO_2CO_3. *Science* **1955**, *121*, 472–473. [CrossRef]
82. Finch, R.J.; Cooper, M.A.; Hawthorne, F.C.; Ewing, R.C. Refinement of the crystal structure of rutherfordine. *Can. Mineral.* **1999**, *37*, 929–938.
83. Demartin, F.; Diella, V.; Donzelli, S.; Gramaccioli, C.M.; Pilati, T. The importance of accurate crystal structure determination of uranium minerals. I. Phosphuranylite $KCa(H_3O)_3(UO_2)_7(PO_4)_4O_4 \cdot 8H_2O$. *Acta Crystallogr.* **1991**, *B47*, 439–446. [CrossRef]
84. Shvareva, T.Y.; Albrecht-Schmitt, T.E. General route to three-dimensional framework uranyl transition metal phosphates with atypical structural motifs: The case examples of $Cs_2\{(UO_2)_4[Co(H_2O)_2]_2(HPO_4)(PO_4)_4\}$ and $Cs_{3+x}[(UO_2)_3CuH_{4-x}(PO_4)_5] \cdot H_2O$. *Inorg. Chem.* **2006**, *45*, 1900–1902. [CrossRef] [PubMed]
85. Plášil, J.; Hauser, J.; Petříček, V.; Meisser, N.; Mills, S.J.; Škoda, R.; Fejfarová, K.; Čejka, J.; Sejkora, J.; Hloušek, J.; et al. Crystal structure and formula revision of deliensite, $Fe[(UO_2)_2(SO_4)_2(OH)_2](H_2O)_7$. *Mineral. Mag.* **2012**, *76*, 2837–2860. [CrossRef]
86. Kampf, A.R.; Kasatkin, A.V.; Čejka, J.; Marty, J. Plášilite, $Na(UO_2)(SO_4)(OH) \cdot 2H_2O$, a new uranyl sulfate mineral from the Blue Lizard mine, San Juan County, Utah, USA. *J. Geosci.* **2015**, *60*, 1–10. [CrossRef]

87. Blatov, V.A.; Shevchenko, A.P.; Proserpio, D.M. Applied topological analysis of crystal structures with the program package ToposPro. *Cryst. Growth Des.* **2014**, *14*, 3576–3586. [CrossRef]
88. Guesdon, A.; Chardon, J.; Provost, J.; Raveau, B. A copper uranyl monophosphate built up from CuO_2 infinity chains: $Cu_2UO_2(PO_4)_2$. *J. Solid State Chem.* **2002**, *165*, 89–93. [CrossRef]
89. Krivovichev, S.V.; Burns, P.C. Synthesis and crystal structure of $Li_2[(UO_2)(MoO_4)_2]$, a uranyl molybdate with chains of corner-sharing uranyl square bipyramids and MoO4 tetrahedra. *Solid State Sci.* **2003**, *5*, 481–485. [CrossRef]
90. Almond, P.M.; Albrecht-Schmitt, T.E. Expanding the remarkable structural diversity of uranyl tellurites: Hydrothermal preparation and structures of $KUO_2Te_2O_5(OH)$, $Tl_3\{(UO_2)_2Te_2O_5(OH)(Te_2O_6)\}\cdot 2H_2O$, $\beta\text{-}Tl_2(UO_2(TeO_3))_2$, and $Sr_3((UO_2)(TeO_3))_2(TeO_3)_2$. *Inorg. Chem.* **2002**, *41*, 5495–5501. [CrossRef]
91. Charykova, M.V.; Krivovichev, V.G. Mineral systems and the thermodynamics of selenites and selenates in the oxidation zone of sulfide ores—A review. *Mineral. Petrol.* **2017**, *111*, 121–134. [CrossRef]
92. Krivovichev, V.G.; Charykova, M.V.; Vishnevsky, A.V. The thermodynamics of selenium minerals in near-surface environments. *Minerals* **2017**, *7*, 188. [CrossRef]

© 2019 by the authors. Licensee MDPI, Basel, Switzerland. This article is an open access article distributed under the terms and conditions of the Creative Commons Attribution (CC BY) license (http://creativecommons.org/licenses/by/4.0/).

Article

Crystallographic Insights into Uranyl Sulfate Minerals Formation: Synthesis and Crystal Structures of Three Novel Cesium Uranyl Sulfates

Olga S. Tyumentseva [1], Ilya V. Kornyakov [1,2], Sergey N. Britvin [1,2], Andrey A. Zolotarev [1] and Vladislav V. Gurzhiy [1,*]

[1] Department of Crystallography, Institute of Earth Sciences, St. Petersburg State University, University Emb. 7/9, St. 199034 Petersburg, Russian; o-tyumentseva@mail.ru (O.S.T.); ikornyakov@mail.ru (I.V.K.); sbritvin@gmail.com (S.N.B.); a.zolotarev@spbu.ru or aazolotarev@gmail.com (A.A.Z.)
[2] Kola Science Centre, Russian Academy of Sciences, Fersmana 14, 184209 Apatity, Russia
* Correspondence: vladislav.gurzhiy@spbu.ru or vladgeo17@mail.ru

Received: 16 November 2019; Accepted: 5 December 2019; Published: 9 December 2019

Abstract: An alteration of the uranyl oxide hydroxy-hydrate mineral schoepite [$(UO_2)_8O_2(OH)_{12}$] $(H_2O)_{12}$ at mild hydrothermal conditions was studied. As the result, four different crystalline phases $Cs[(UO_2)(SO_4)(OH)](H_2O)_{0.25}$ (**1**), $Cs_3[(UO_2)_4(SO_4)_2O_3(OH)](H_2O)_3$ (**2**), $Cs_6[(UO_2)_2(SO_4)_5](H_2O)_3$ (**3**), and $Cs_2[(UO_2)(SO_4)_2]$ (**4**) were obtained, including three novel compounds. The obtained Cs uranyl sulfate compounds **1**, **3**, and **4** were analyzed using single-crystal XRD, EDX, as well as topological analysis and information-based structural complexity measures. The crystal structure of **3** was based on the 1D complex, the topology of which was unprecedented for the structural chemistry of inorganic oxysalts. Crystal chemical analysis performed herein suggested that the majority of the uranyl sulfates minerals were grown from heated solutions, and the temperature range could be assumed from the manner of interpolyhedral linkage. The presence of edge-sharing uranyl bipyramids most likely pointed to the temperatures of higher than 100 °C. The linkage of sulfate tetrahedra with uranyl polyhedra through the common edges involved elevated temperatures but of lower values (~70–100 °C). Complexity parameters of the synthetic compounds were generally lower than that of uranyl sulfate minerals, whose structures were based on the complexes with the same or genetically similar topologies. The topological complexity of the uranyl sulfate structural units contributed the major portion to the overall complexity of the synthesized compounds, while the complexity of the respective minerals was largely governed by the interstitial structure and H-bonding system.

Keywords: uranyl; hydroxy-hydrate; sulfate; cesium; schoepite; crystal structure; topology; structural complexity

1. Introduction

Uranyl-oxide hydroxy-hydrate minerals are regarded to be the products of the first stages of uraninite alteration under oxidizing conditions [1–5]. Being formed as the result of the reaction of bedrock with aqueous fluids, these natural compounds obviously play an important, if not a key, role in the uranium transfer to the environment. In addition, uranyl-oxide hydroxy-hydrate phases can be regarded as the precursors of the formation of other secondary uranium-bearing minerals, under reaction with waters enriched by various cations (usually mono- and divalent) and oxyanions (CO_3^{2-}, SO_4^{2-}, PO_4^{2-}, etc.). The description of the new mineral species is rarely followed by the experiments, which could shed some light on the conditions of their genesis. Due to the complexity of uranyl-bearing complexes forming in aqueous solutions in the presence of a certain cation/oxyanion combination as a function of pH, the exact formation mechanisms of uranyl minerals are not fully understood. In this study, we presented the results of uranyl sulfate synthesis experiments that might elucidate some of the formation behavior of natural uranyl sulfates.

Herein, we reported on the alteration experiment of the synthetic analog of uranyl-oxide hydroxy-hydrate mineral schoepite, $[(UO_2)_8O_2(OH)_{12}](H_2O)_{12}$ [6,7]. As the result, four different crystalline phases $Cs[(UO_2)(SO_4)(OH)](H_2O)_{0.25}$ (**1**), $Cs_3[(UO_2)_4(SO_4)_2O_3(OH)](H_2O)_3$ (**2**) [8], $Cs_6[(UO_2)_2(SO_4)_5]$$(H_2O)_3$ (**3**), and $Cs_2[(UO_2)(SO_4)_2]$ (**4**) were obtained, including three novel compounds. The obtained Cs uranyl sulfate compounds were analyzed using single-crystal X-ray diffraction (SC XRD), energy-dispersive X-ray analysis (EDX), as well as topological analysis and information-based structural complexity measures.

2. Materials and Methods

2.1. Synthesis

An analog of the hydroxy-hydrated uranyl oxide mineral schoepite ($[(UO_2)_8O_2(OH)_{12}](H_2O)_{12}$ [6,7]) was synthesized according to the procedure discussed in [9], and its purity was checked using powder XRD; H_2SO_4 (Sigma-Aldrich, 98%) and Cs_2SO_4 (Vekton, Russia, 99%) were used as received. The 0.2 g of synthetic schoepite (0.03 mmol), 0.12 g of cesium sulfate (0.33 mmol), and 0.01 (0.19 mmol) ml of sulfuric acid were dissolved in 10 mL of deionized water. The solution was stirred and loaded in a 23 mL Teflon-lined steel autoclave, which was placed in a box furnace and heated to 110 °C. After 24 h, the furnace was cooled at ~20 °C/h to room temperature. A fine crystalline precipitate covered by a translucent light-yellow solution was found at the bottom of the Teflon capsule. Afterward, the product was poured into a watch glass. The detailed examination under the optical microscope revealed the presence of two types of crystals (Figure 1): The bulk of the precipitant was tiny yellowish isometric crystals with nearly diamond luster (compound **1**), interspersed with larger in size, but of much worse quality orange rhombus lamina crystals (compound **2**). After picking several crystals for further diagnostics, the product was left in a watch glass to evaporate at room temperature. In a few days, two more kinds of crystals were detected in a small amount at the edge of the solution (Figure 1): green rosettes of thin plates (compound **3**) and light green flattened rhombic crystals (compound **4**).

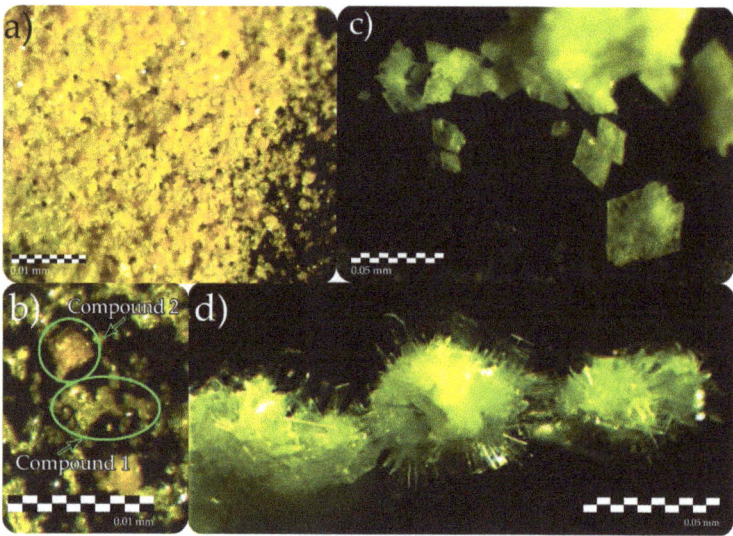

Figure 1. Crystals of **1** (a,b), **2** (a,b), **3** (c) and **4** (d) formed in the described synthetic experiment.

2.2. Chemical Analysis

Small pieces of single crystals of **1–4** verified on the diffractometer were crushed, pelletized, and carbon-coated. The chemical composition of the samples was determined using a TM 3000 scanning electron microscope equipped with an Oxford EDX spectrometer, with an acquisition time of 15 s per point in energy-dispersive mode (acceleration voltage 15 kV). The following analytical standards were used: CsBr (CsK), barite (SK), and U_3O_8 (UK).

Analytical calculations. Compound **1**: atomic ratio from structural data Cs 1.00, U 1.00, S 1.00; found by EDX: Cs 1.03, U 1.05, S 0.92. Compound **2**: atomic ratio from structural data Cs 3.00, U 4.00, S 2.00; found by EDX: Cs 2.93, U 4.06, S 2.01. Compound **3**: atomic ratio from structural data Cs 6.00, U 2.00, S 5.00; found by EDX: Cs 5.96, U 2.07, S 4.97. Compound **4**: atomic ratio from structural data Cs 2.00, U 1.00, S 2.00; found by EDX: Cs 1.95, U 1.05, S 2.00.

2.3. Single-Crystal X-Ray Diffraction Study

Single crystals of **1–4** were selected under binocular, coated in oil-based cryoprotectant, and mounted on cryoloops. Diffraction data were collected using a Bruker Kappa Duo diffractometer (Bruker AXS, Madison, WI, USA) equipped with a CCD (charge-coupled device) Apex II detector operated with monochromated microfocused MoKα radiation (λ[MoKα] = 0.71073 Å) at 45 kV and 0.6 mA. Diffraction data were collected at room temperature with frame widths of 0.5° in ω and φ, and exposures of 40 to 120 s per frame. Diffraction data were integrated and corrected for polarization, background, and Lorentz effects. An empirical absorption correction based on spherical harmonics implemented in the SCALE3 ABSPACK algorithm was applied in the CrysAlisPro program [10]. The unit-cell parameters (Table 1) were refined using the least-squares techniques. The structure was solved by a dual-space algorithm and refined using the SHELX programs [11,12] incorporated in the OLEX2 program package [13]. The final models include coordinates and anisotropic displacement parameters for all non-H atoms. The H atoms of OH groups and H_2O molecules were localized from difference Fourier maps and were included in the refinement with U_{iso}(H) set to 1.5U_{eq}(O) and O–H bond-length restraints to 0.95 Å. Selected bond lengths and angles are listed in Tables 2–4. Checking of the unit-cell parameters along with the results of the chemical analyses showed that the crystals of **2** were the cesium uranyl sulfate phase, which was reported previously [8]. Due to the small size and low quality of the crystals of **3**, refinement parameters were rather high, but the structural model was quite reliable.

Table 1. Crystallographic data for **1, 3**, and **4**.

Compound	1	3	4
Crystal System	Orthorhombic	Triclinic	Monoclinic
a (Å)	9.2021(3)	7.5829(3)	10.8351(5)
b (Å)	13.2434(5)	14.4441(11)	9.0317(5)
c (Å)	12.5610(3)	14.6458(14)	11.8494(6)
α (°)	90	93.737(7)	90
β (°)	90	99.535(5)	110.7510(10)
γ (°)	90	99.614(5)	90
V (Å^3)	1530.77(8)	1552.6(2)	1084.35(10)
Molecular weight	520.51	1871.87	727.97
Space group	$Pnma$	P–1	$P2_1/n$
μ (mm^{-1})	26.156	17.779	22.004
Temperature (K)		293(2)	
Z	8	2	4
D_{calc} (g/cm^3)	4.517	4.004	4.459
Crystal size (mm^3)	0.06 × 0.04 × 0.02	0.04 × 0.03 × 0.005	0.08 × 0.05 × 0.02
Diffractometer		Bruker Kappa Apex II Duo	
Radiation		MoKα	

Table 1. Cont.

Compound	1	3	4
Total reflections	13617	8814	27773
Unique reflections	1824	5151	2493
Angle range 2θ (°)	4.47–55.00	3.84–50.00	4.38–55.00
Reflections with $|F_o| \geq 4\sigma_F$	1560	3806	2308
R_{int}	0.0503	0.0784	0.0358
R_σ	0.0201	0.0949	0.0154
R_1 ($|F_o| \geq 4\sigma_F$)	0.0261	0.0829	0.0131
wR_2 ($|F_o| \geq 4\sigma_F$)	0.0607	0.2050	0.0278
R_1 (all data)	0.0336	0.1077	0.0157
wR_2 (all data)	0.0648	0.2215	0.0287
S	1.054	1.068	1.040
$\rho_{min}, \rho_{max}, e/\text{Å}^3$	−1.097, 2.145	−4.179, 3.819	−0.604, 0.692
CSD	1965819	1965817	1965818

$R_1 = \Sigma||F_o| - |F_c||/\Sigma|F_o|$; $wR_2 = \{\Sigma[w(F_o^2 - F_c^2)^2]/\Sigma[w(F_o^2)^2]\}^{1/2}$; $w = 1/[\sigma^2(F_o^2) + (aP)^2 + bP]$, where $P = (F_o^2 + 2F_c^2)/3$; $s = \{\Sigma[w(F_o^2 - F_c^2)]/(n - p)\}^{1/2}$ where n is the number of reflections, and p is the number of refinement parameters.

Table 2. Selected bond lengths (Å) and angles (°) in the structure of 1.

Bond		Bond	
U1–O1	1.768(5)	S1–O6	1.481(5)
U1–O2	1.768(5)	S1–O7	1.426(6)
< U1–O$_{Ur}$ >	1.768	< S1–O >	1.465
U1–O3	2.351(4)		
U1–OH4	2.330(5)	< Cs1–O > CN * = 8	3.313
U1–OH4	2.322(5)	< Cs2–O > CN * = 10	3.302
U1–O5	2.421(5)		
U1–O6	2.358(5)	Angle	
< U1–O$_{eq}$ >	2.356	U1–O4–U1	113.0(2)
		S1–O5–U1	140.3(3)
S1–O3	1.478(4)	S1–O6–U1	140.3(3)
S1–O5	1.473(5)		

*-Coordination numbers (CN) at the 3.6 Å limit, for the average bond length value.

Table 3. Selected bond lengths (Å) and angles (°) in the structure of 3.

Bond		Bond	
U1–O1	1.770(19)	S3–O14	1.50(2)
U1–O2	1.80(2)	< S3–O >	1.47
< U1–O$_{Ur}$ >	1.785		
U1–O3	2.336(19)	S4–O17	1.51(2)
U1–O4	2.40(2)	S4–O18	1.48(2)
U1–O5	2.428(17)	S4–O19	1.43(2)
U1–O6	2.347(19)	S4–O20	1.42(3)
U1–O7	2.35(2)	< S4–O >	1.46
< U1–O$_{eq}$ >	2.37		
		S5–O21	1.48(3)
U2–O15	1.76(2)	S5–O22	1.43(3)
U2–O16	1.77(2)	S5–O23	1.46(3)
< U2–O$_{Ur}$ >	1.765	S5–O24	1.43(3)
U2–O14	2.349(19)	< S5–O >	1.45
U2–O17	2.42(2)		
U2–O21	2.35(3)	< Cs1–O > CN * = 12	3.26
U2–O22	2.38(3)	< Cs2–O > CN * = 10	3.32

Table 3. Cont.

Bond		Bond	
U2–O22A	2.26(10)	< Cs3–O > CN * = 9	3.33
< U2–O$_{eq}$ >	2.35	< Cs4–O > CN * = 10	3.33
		< Cs5–O > CN * = 9	3.27
S1–O3	1.492(19)	< Cs6–O > CN * = 10	3.28
S1–O7	1.50(2)		
S1–O8	1.42(2)	Angle	
S1–O9	1.43(3)	U1–O3–S1	138.4(13)
< S1–O >	1.46	U1–O4–S2	100.9(9)
		U1–O5–S2	99.6(9)
S2–O4	1.505(19)	U1–O6–S3	141.8(12)
S2–O5	1.510(19)	U1–O7–S1	140.0(14)
S2–O10	1.42(2)	U2–O14–S3	135.7(12)
S2–O11	1.42(2)	U2–O17–S4	100.5(10)
< S2–O >	1.46	U2–O18–S4	101.3(11)
		U2–O21–S5	146.0(16)
S3–O6	1.497(19)	U2–O22–S5	146(2)
S3–O12	1.494(19)	U2–O22A–S5	157(7)
S3–O13	1.39(2)		

*-Coordination numbers (CN) at the 3.6 Å limit, for the average bond length value.

Table 4. Selected bond lengths (Å) and angles (°) in the structure of **4**.

Bond		Bond	
U1–O1	1.775(2)	S2–O7	1.477(2)
U1–O2	1.768(2)	S2–O8	1.430(3)
< U1–O$_{Ur}$ >	1.772	S2–O9	1.486(2)
U1–O4	2.301(2)	S2–O10	1.492(2)
U1–O5	2.319(2)	< S2–O >	1.471
U1–O7	2.322(2)		
U1–O9	2.478(2)	< Cs1–O > CN * = 10	3.286
U1–O10	2.482(2)	< Cs2–O > CN * = 10	3.311
< U1–O$_{eq}$ >	2.380		
		Angle	
S1–O3	1.446(2)	S1–O4–U1	142.45(14)
S1–O4	1.502(2)	S1–O5–U1	133.36(14)
S1–O5	1.502(2)	S2–O7–U1	148.09(16)
S1–O6	1.440(2)	S2–O9–U1	99.54(11)
< S1–O >	1.473	S2–O10–U1	99.22(11)

*-Coordination numbers (CN) at the 3.6 Å limit, for the average bond length value.

Supplementary crystallographic data (see online Supplementary Materials) for **1**, **3**, and **4** were deposited in the Inorganic Crystal Structure Database and could be obtained from Fachinformationszentrum Karlsruhe via https://www.ccdc.cam.ac.uk/structures/.

3. Results

3.1. Structure Descriptions

The crystal structure of **1** contained one crystallographically nonequivalent U^{6+} atom with two short $U^{6+} \equiv O^{2-}$ bonds (1.768(5) Å), forming nearly linear UO_2^{2+} uranyl ion (Ur), which was coordinated by another five oxygen atoms < U1–O$_{eq}$ > = 2.356 Å that were arranged in the equatorial plane of the UO_7 pentagonal bipyramid. Three of those O$_{eq}$ atoms belonged to sulfate tetrahedra, while the other two *cis*-O atoms made an edge shared with the neighbor pentagonal bipyramid, thus forming a dimer. In addition, both O atoms from the shared edge were protonated to form OH$^-$ groups. There was one crystallographically nonequivalent S^{6+} atom tetrahedrally coordinated by four O^{2-} atoms. U dimers and S-centered tetrahedra shared common vertices to form a layered $[(UO_2)(SO_4)(OH)]^-$ complexes (Figure 2a). Being 3-connected, sulfate tetrahedra had the fourth non-shared vertex oriented either *up* or *down* relative to the plane of the layer, which gave rise to geometric isomerism with various orientations of the sulfate tetrahedra. To distinguish the isomers, their orientation matrices were assigned using symbols **u** (*up*) and **d** (*down*). There were two nonequivalent Cs^+ atoms in the structure of **1**, occupying special positions on a mirror plane and arranged in the interlayer space along with substantially vacant and disordered over two sites H_2O molecule.

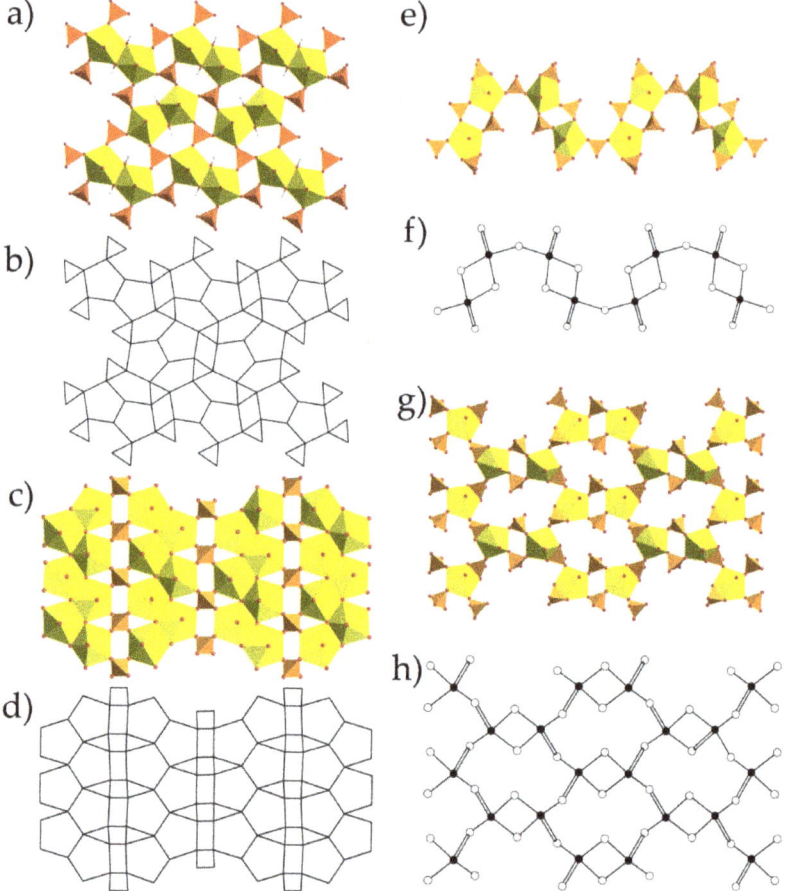

Figure 2. 1D and 2D complexes and their respective topologies in the structures of **1** (a,b), **2** (c,d), **3** (e,f), and **4** (g,h). Legend: U-bearing coordination polyhedra = yellow; S-centered tetrahedra = orange; O atoms = red; H atoms = light grey; black nodes = U atoms, white nodes = S atoms.

There were two crystallographically nonequivalent U^{6+} atoms in the structure of **3** with two short $U^{6+} \equiv O^{2-}$ bonds (1.76(2)–1.80(2) Å), forming *Ur*. The *Ur* cations were coordinated by five oxygen atoms < Ur-O_{eq} > = 2.35 and 2.37 Å (for U1 and U2, respectively), which belonged to sulfate tetrahedra that were arranged in the equatorial plane of the UO_7 pentagonal bipyramid. There were five S^{6+} atoms in the structure of **3**, tetrahedrally coordinated by four O^{2-} atoms each. All sulfate tetrahedra were 2-connected, having only two vertices shared with the uranyl bipyramids, while the other two vertices were left non-shared. But if S1-, S3-, and S5-centered tetrahedra shared two of their vertices with two

Ur polyhedra, S2- and S4-centered tetrahedra linked to the single bipyramid via sharing a common edge. U and S coordination polyhedra shared common vertices and edges to form wave-like infinite chains of $[(UO_2)_2(SO_4)_5]^{6-}$ composition (Figure 2e). Besides, the linkage of coordination polyhedra inside the chain occurred only via corner-sharing, while edge-shared sulfate tetrahedra decorated the exterior of the chain. Uranyl sulfate chains were interlinked into the pseudo layered structure via an H-bonding system involving H_2O molecules. The negative charge of the uranyl sulfate complex was compensated by six Cs^+ cations arranged in between the pseudo layers. It should be noted that three sites occupied by H_2O molecules were localized only in the voids arranged in the plane of the chains but not between the pseudo-2D complexes.

There was one crystallographically nonequivalent U^{6+} atom in the structure of **4** with two short bonds ($< U1–O_{Ur} > = 1.772$ Å) forming Ur, which was coordinated by five oxygen atoms $< Ur\text{-}O_{eq} > = 2.38$ Å, belonging to sulfate tetrahedra and arranged in the equatorial plane of the pentagonal bipyramid. Two nonequivalent S^{6+} atoms were tetrahedrally coordinated by four O^{2-} atoms each. There were two types of sulfate oxyanions: $[S1O_4]^{2-}$ were 2-connected, sharing two vertices with two adjacent Ur polyhedra, and $[S2O_4]^{2-}$ were 3-connected (chelating-bridging), sharing vertex with one uranyl bipyramid and an edge with the neighbor one. The structure of **4** was based on the layered complexes of $[(UO_2)(SO_4)_2]^{2-}$ composition (Figure 2g), interlinked by the two non-equivalent Cs^+ cations.

3.2. Topological Analysis

The anion topology of the U-S layer in **1** (Figure 2b) was determined using the approach described in [14]. It belonged to the $5^44^13^2$-I topological type according to the classification suggested in [15], and it has been observed in the structures of several synthetic uranyl sulfates [16,17], chromates [18], and phosphates [19]. Topology consisted of chains of pentagons, half of which were occupied by the Ur edge-sharing dimers, separated by the groups of one square and two triangles. Each triangle corresponded to the 3-connected face of the sulfate tetrahedra, while the squares were vacant. The orientation of non-shared vertex in the structure of **1** alternated by rows (Figure 2a): in the first row, all vertices were oriented *down*, in the second–*up*, then again *down*, etc. Thus, the geometric isomer represented in **1** was described by the (**u**)(**d**) matrix. It should be noted that the layered complex in the structure of **1** was very similar to those found in the structures of uranyl sulfate minerals as deliensite, $Fe[(UO_2)_2(SO_4)_2(OH)_2](H_2O)_7$ [20], plášilite, $Na(UO_2)(SO_4)(OH)\cdot 2H_2O$ [21], and others [22]. But their topologies were significantly distinct due to various arrangements of the uranyl dimers within the layer. In the structure of **1**, dimers were stacked in a ladder fashion, while in the structures of aforementioned minerals, dimers were arranged parallel to each other, thus forming, typical for minerals, so-called phosphuranylite topology [23]. Differences between these topological isomers have been recently described in [17].

The crystal structure of **2** [8] was based on the layered complexes of the $[(UO_2)_4(SO_4)_2O_3(OH)]^{6-}$ composition (Figure 2c,d). Its topology was described by the $5^24^33^2$ ring symbol and was related to one of the most common topological types among the natural uranyl sulfates [22], the so-called zippeite topology. Topology consisted of zig-zag infinite chains of edge-sharing pentagons separated by chains of squares and triangles.

The topology of the uranyl sulfate chain in the structure of **3** (Figure 2f) could be visualized using the theory of graphical representation [24]. Double links between the black and white vertices in a graph indicate the sharing of an edge between uranyl coordination polyhedra and sulfate oxyanion. The chain topology in **3** was unprecedented for the structural chemistry of inorganic oxysalts and belonged to the novel *cc*1-2:5-1 type.

The graph of the uranyl sulfate layered complex in the structure of **4** belonged to the *cc*2–1:2–21 topological type (Figure 2h) and consisted of dense 4-membered and large 12-membered rings. This topology is rather rare but has been observed in the structures of a few actinide-bearing compounds. It has been described at first in the structure of isotypic Cs neptunyl sulfate compound [25], and later in the structures of Ba uranyl selenite [26] and organically templated uranyl sulfate [27].

3.3. Structural Complexity

The information-based complexity parameters for **1–4** are given in Table 5. This approach, recently developed by S.V. Krivovichev [28–30] and successfully implemented in [22,31,32], allowed to compare the structures in terms of their information content and to analyze contributions of various substructural building blocks into the complexity of the whole structure. Structural complexity is a negative contribution to the configurational entropy of a crystalline compound and thus could help to understand and describe the processes of various structures formation in the laboratory and, what is of special interest, in nature.

Table 5. Structural and topological complexity parameters for the uranyl sulfate compounds.

Compound	Complexity Parameters of the Crystal Structure				Structural Complexity of the U-S Unit				Topological Complexity of the U-S Unit			
	Sp. Gr.	v	I_G	$I_{G,total}$	Layer/Rod Gr.	v	I_G	$I_{G,total}$	Layer/Rod Gr.	v	I_G	$I_{G,total}$
1	$Pnma$	100	3.844	384.386	$p2_1/a$	40	3.322	132.877	$p2_1/a$	40	3.322	132.877
2	P-1	78	5.285	412.261	p-1	54	4.755	256.764	p-1	54	4.755	256.764
3	P-1	92	5.524	508.168	$\not{t}_{-1}$	62	4.954	307.160	$\not{t}_a mam$	62	3.599	223.160
4	P-1	141	6.161	868.677	p-1	54	4.755	256.764	p-1	54	4.755	256.764
$K_4[UO_2(SO_4)_3]$	$Pnma$	176	4.641	816.860	pm	36	4.337	156.117	$pmm2$	36	3.892	140.117
Marecottite	$P2_1/n$	60	3.907	234.413	$p2_1/a$	52	3.700	192.423	$p2_1/a$	52	3.700	192.423
Peligotite	P-1	82	5.382	441.319								
Lussierite	Cc	104	5.700	592.846	p-1	23	4.524	104.042	pm	23	3.915	90.042
Klaprothite	$P2_1/c$	164	5.382	882.639								
Ottohahnite	P-1	126	6.000	768.000	p-1	62	4.954	307.160	p-1	62	4.954	307.160

The general trend in evolution of crystallization was recently summarized as follows [33–36]: complexities of structures formed on the latter stages of crystallization are higher than those for the phases growing on the primary stages, wherein very complex structures may form as transitional architectures prior or between phases with relatively small amounts of structural information.

The evaluation was performed in several steps (Figure 3). First, the topological complexity (**TI**), according to the maximal rod (for chains) or layer symmetry group, was calculated since these are the basic structural units. Second, the structural complexity (**SI**) of the units was analyzed, taking into account its real symmetry. The next contribution to information came from the stacking (**LS**) of chained and layered complexes (if more than one layer or chain is in the unit cell). The fourth contribution to the total structural complexity was given by the interstitial structure (**IS**). And the last portion of information came from the interstitial H bonding system (**H**). It should be noted that the H atoms related to the U-bearing chains and layers were considered as a part of those complexes, but not within the contribution of the H-bonding system. For instance, contribution of **H** was equal to the difference between complexity parameters for the whole structure and those for the structural model with no H atoms; contribution of **IS** was equal to the difference between complexity parameters for the structural model with no H atoms and those for the structural model with no interstitial substructure at all; etc. Complexity parameters for the aforementioned contributions were calculated manually using the general formulae [28–30]. Complexity parameters for the whole structures were calculated using the *ToposPro* package [37]. Complexity calculations showed that the crystal structures of **1**, **2**, and **4** should be described as intermediate, possessing the values below 500 bits/cell, while compound **3** just slightly passed through this border (508.168 bits/cell) and should be described as complex.

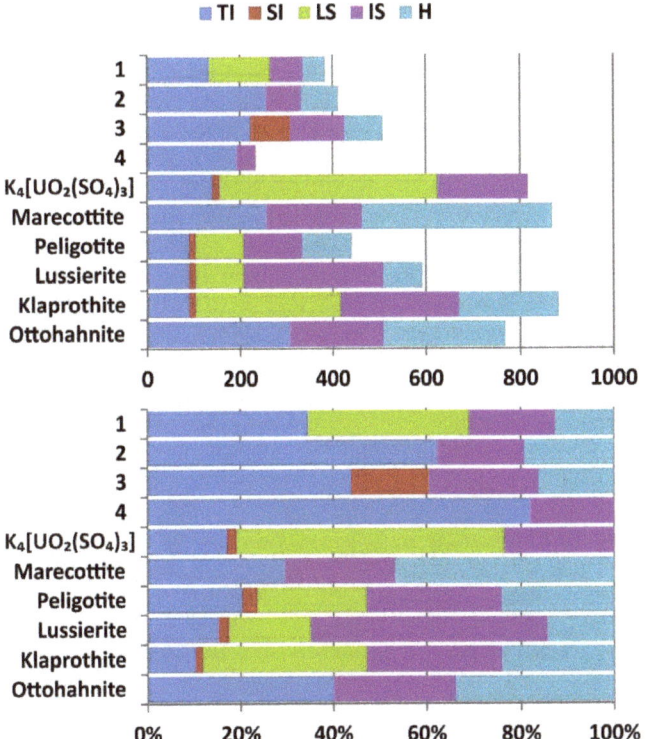

Figure 3. Ladder diagrams showing contributions and normalized contributions (in %) of various factors to structural complexity in terms of bits per unit cell. Legend: TI = topological information; CI = cluster information; SI = structural information; LS = layer stacking; IS = interstitial structure; HB = hydrogen bonding. See Table 5 and text for details.

4. Discussion

The crystals of **1–4** could be distributed over two genetically distinct groups. The crystals of **1** and **2** were formed during the first stage of synthetic schoepite alteration. Their structures were based on the layered complexes with the edge-sharing linkage of uranyl pentagonal bipyramids, which reflected the heating of the reaction solution during the growth processes. It should be noted that original schoepite was obtained from nearly neutral solutions, whereas the aqueous medium in our experiment was significantly more acidic (pH ~ 2). Acidic conditions and the presence of additional Cs^+ cations destroyed the dense layer in the structure of schoepite, but high temperature allowed preserving an edge-sharing complexation of *Ur* coordination polyhedra. An initial solution in the experiment contained the Cs:U:S molar ratio ~ 1:1:1, which explained the predominance of the $Cs[(UO_2)(SO_4)(OH)](H_2O)_{0.25}$ (**1**) phase in the precipitate. The lower amount of crystals of **2** could be explained by the lower temperature of the experiment that is preferable for the formation of the zippeite-type structures.

The crystals of **3** and **4** could be attributed to the later genetic type because they were grown after cooling the system at room temperature conditions. It is of interest that both phases had in their structures uranyl pentagonal bipyramids that shared an edge with the sulfate tetrahedra. Similar arrangement of *Ur* and sulfate oxyanions were found in the structures of four natural uranyl sulfates: klaprothite, $Na_6(UO_2)(SO_4)_4(H_2O)_4$, its polymorph peligotite, $Na_6(UO_2)(SO_4)_4(H_2O)_4$, ottohahnite, $Na_6(UO_2)_2(SO_4)_5(H_2O)_{8.5}$ [38], and lussierite $Na_{10}[(UO_2)(SO_4)_4](SO_4)_2 \cdot 3H_2O$ [39]. The same clusters that were observed in the structures of klaprothite and peligotite (Figure 4) were previously described in a few synthetic compounds [40–43], which were grown using low temperature (70 °C) hydrothermal experiments. The presence of such unusual arrangements of edge-sharing uranyl bipyramid and sulfate tetrahedra have never been observed during regular evaporation experiments at room temperature. Thus, we could assume that these clusters were formed on the first, hydrothermal stage of our experiment. Moreover, the presence of such S-enriched [(UO$_2$)(SO$_4$)$_4$]$^{6-}$ clusters in the heated solution could explain the local disturbance of the Cs:U:S ~ 1:1:1 concentration, which induced the formation of S-"depleted" zippeite-like compound 2. There is one more known synthetic K-bearing uranyl sulfate, whose structure is based on the double klaprothite-type clusters [44]. This row could be continued by the further doubling of the cluster in K$_4$(UO$_2$(SO$_4$)$_3$) [44] to get the quadruple 0D unit in the structure of ottohahnite (Figure 4). Further increase of the cluster size led to the arrangement of the infinite chains in the structure of 3, which in turn, during the dehydration process [34], would transform into the layer in the structure of 4. The absence of the structures based on the 0D structural units in our experiment might come from the requirements for longer storage at elevated, but not high, temperatures, and higher concentrations of Cs$^+$ cations and [SO$_4$]$^{2-}$ oxyanions in the initial solution.

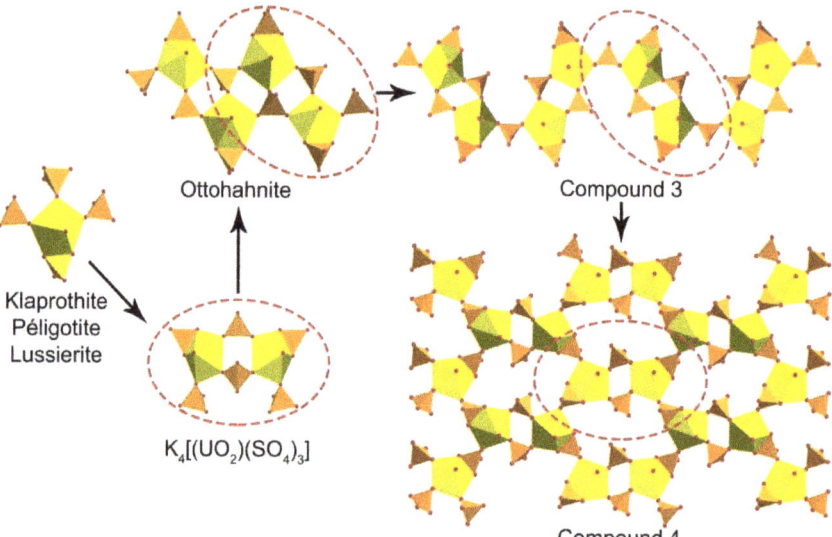

Figure 4. Scheme of structural evolution for the 1D and 2D uranyl sulfate complexes, which have an edge-sharing uranyl bipyramid and sulfate tetrahedron.

5. Conclusions

Summarizing our crystal's chemical observations, we could suggest that the majority of the discovered natural uranyl sulfates were grown from heated solutions, and the temperature range could be assumed from the manner of interpolyhedral linkage. The presence of edge-sharing uranyl bipyramids (phosphuranylite anion topology [22]), most likely pointed to the temperatures of higher than 100 °C, and the crystal growth should apparently occur directly in hydrothermal conditions. The linkage of sulfate tetrahedra with *Ur* through the common edges also involved elevated temperatures, but of less values (~70–100 °C), which could be achieved by cooling the system. Moreover, in the second case, crystallization might start much later at environmental conditions but from the initially heated solutions. The enriched solution might pass some way along the cracks in the bedrock, transferring klaprothite-like clusters in the dissolved form. It is of interest that complexity parameters of the synthetic compounds were generally lower than that of minerals, whose structures were based on the complexes with the same or genetically similar topologies. Furthermore, the topological complexity of the uranyl sulfate structural units contributed the major portion to the overall complexity of the synthesized compounds, while the complexity of the respective minerals was largely governed by the interstitial structure and H-bonding system.

Supplementary Materials: The following are available online at http://www.mdpi.com/2073-4352/9/12/660/s1: Cif files for 1, 3, and 4.

Author Contributions: Conceptualization, O.S.T. and V.V.G.; Methodology, O.S.T. and I.V.K.; Investigation, O.S.T., I.V.K., S.N.B., and A.A.Z.; Writing-Original Draft Preparation, O.S.T. and V.V.G.; Writing-Review and Editing, V.V.G.; Visualization, V.V.G. and I.V.K.

Funding: This research was funded by the Russian Science Foundation (grant 18-17-00018).

Acknowledgments: The XRD and EDX measurements have been performed at the X-ray Diffraction Center and Center for Microscopy and Microanalysis of the St. Petersburg State University.

Conflicts of Interest: The authors declare no conflict of interest.

References

1. Finch, R.J.; Ewing, R.C. The corrosion of uraninite under oxidizing conditions. *J. Nucl. Mater.* **1992**, *190*, 133–156.
2. Finch, R.J.; Murakami, T. Systematics and paragenesis of uranium minerals. In *Uranium: Mineralogy, Geochemistry and the Environment*; Burns, P.C., Ewing, R.C., Eds. *Rev. Mineral. Geochem.* **1999**, *38*, 91–179.
3. Plášil, J. Oxidation-hydration weathering of uraninite: The current state-of-knowledge. *J. Geosci.* **2014**, *59*, 99–114.
4. Krivovichev, S.V.; Plášil, J. Mineralogy and crystallography of uranium. In *Uranium: From Cradle to Grave*; Burns, P.C., Sigmon, G.E., Eds. *MAC Short Courses* **2013**, *43*, 15–119.
5. Plášil, J. Uranyl-oxide hydroxy-hydrate minerals: Their structural complexity and evolution trends. *Eur. J. Mineral.* **2018**, *30*, 237–251.
6. Walker, T.L. Schoepite, a new uranium mineral from Kasolo, Belgian Congo. *Amer. Miner.* **1923**, *8*, 67–69.
7. Plášil, J. The crystal structure of uranyl-oxide mineral schoepite, $[(UO_2)_4O(OH)_6](H_2O)_6$, revisited. *J. Geosci.* **2018**, *63*, 65–73.
8. Serezhkina, L.B.; Grigor'ev, M.S.; Makarov, A.S.; Serezhkin, V.N. Synthesis and Structure of Cesium-Containing Zippeite. *Radiochemistry* **2015**, *57*, 20–25.
9. Nipruk, O.V.; Knyazev, A.V.; Chernorukov, G.N.; Pykhova, Y.P. Synthesis and study of hydrated uranium(VI) oxides, $UO_3 \cdot nH_2O$. *Radiochemistry* **2011**, *53*, 146–150.
10. *CrysAlisPro Software System*, version 1.171.38.46; Rigaku Oxford Diffraction: Oxford, UK, 2015.
11. Sheldrick, G.M. SHELXT—Integrated space-group and crystal structure determination. *Acta Crystallogr.* **2015**, *A71*, 3–8.
12. Sheldrick, G.M. Crystal structure refinement with SHELXL. *Acta Crystallogr.* **2015**, *C71*, 3–8.
13. Dolomanov, O.V.; Bourhis, L.J.; Gildea, R.J.; Howard, J.A.K.; Puschmann, H. OLEX2: A complete structure solution, refinement and analysis program. *J. Appl. Cryst.* **2009**, *42*, 339–341.
14. Burns, P.C.; Miller, M.L.; Ewing, R.C. U^{6+} minerals and inorganic phases: A comparison and hierarchy of structures. *Can. Mineral.* **1996**, *34*, 845–880.
15. Krivovichev, S.V. *Structural Crystallography of Inorganic Oxysalts*; Oxford University Press: Oxford, UK, 2008.
16. Grechishnikova, E.V.; Virovets, A.V.; Peresypkina, E.V.; Serezhkina, L.B. Synthesis and crystal structure of the $(C_2N_4H_7O)[UO_2(SO_4)(OH)] \cdot 0.5H_2O$. *Zh. Neorg. Khim. (Russ.)* **2005**, *50*, 1800–1805.
17. Nazarchuk, E.V.; Charkin, D.O.; Siidra, O.I.; Gurzhiy, V.V. Synthesis and Crystal Structures of New Layered Uranyl Compounds Containing Dimers $[(UO_2)_2O_8]$ of Edge-Linked Pentagonal Bipyramids. *Radiochemistry* **2018**, *60*, 498–506.
18. Unruh, D.K.; Baranay, M.; Pressprich, L.; Stoffer, M.; Burns, P.C. Synthesis and characterization of uranyl chromate sheet compounds containing edge-sharing dimers of uranyl pentagonal bipyramids. *J. Solid State Chem.* **2012**, *186*, 158–164.

19. Ok, K.M.; Baek, J.; Halasyamani, P.S.; O'Hare, D. New Layered Uranium Phosphate Fluorides: Syntheses, Structures, Characterizations, and Ion-Exchange Properties of $A(UO_2)F(HPO_4) \cdot xH_2O$ (A = Cs$^+$, Rb$^+$, K$^+$; x = 0–1). *Inorg. Chem.* **2006**, *45*, 10207–10214.
20. Plášil, J.; Hauser, J.; Petříček, V.; Meisser, N.; Mills, S.J.; Škoda, R.; Fejfarová, K.; Čejka, J.; Sejkora, J.; Hloušek, J.; et al. Crystal structure and formula revision of deliensite, $Fe[(UO_2)_2(SO_4)_2(OH)_2](H_2O)_7$. *Mineral. Mag.* **2012**, *76*, 2837–2860.
21. Kampf, A.R.; Kasatkin, A.V.; Čejka, J.; Marty, J. Plášilite, $Na(UO_2)(SO_4)(OH) \cdot 2H_2O$, a new uranyl sulfate mineral from the Blue Lizard mine, San Juan County, Utah, USA. *J. Geosci.* **2015**, *60*, 1–10.
22. Gurzhiy, V.V.; Plášil, J. Structural complexity of natural uranyl sulfates. *Acta Crystallogr.* **2019**, *B75*, 39–48.
23. Demartin, F.; Diella, V.; Donzelli, S.; Gramaccioli, C.M.; Pilati, T. The importance of accurate crystal structure determination of uranium minerals. I. Phosphuranylite $KCa(H_3O)_3(UO_2)_7(PO_4)_4O_4 \cdot 8H_2O$. *Acta Crystallogr.* **1991**, *B47*, 439–446.
24. Krivovichev, S.V. Combinatorial topology of salts of inorganic oxoacids: Zero-, one- and two-dimensional units with corner-sharing between coordination polyhedra. *Crystallogr. Rev.* **2004**, *10*, 185–232.
25. Fedoseev, A.M.; Budantseva, N.A.; Grigor'ev, M.S.; Bessonov, A.A.; Astafurova, L.N.; Lapitskaya, T.S.; Krupa, J.C. Sulfate Compounds of Hexavalent Neptunium and Plutonium. *Radiochim. Acta* **1999**, *86*, 17–22.
26. Almond, P.M.; Peper, S.; Bakker, E.; Albrecht-Schmitt, T.E. Variable Dimensionality and New Uranium Oxide Topologies in the Alkaline-Earth Metal Uranyl Selenites $AE[(UO_2)(SeO_3)_2]$ (AE=Ca, Ba) and $Sr[(UO_2)(SeO_3)_2] \cdot 2H_2O$. *J. Solid State Chem.* **2002**, *168*, 358–366.
27. Norquist, A.J.; Doran, M.B.; O'Hare, D. The effects of linear diamine chain length in uranium sulfates. *Solid State Sci.* **2003**, *5*, 1149–1158.
28. Krivovichev, S.V. Topological complexity of crystal structures: Quantitative approach. *Acta Crystallogr. A* **2012**, *68*, 393–398.
29. Krivovichev, S.V. Structural complexity and configurational entropy of crystalline solids. *Acta Crystallogr. B* **2016**, *72*, 274–276.
30. Krivovichev, S.V. Ladders of information: What contributes to the structural complexity in inorganic crystals. *Z. Kristallogr.* **2018**, *233*, 155–161.
31. Krivovichev, V.G.; Krivovichev, S.V.; Charykova, M.V. Selenium Minerals: Structural and Chemical Diversity and Complexity. *Minerals* **2019**, *9*, 455.
32. Gurzhiy, V.V.; Kuporev, I.V.; Kovrugin, V.M.; Murashko, M.N.; Kasatkin, A.V.; Plášil, J. Crystal Chemistry and Structural Complexity of Natural and Synthetic Uranyl Selenites. *Crystals* **2019**, *9*, 639.
33. Gurzhiy, V.V.; Tyumentseva, O.S.; Krivovichev, S.V.; Krivovichev, V.G.; Tananaev, I.G. Mixed uranyl sulfate-selenates: variable composition and crystal structures. *Cryst. Growth Des.* **2016**, *16*, 4482–4492.
34. Gurzhiy, V.V.; Krivovichev, S.V.; Tananaev, I.G. Dehydration-driven evolution of topological complexity in ethylamonium uranyl selenates. *J. Solid State Chem.* **2017**, *247*, 105–112.
35. Gurzhiy, V.V.; Tyumentseva, O.S.; Britvin, S.N.; Krivovichev, S.V.; Tananaev, I.G. Ring opening of azetidine cycle: First examples of 1-azetidinepropanamine molecules as a template in hybrid organic-inorganic compounds. *J. Molec. Struct.* **2018**, *1151*, 88–96.
36. Gurzhiy, V.V.; Tyumentseva, O.S.; Izatulina, A.R.; Krivovichev, S.V.; Tananaev, I.G. Chemically Induced Polytypic Phase Transitions in the $Mg[(UO_2)(TO_4)_2(H_2O)](H_2O)_4$ (T = S, Se) System. *Inorg. Chem.* **2019**, *58*, 14760–14768.
37. Blatov, V.A.; Shevchenko, A.P.; Proserpio, D.M. Applied topological analysis of crystal structures with the program package TopoPro. *Cryst. Growth Des.* **2014**, *14*, 3576–3586.
38. Kampf, A.R.; Plášil, J.; Kasatkin, A.V.; Marty, J.; Čejka, J. Klaprothite, péligotite and ottohahnite, three new minerals with bidentate UO_7–SO_4 linkages from the Blue Lizard mine, San Juan County, Utah, USA. *Miner. Mag.* **2017**, *81*, 753–779.
39. Kampf, A.R.; Olds, T.A.; Plášil, J.; Nash, B.P.; Marty, J. Lussierite, a new sodium uranyl sulfate mineral with bidentate UO_7–SO_4 linkage from the Blue Lizard mine, San Juan County, Utah, USA. *Miner. Mag.* **2019**, 1–25. [CrossRef]
40. Plášil, J.; Meisser, N.; Čejka, J. The crystal structure of $Na_6[(UO_2)(SO_4)_4](H_2O)_4$: X-ray and Raman spectroscopy study. *Canad. Miner.* **2015**, *54*, 5–20.
41. Burns, P.C.; Hayden, L.A. A uranyl sulfate cluster in $Na_{10}[(UO_2)(SO_4)_4](SO_4)_2 \cdot 3H_2O$. *Acta Crystallogr. C* **2002**, *58*, i121–i123.

42. Hayden, L.A.; Burns, P.C. The sharing of an edge between a uranyl pentagonal bipyramid and sulfate tetrahedron in the structure of $KNa_5[(UO_2)(SO_4)_4](H_2O)$. *Canad. Miner.* **2002**, *40*, 211–216.
43. Hayden, L.A.; Burns, P.C. A novel uranyl sulfate cluster in the structure of $Na_6(UO_2)(SO_4)_4(H_2O)_2$. *J. Solid State Chem.* **2002**, *163*, 313–318.
44. Mikhailov, Y.N.; Kokh, L.A.; Kuznetsov, V.G.; Grevtseva, T.G.; Sokol, S.K.; Ellert, G.V. Synthesis and crystal structure of potassium trisulfatouranylate $K_4(UO_2(SO_4)_3$. *Koord. Khimiya* **1977**, *3*, 500–513. (In Russian)

© 2019 by the authors. Licensee MDPI, Basel, Switzerland. This article is an open access article distributed under the terms and conditions of the Creative Commons Attribution (CC BY) license (http://creativecommons.org/licenses/by/4.0/).

Article

Krasnoshteinite, Al$_8$[B$_2$O$_4$(OH)$_2$](OH)$_{16}$Cl$_4$·7H$_2$O, a New Microporous Mineral with a Novel Type of Borate Polyanion

Igor V. Pekov [1,*], Natalia V. Zubkova [1], Ilya I. Chaikovskiy [2], Elena P. Chirkova [2], Dmitry I. Belakovskiy [3], Vasiliy O. Yapaskurt [1], Yana V. Bychkova [1], Inna Lykova [4], Sergey N. Britvin [5,6] and Dmitry Yu. Pushcharovsky [1]

[1] Faculty of Geology, Moscow State University, Vorobievy Gory, 119991 Moscow, Russia; n.v.zubkova@gmail.com (N.V.Z.); yvo72@geol.msu.ru (V.O.Y.); yanab66@yandex.ru (Y.V.B.); dmitp@geol.msu.ru (D.Y.P.)
[2] Mining Institute, Ural Branch of the Russian Academy of Sciences, Sibirskaya str., 78a, 614007 Perm, Russia; ilya@mi-perm.ru (I.I.C.); zaitseva59@mail.ru (E.P.C.)
[3] Fersman Mineralogical Museum, Russian Academy of Sciences, Leninsky Prospekt 18-2, 119071 Moscow, Russia; dmzvr@mail.ru
[4] Canadian Museum of Nature, 240 McLeod St, Ottawa, ON K2P 2R1, Canada; ilykova@nature.ca
[5] Department of Crystallography, St. Petersburg State University, Universitetskaya nab. 7/9, 199034 St. Petersburg, Russia; sbritvin@gmail.com
[6] Kola Science Centre, Russian Academy of Sciences, Fersman Street 14, 184209 Apatity, Russia
* Correspondence: igorpekov@mail.ru

Received: 6 April 2020; Accepted: 12 April 2020; Published: 15 April 2020

Abstract: A new mineral, krasnoshteinite (Al$_8$[B$_2$O$_4$(OH)$_2$](OH)$_{16}$Cl$_4$·7H$_2$O), was found in the Verkhnekamskoe potassium salt deposit, Perm Krai, Western Urals, Russia. It occurs as transparent colourless tabular to lamellar crystals embedded up to 0.06 × 0.25 × 0.3 mm in halite-carnallite rock and is associated with dritsite, dolomite, magnesite, quartz, baryte, kaolinite, potassic feldspar, congolite, members of the goyazite–woodhouseite series, fluorite, hematite, and anatase. D_{meas} = 2.11 (1) and D_{calc} = 2.115 g/cm^3. Krasnoshteinite is optically biaxial (+), α = 1.563 (2), β = 1.565 (2), γ = 1.574 (2), and $2V_{meas}$ = 50 (10)°. The chemical composition (wt.%; by combination of electron microprobe and ICP-MS; H$_2$O calculated from structure data) is: B$_2$O$_3$ 8.15, Al$_2$O$_3$ 46.27, SiO$_2$ 0.06, Cl 15.48, H$_2$O$_{calc.}$ 33.74, –O=Cl –3.50, totalling 100.20. The empirical formula calculated based on O + Cl = 33 apfu is (Al$_{7.87}$Si$_{0.01}$)$_{\Sigma 7.88}$[B$_{2.03}$O$_4$(OH)$_2$][(OH)$_{15.74}$(H$_2$O)$_{0.26}$]$_{\Sigma 16}$[(Cl$_{3.79}$(OH)$_{0.21}$]$_{\Sigma 4}$·7H$_2$O. The mineral is monoclinic, P2$_1$, a = 8.73980 (19), b = 14.4129 (3), c = 11.3060 (3) Å, β = 106.665 (2)°, V = 1364.35 (5) Å^3, and Z = 2. The crystal structure of krasnoshteinite (solved using single-crystal data, R_1 = 0.0557) is unique. It is based upon corrugated layers of Al-centered octahedra connected via common vertices. BO$_3$ triangles and BO$_2$(OH)$_2$ tetrahedra share a common vertex, forming insular [B$_2$O$_4$(OH)$_2$]$^{4-}$ groups (this is a novel borate polyanion) which are connected with Al-centered octahedra via common vertices to form the aluminoborate pseudo-framework. The structure is microporous, zeolite-like, with a three-dimensional system of wide channels containing Cl$^-$ anions and weakly bonded H$_2$O molecules. The mineral is named in honour of the Russian mining engineer and scientist Arkadiy Evgenievich Krasnoshtein (1937–2009). The differences in crystal chemistry and properties between high-temperature and low-temperature natural Al borates are discussed.

Keywords: krasnoshteinite; zeolite-like borate; hydrous aluminum chloroborate; new mineral; crystal structure; microporous crystalline material; evaporitic salt rock; Verkhnekamskoe potassium salt deposit; Perm Krai

1. Introduction

Boron is a rare chemical element in nature; its average content in the upper continental crust of the Earth is 0.0011 wt.% [1]. Despite its rarity, boron demonstrates diverse and complicated mineralogy and mineral crystal chemistry. Three hundred minerals with species-defining B are known, including 160 borates and oxoborates [2], and some of these minerals form huge and sometimes extremely rich deposits. Unusual geochemical and mineralogical features of boron are due to its very bright crystal chemical individuality which causes strong ability to separate from other elements in crystal structures and form very specific, unique structural units [3,4]. Unlike boron, aluminum is one of the most abundant elements in the lithosphere, however, natural Al borates are not numerous (only twelve borate and oxoborate minerals with species-defining Al are known: see Discussion) and are classified as rare minerals.

In the present article, we characterize the new mineral species krasnoshteinite (Cyrillic: красноштейнит), a hydrous aluminum chloroborate, and its unusual crystal structure. The mineral is named in honour of the Russian mining engineer and scientist, corresponding member of the Russian Academy of Sciences, Arkadiy Evgenievich Krasnoshtein (1937–2009), an outstanding specialist in the mining of potassium salts who made a great contribution to the exploitation of underground mines at the Verkhnekamskoe deposit. Dr. Krasnoshtein was the founder (1988) and first director of the Mining Institute of the Ural Branch of the Russian Academy of Sciences in Perm. Both the new mineral and its name have been approved by the Commission on New Minerals, Nomenclature and Classification of the International Mineralogical Association, IMA No. 2018-077.

The type specimen of krasnoshteinite was deposited in the systematic collection of the Fersman Mineralogical Museum of the Russian Academy of Sciences (Moscow, Russia), under the catalogue number 96274.

2. Materials and Methods

2.1. Occurrence, General Appearance, Physical Properties and Optical Data

Krasnoshteinite was found in the core of the borehole #2001, with a depth of 247.6–248 m, drilled in the Romanovskiy area (30 km south of the city of Berezniki) of the Verkhnekamskoe potassium salt deposit, Perm Krai, Western Urals, Russia. The general data on this well-known. Huge deposits are given in monographs [5,6]. Krasnoshteinite occurs in halite-carnallite rock and is associated with dritsite ($Li_2Al_4(OH)_{12}Cl_2 \cdot 3H_2O$) [7], dolomite, magnesite, quartz, Sr-bearing baryte, kaolinite, potassic feldspar, congolite, members of the goyazite $SrAl_3(PO_4)(PO_3OH)(OH)_6$–woodhouseite $CaAl_3(PO_4)(SO_4)(OH)_6$ series, fluorite, hematite, and anatase. The new mineral was probably formed as a result of diagenetic or post-diagenetic processes in halite-carnallite evaporitic rock of the Layer E of the Verkhnekamskoe deposit.

Krasnoshteinite occurs as separate tabular to lamellar crystals of up to 0.06 × 0.25 × 0.3 mm (Figure 1a,b) and their parallel intergrowths (Figure 1c) embedded in carnallite and halite. In some cases, tiny crystals of krasnoshteinite overgrow its larger crystal in random orientations to form a crystal cluster (Figure 1d). Samples shown in Figure 1 were separated after dissolution of a host halite-carnallite rock in water.

Crystals of krasnoshteinite are flattened on the *ab* plane. The pedions {010} and {0-10} and the pinacoid {100} are major lateral faces of the tabular crystals. The surface of the most developed "face" of a crystal is typically complicated, rough, and demonstrating coarse or/and fine striation along {100} (Figure 1); it is usually composed by several poorly formed faces belonging to the 0*kl* zone.

Krasnoshteinite is a transparent colorless mineral with a white streak and vitreous luster. It is brittle, with a Mohs hardness is ca. of 3. Krasnoshteinite demonstrates perfect cleavage on {010} and an imperfect cleavage on {100}. A fracture is stepped (observed under the microscope). The mineral is non-fluorescent in the ultraviolet light. The density measured by flotation in heavy liquids (bromoform

+ dimethylformamide) is 2.11 (1) g/cm^3, and the density calculated using the empirical formula and the unit-cell parameters determined from single-crystal X-ray diffraction data is 2.115 g/cm^3.

In plane polarized light, krasnoshteinite is colorless and non-pleochroic. It is optically biaxial (+), α = 1.563 (2), β = 1.565 (2), γ = 1.574 (2) (589 nm). 2V (meas.) = 50 (10)° and 2V (calc.) = 51°. Dispersion of optical axes is distinct, $r > v$. Optical orientation is: $Y = b$, and $X = a$.

Figure 1. Separate crystals (**a**,**b**) and crystal clusters ((**c**,**d**): small crystals overgrow large crystal) of krasnoshteinite. Blocky crystals of dolomite and distorted quartz crystal are observed in (**b**).

2.2. Chemical Composition

The chemical composition of krasnoshteinite was studied using a Jeol JSM-6480LV scanning electron microscope equipped with an INCA-Wave 500 wavelength-dispersive spectrometer (Laboratory of Analytical Techniques of High Spatial Resolution, Dept. of Petrology, Moscow State University). Electron microprobe analyses were obtained in the wavelength-dispersive spectroscopy mode (20 kV and 20 nA; the electron beam was rastered to the 5 × 5 µm area to avoid damage of the highly hydrated mineral) and gave contents of Al, Si, O, and Cl. The standards used were: Al$_2$O$_3$ (Al), wollastonite (Si), YAl$_3$(BO$_3$)$_4$ (O), and NaCl (Cl). The contents of other elements with atomic numbers higher than 8 are below detection limits.

The presence of significant amount of chlorine in krasnoshteinite prevents the quantitative determination of boron by electron microprobe, due to the overlap of X-ray emission lines of the *K* series of B with *L* lines of Cl. The boron content was determined using ICP-MS. The measurements were carried out with the Element-2 (Thermo Fisher Scientific) instrument which has high resolution (that avoids interference of components) and sensitivity. Several crystals of the mineral were dissolved in 10 cm^3 of 3% HNO$_3$ solution (Merck, Suprapur®) in deionized water (EasyPure). Since the mass of the mineral was too small for accurate weighing, we have determined contents of B and Al in relative units and further used averaged Al content, obtained by electron microprobe, for B content calculation. The obtained value is in good agreement with the boron content determined from the crystal structure refinement. Contents of Li and Be in krasnoshteinite are below detection limits.

H$_2$O was not analysed because of the paucity of material. Hydrogen (H$_2$O) content was calculated based on the structure data (see below) and taking into account the charge balance requirement. The analytical total is close to 100 wt.% (Table 1) that demonstrates a good agreement between electron microprobe data for Al, Si, O, and Cl, ICP-MS data for B and calculated value for H. The correctness of the obtained chemical data was also confirmed by the superior value of the Gladstone–Dale compatibility index [8]: 1 − (K$_p$/K$_c$) = 0.003 (superior) with measured density value, or 0.006 (superior) with calculated density value.

CO$_2$ was not analysed because the structure data showed the absence of this constituent. The absence of gas release in HCl (see below) also indicated that krasnoshteinite does not contain carbonate groups.

2.3. Single-Crystal X-ray Diffraction and Crystal Structure Determination

Single-crystal X-ray diffraction data were collected by means of an Xcalibur S CCD diffractometer (Dept. of Crystallography and Crystal Chemistry, Faculty of Geology, Moscow State University) operated at 40 kV and 50 mA using MoKα radiation. A full sphere of three-dimensional data was collected. Data reduction was performed using CrysAlisPro Version 1.171.37.35 [9]. The data were corrected for Lorentz factor and polarization effects. The crystal structure was solved and refined with the ShelX program package using direct methods [10].

2.4. Powder X-ray Diffraction

Powder X-ray diffraction data were collected by means of a Rigaku R-Axis Rapid II diffractometer (XRD Resource Center, St. Petersburg State University) equipped with a rotating anode X-ray source and a curved image plate detector (Debye-Scherrer geometry, d = 127.4 mm, CoKα, λ = 1.79021 Å). The data were integrated using the software package Osc2Tab/SQRay [11]. The unit-cell parameters were refined from the powder data using the Pawley method and Topas software [12].

3. Results

3.1. Chemical Data

Chemical composition of krasnoshteinite is given in Table 1. The empirical formula calculated on the basis of O + Cl = 33 atoms per formula unit is H$_{32.47}$Al$_{7.87}$Si$_{0.01}$B$_{2.03}$Cl$_{3.79}$O$_{29.71}$ or, after recalculation of the anionic part and taking into account crystal-structure data, (Al$_{7.87}$Si$_{0.01}$)$_{\Sigma 7.88}$[B$_{2.03}$O$_4$(OH)$_2$][(OH)$_{15.74}$(H$_2$O)$_{0.26}$]$_{\Sigma 16}$[(Cl$_{3.79}$(OH)$_{0.21}$]$_{\Sigma 4}$·7H$_2$O. The ideal formula is Al$_8$[B$_2$O$_4$(OH)$_2$](OH)$_{16}$Cl$_4$·7H$_2$O, which requires Al 24.65, B 2.47, Cl 16.19, H 3.69, O 53.00, total 100 wt.%, or, in oxides, Al$_2$O$_3$ 46.58, B$_2$O$_3$ 7.95, H$_2$O 32.93, Cl 16.19, –O=Cl –3.65, total 100 wt.%.

Table 1. Chemical composition (in wt.%) of krasnoshteinite in elements and in oxides (Al, Si, Cl, and O: average data for 7 spot electron-microprobe analyses).

Data in Elements, with Measured O Content				Data Recalculated in Oxides	
Constituent	Wt.%	Range	Stand. Dev.	Constituent	Wt.%
B	2.53			B$_2$O$_3$	8.15
Al	24.49	23.79–24.96	0.41	Al$_2$O$_3$	46.27
Si	0.03	0.02–0.05	0.01	SiO$_2$	0.06
Cl	15.48	15.01–16.69	0.59	Cl	15.48
H(calc.)	3.75			H$_2$O (calc.)	33.74
O	53.92	52.43–56.62	1.42	–O=Cl	−3.50
Total	100.20			Total	100.20

Krasnoshteinite is insoluble in water and slowly dissolves in cold diluted HCl without effervescence. The obtained solution shows characteristic color reaction, with quinalizarin clearly indicating boron presence.

3.2. Single-Crystal X-ray Diffraction and Crystal Structure Determination

The single-crystal X-ray diffraction data were indexed in the $P2_1$ space group with the following unit-cell parameters: a = 8.73980 (19), b = 14.4129 (3), c = 11.3060 (3) Å, β = 106.665 (2)°, and V = 1364.35 (5) Å^3 (Table 2). Details on data collection and structure refinement are also given in Table 2. The final structure refinement converged to R_1 = 0.0557 for 6142 unique observed reflections with $I > 2\sigma(I)$. The H atoms of OH groups and H_2O molecules were located from the difference Fourier synthesis. The studied crystal is microtwinned with the inversion center as a twin operation: twinning by merohedry Class I [13] with the twin domain ratio of 68/32. Coordinates and equivalent thermal displacement parameters of atoms are given in Table 3, selected interatomic distances in Table 4, and H-bonding in Table 5. Other crystal structure information for krasnoshteinite has been deposited with the Editors and is available as Supplementary Materials (see below): anisotropic displacement parameters of non-hydrogen atoms in the structure are presented in Table S1 and bond valence calculations in Table S2; crystallographic information file (CIF) is given as a separate Supplementary Material. Bond-valence parameters for Al-O and B-O were taken from [14] and for H-bonding from [15,16].

Table 2. Crystal data, data collection information and structure refinement details for krasnoshteinite.

Formula	$Al_8[B_2O_4(OH)_2](OH)_{16}Cl_4 \cdot 7H_2O$
Formula weight	875.52
Temperature, K	293 (2)
Radiation and wavelength, Å	MoKα; 0.71073
Crystal system, space group, Z	Monoclinic, $P2_1$, 2
Unit cell dimensions, Å/°	a = 8.73980 (19)
	b = 14.4129 (3) β = 106.665 (2)
	c = 11.3060 (3)
V, Å^3	1364.35 (5)
Absorption coefficient μ, mm^{-1}	0.809
F_{000}	892
Crystal size, mm	0.06 × 0.16 × 0.17
Diffractometer	Xcalibur S CCD
θ range for data collection, °/Collection mode	2.81 – 28.28/full sphere
Index ranges	$-11 \leq h \leq 11, -19 \leq k \leq 19, -15 \leq l \leq 15$
Reflections collected	23,807
Independent reflections	6773 (R_{int} = 0.0759)
Independent reflections with $I > 2\sigma(I)$	6142
Data reduction	CrysAlisPro, Agilent Technologies, v. 1.171.37.35 [9]
Absorption correction	multi-scan
	Empirical absorption correction using spherical harmonics, implemented in SCALE3 ABSPACK scaling algorithm
Structure solution	direct methods
Refinement method	full-matrix least-squares on F^2
Number of refined parameters	485
Final R indices [$I > 2\sigma(I)$]	R1 = 0.0557, wR2 = 0.1157
R indices (all data)	R1 = 0.0633, wR2 = 0.1196
GooF	1.107
Largest diff. peak and hole, e/Å^3	0.60 and −0.56

Table 3. Coordinates and equivalent displacement parameters (U_{eq}, in Å^2) of atoms in krasnoshteinite.

Site	X	Y	Z	U_{eq} *
Al1	0.16497(17)	0.29061(10)	0.29761(12)	0.0100(3)
Al2	−0.15081(17)	0.71563(9)	−0.04927(12)	0.0097(3)
Al3	0.11906(16)	0.67894(10)	−0.16964(12)	0.0098(3)
Al4	−0.15717(16)	0.11133(9)	−0.07805(12)	0.0089(3)
Al5	0.50122(18)	0.16162(11)	−0.01146(13)	0.0101(2)
Al6	0.11383(16)	0.13955(10)	−0.19674(12)	0.0103(3)
Al7	0.16655(16)	0.52591(10)	0.31878(12)	0.0094(3)
Al8	0.00144(18)	0.65411(11)	0.51556(12)	0.0112(3)
B1	0.2145(5)	0.4153(4)	0.0964(4)	0.0102(9)
B2	0.7283(6)	−0.0017(4)	0.0864(5)	0.0123(10)
Cl1	0.62010(14)	0.41682(12)	0.28131(13)	0.0305(3)
Cl2	0.24073(14)	0.91650(11)	0.04320(14)	0.0318(3)
Cl3	0.61509(17)	0.82518(11)	0.35183(13)	0.0321(3)
Cl4	0.5240(2)	0.57208(11)	0.64519(13)	0.0371(4)
O1	0.1226(4)	0.1723(2)	0.2021(3)	0.0096(6)
H1	0.182(5)	0.122(2)	0.247(4)	0.012
O2	0.3923(4)	0.2700(3)	0.3509(3)	0.0175(8)
H2A	0.444(5)	0.2119(17)	0.369(5)	0.021
H2B	0.451(5)	0.302(3)	0.305(4)	0.021
O3	0.1966(3)	0.4058(3)	0.3715(3)	0.0145(6)
H3	0.232(5)	0.402(4)	0.4585(10)	0.017
O4	0.0817(4)	0.7234(2)	−0.0141(3)	0.0103(6)
H4	0.128(6)	0.7797(19)	0.022(4)	0.012
O5	0.3314(4)	0.7079(2)	−0.1198(3)	0.0142(7)
H5	0.341(6)	0.7694(14)	−0.144(4)	0.017
O6	−0.1336(4)	0.7504(2)	0.1100(3)	0.0112(6)
H6	−0.149(6)	0.8122(14)	0.133(4)	0.013
O7	0.0760(4)	0.1003(2)	−0.0385(3)	0.0109(7)
H7	0.137(5)	0.047(2)	−0.008(4)	0.013
O8	0.6348(4)	0.1519(3)	−0.1168(3)	0.0143(7)
H8	0.588(6)	0.165(4)	−0.2011(16)	0.017
O9	0.3603(4)	0.1780(2)	0.0857(3)	0.0104(6)
H9	0.396(6)	0.146(3)	0.161(2)	0.012
O10	−0.0590(4)	0.2960(2)	0.2401(3)	0.0119(7)
H10	−0.113(5)	0.3394(7)	0.179(2)	0.014
O11	−0.1825(4)	0.8288(2)	−0.1385(3)	0.0124(7)
O12	0.5632(4)	0.0394(2)	0.0507(3)	0.0130(7)
H12	0.479(4)	−0.003(3)	0.043(5)	0.016
O13	0.8084(4)	−0.0053(2)	−0.1569(3)	0.0092(6)
O14	0.1514(4)	0.2235(2)	0.4336(3)	0.0124(7)

Table 3. Cont.

Site	X	Y	Z	U_{eq} *
H14	0.223(4)	0.246(3)	0.5072(14)	0.015
O15	0.7339(3)	−0.0820(3)	0.0047(3)	0.0143(6)
O16	0.5396(4)	0.7888(2)	0.0675(3)	0.0165(7)
H16A	0.625(4)	0.829(3)	0.065(5)	0.020
H16B	0.478(5)	0.824(3)	0.108(4)	0.020
O17	0.0990(4)	0.6243(2)	0.6804(3)	0.0120(7)
H17	0.144(5)	0.5646(16)	0.6811(15)	0.014
O18	0.7769(4)	−0.0354(3)	0.2134(3)	0.0201(7)
H18	0.706(5)	−0.070(3)	0.245(4)	0.024
O19	−0.1261(4)	0.1448(2)	0.7681(3)	0.0092(6)
H19	−0.172(6)	0.198(2)	0.723(4)	0.011
O20	−0.1619(4)	0.0707(2)	0.0752(3)	0.0088(6)
O21	0.3296(4)	0.1147(2)	−0.1411(3)	0.0121(7)
H21	0.364(6)	0.0579(19)	−0.166(4)	0.015
O22	0.1002(4)	0.1859(2)	0.6494(3)	0.0147(7)
H22	0.180(4)	0.228(3)	0.642(5)	0.018
O23	0.0542(4)	0.0215(2)	0.7411(3)	0.0123(7)
H23	0.106(5)	−0.030(2)	0.717(4)	0.015
O24	0.1359(5)	0.0438(3)	0.4965(3)	0.0214(8)
H24A	0.214(5)	0.022(4)	0.567(2)	0.026
H24B	0.150(6)	0.010(3)	0.429(3)	0.026
O25	−0.1540(4)	0.0819(2)	0.5351(3)	0.0145(7)
H25	−0.248(4)	0.088(4)	0.470(3)	0.017
O26	0.3966(4)	0.5521(2)	0.3674(3)	0.0173(8)
H26A	0.458(5)	0.502(2)	0.352(4)	0.021
H26B	0.444(5)	0.566(3)	0.4523(16)	0.021
O27	−0.1362(4)	0.2617(3)	0.4649(3)	0.0230(8)
H27A	−0.134(6)	0.308(3)	0.408(4)	0.028
H27B	−0.223(4)	0.274(3)	0.496(4)	0.028
O28	0.7250(5)	0.2946(3)	0.6403(4)	0.0354(10)
H28A	0.617(3)	0.289(4)	0.641(6)	0.043
H28B	0.760(6)	0.351(3)	0.684(5)	0.043
O29	0.1131(6)	0.9559(4)	0.2765(5)	0.0571(15)
H29A	0.127(8)	0.952(5)	0.197(3)	0.069
H29B	0.010(5)	0.930(6)	0.269(6)	0.069

* The positions of H atoms were located from the difference Fourier map and refined with O-H and H-H distances softly restrained to 0.95(1) and 1.50(1) Å, respectively, to hold near-optimal geometry. U_{iso} (H) = 1.2 U_{eq} (O).

Table 4. Selected interatomic distances (Å) in the structure of krasnoshteinite.

Al1 - O3	1.843(4)		Al5 - O5	1.882(4)
- O14	1.849(3)		- O9	1.885(3)
- O10	1.879(3)		- O21	1.896(3)
- O2	1.927(4)		- O8	1.897(3)
- O11	1.929(3)		- O12	1.916(4)
- O1	1.996(3)		- O16	1.939(4)
<Al1 – O>	1.904		<Al5 – O>	1.903
Al2 - O6	1.833(3)		Al6 - O22	1.835(3)
- O9	1.839(3)		- O21	1.844(3)
- O11	1.896(3)		- O6	1.857(3)
- O1	1.918(3)		- O23	1.858(4)
- O7	1.951(3)		- O7	1.992(3)
- O4	1.959(3)		- O19	2.021(3)
<Al2 – O>	1.899		<Al6 – O>	1.901
Al3 - O5	1.827(4)		Al7 - O3	1.825(4)
- O17	1.831(3)		- O23	1.853(3)
- O20	1.867(3)		- O25	1.869(3)
- O10	1.875(4)		- O19	1.956(3)
- O4	1.984(3)		- O13	1.956(3)
- O1	2.040(3)		- O26	1.963(4)
<Al3 – O>	1.904		<Al7 – O>	1.904
Al4 - O8	1.839(3)		Al8 - O17	1.864(3)
- O20	1.841(3)		- O22	1.878(3)
- O13	1.886(3)		- O14	1.886(3)
- O19	1.898(3)		- O25	1.905(4)
- O4	1.932(3)		- O27	1.921(4)
- O7	1.963(3)		- O24	1.973(4)
<Al4 - O>	1.893		<Al8 – O>	1.905
B1 - O15	1.344(5)		B2 - O20	1.448(6)
- O13	1.377(7)		- O18	1.459(6)
- O11	1.392(6)		- O15	1.490(6)
<B1 – O>	1.371		- O12	1.504(6)
			<B2 – O>	1.475

Table 5. Hydrogen-bond geometry (Å,°) in the structure of krasnoshteinite.

D – H···A	D – H	H···A	D···A	∠(D – H···A)
O1 - H1···O29	0.946(10)	2.52(4)	3.237(7)	133(4)
O1 - H1···Cl4		2.61(3)	3.398(3)	141(4)
O2 - H2A···Cl4	0.947(10)	2.05(2)	2.942(4)	157(5)
O2 - H2B···Cl1	0.947(10)	2.28(2)	3.156(4)	153(4)
O3 - H3···Cl3	0.945(10)	2.44(3)	3.300(3)	151(5)
O4 - H4···Cl2	0.947(10)	2.185(19)	3.096(4)	161(4)
O5 - H5···Cl1	0.940(10)	2.71(2)	3.609(4)	161(4)
O6 - H6···O15	0.946(10)	2.15(4)	2.794(5)	124(4)
O6 - H6···O18		2.533(15)	3.468(5)	170(4)
O7 - H7···Cl2	0.945(10)	2.093(14)	3.030(4)	171(5)
O8 - H8···O26	0.942(10)	2.52(5)	3.119(5)	121(4)
O9 - H9···Cl4	0.943(10)	2.352(11)	3.294(3)	177(4)
O10 - H10···Cl2	0.948(10)	2.684(17)	3.589(3)	160(3)
O12 - H12···Cl2	0.946(10)	2.380(15)	3.309(3)	168(4)
O14 - H14···Cl3	0.943(10)	2.134(12)	3.059(3)	167(3)
O16 - H16A···O15	0.949(10)	1.84(2)	2.747(5)	158(5)

Table 5. Cont.

D – H···A	D – H	H···A	D···A	∠(D – H···A)
O16 - H16A···O6	0.949(10)	2.32(4)	2.812(5)	111(3)
O16 - H16B···Cl2	0.952(10)	2.40(4)	3.143(4)	135(4)
O16 - H16B···Cl3	0.952(10)	2.67(5)	3.135(4)	111(3)
O17 - H17···O18	0.944(10)	1.873(13)	2.677(5)	141.5(19)
O18 - H18···Cl3	0.945(10)	2.22(2)	3.123(4)	160(5)
O19 - H19···O28	0.946(10)	1.770(12)	2.714(5)	175(5)
O21 - H21···Cl1	0.941(10)	2.441(18)	3.353(4)	163(4)
O22 - H22···Cl3	0.946(10)	2.259(12)	3.200(4)	173(4)
O23 - H23···Cl1	0.944(10)	2.51(3)	3.295(3)	140(4)
O23 - H23···O10		2.61(4)	3.257(4)	126(4)
O24 - H24A···Cl1	0.940(10)	2.43(2)	3.337(4)	161(4)
O24 - H24B···O29	0.942(10)	1.835(16)	2.747(6)	162(4)
O25 - H25···Cl4	0.941(10)	2.40(2)	3.294(4)	160(4)
O26 - H26A···Cl1	0.951(10)	2.187(18)	3.107(4)	163(5)
O26 - H26B···Cl4	0.952(10)	2.092(16)	3.030(4)	168(5)
O27 - H27A···O10	0.935(10)	2.19(4)	2.853(5)	128(4)
O27 - H27A···Cl1		2.71(4)	3.363(4)	128(3)
O27 - H27B···O28	0.940(10)	1.84(3)	2.648(5)	142(4)
O28 - H28A···Cl3	0.951(10)	2.12(2)	3.031(4)	161(5)
O28 - H28B···O29	0.951(10)	1.86(3)	2.745(7)	154(5)
O29 - H29A···Cl2	0.946(10)	2.29(3)	3.196(6)	160(6)
O29 - H29B···O18	0.951(10)	2.02(6)	2.820(6)	141(7)

3.3. Powder X-ray Diffraction

The indexed powder X-ray diffraction data are given in Table S3 in Supplementary Materials (see below). The powder X-ray diffraction pattern of krasnoshteinite is unique and can be used as a good diagnostic tool of the mineral. The parameters of a monoclinic unit cell refined from the powder data are as follows: $a = 8.740\ (4)$, $b = 14.409\ (4)$, $c = 11.316\ (4)$ Å, $\beta = 106.58\ (3)°$, and $V = 1366\ (1)$ Å^3.

4. Discussion

The crystal structure of krasnoshteinite (Figure 2) is unique. It is based upon the (010) corrugated layers of Al-centered octahedra connected via common vertices to form a pseudo-framework. There are eight crystallographically non-equivalent octahedrally coordinated Al sites: Al(1) and Al(7) cations center octahedra AlO(OH)$_4$(OH$_2$), Al(2,3) – AlO(OH)$_5$, Al(4) – AlO$_2$(OH)$_4$, Al(5) – Al(OH)$_5$(OH$_2$), Al(6) – Al(OH)$_6$, and Al(8) – Al(OH)$_4$(OH$_2$)$_2$. These Al-centered octahedra play different structural roles. Al(1–4)- and Al(6,7)-centered octahedra share edges to form six-membered clusters. Al(8)-centered octahedra link adjacent clusters along the c axis sharing two corners with each cluster, while Al(5)-centered octahedra play the same role linking the clusters along the a axis to form octahedral layers (Figure 3a). Adjacent layers are connected via the common O(3) vertex of Al(7)- and Al(1)-centered octahedra, forming the three-dimensional octahedral motif.

Boron atoms occupy two crystallographically non-equivalent sites and center B(1)O$_3$ triangles and B(2)O$_2$(OH)$_2$ tetrahedra, which share a common vertex to form insular [B$_2$O$_4$(OH)$_2$]$^{4-}$ groups (Figure 3b). According to the classification of fundamental building blocks (FBB) in borates [17,18], FBB in krasnoshteinite is 1△□:△□, i.e., the block with one triangle and one tetrahedron sharing corner. Krasnoshteinite is the first borate with such FBBs. These groups are connected with clusters of Al-centered octahedra via common vertices. Thus, a BO$_3$ triangle shares one O vertex with a

B-centered tetrahedron, one vertex with two Al-centered octahedra [Al(4) and Al(7)] of the layer, and one vertex with Al(1)- and Al(2)-centered octahedra of an adjacent layer (thus reinforcing the linkage between neighboring octahedral layers). A $BO_2(OH)_2$ tetrahedron shares one O vertex with a BO_3 triangle, one O vertex with two Al-centered octahedra [Al(3) and Al(4)], and one O=OH vertex with a Al(5)-centered octahedron; all Al(3,4,5) octahedra belong to the same layer. The resultant aluminoborate pseudo-framework contains three-membered [2B + Al] rings. Such Al-B-O units are known in the porous aluminoborate frameworks as being crucial to stabilizing them [19–21]. The same configuration of the three-membered [2B + Al] ring was described in the crystal structures of satimolite, $KNa_2(Al_5Mg_2)[B_{12}O_{18}(OH)_{12}](OH)_6Cl_4 \cdot 4H_2O$ [22], and synthetic porous Al borates, PKU-3 $H_{24.3}Al_9B_{18}O_{51}Cl_{3.3} \cdot 6.8H_2O$ [23] and PKU-8 $(H_{18}Al_7B_{12}O_{36})Cl_3(NaCl)_{2.4} \cdot 6.5H_2O$ [19].

The aluminoborate pseudo-framework in krasnoshteinite is microporous and zeolite-like (Figure 2). The three-dimensional system of wide channels contains Cl^- anions and H_2O molecules. Together with OH groups and H_2O molecules belonging to Al- and B-centered polyhedra, they form a complicated system of hydrogen bonds (Table 5).

Among 160 natural borates and oxoborates, known to date as valid mineral species, only twelve minerals contain species-defining Al, namely aluminomagnesiohulsite, $(Mg,Fe^{2+})_2$ $(Al,Mg,Sn)(BO_3)O_2$; jeremejevite, $Al_6(BO_3)_5(F,OH)_3$; johachidolite, $CaAlB_3O_7$; krasnoshteinite, $Al_8[B_2O_4(OH)_2](OH)_{16}Cl_4 \cdot 7H_2O$; londonite, $CsAl_4Be_4B_{12}O_{28}$; mengxianminite, $(Ca,Na)_2Sn_2(Mg,Fe)_3$ $Al_8[(BO_3)(BeO_4)O_6]_2$; painite, $CaZrAl_9(BO_3)O_{15}$; peprossiite-(Ce), $CeAl_2B_3O_9$; pseudosinhalite, $Mg_2Al_3O(BO_4)_2(OH)$; rhodizite, $KAl_4Be_4B_{12}O_{28}$; satimolite, $KNa_2(Al_5Mg_2)$ $[B_{12}O_{18}(OH)_{12}]$ $(OH)_6Cl_4 \cdot 4H_2O$; and sinhalite, $MgAl(BO_4)$ [2]. Satimolite and krasnoshteinite are low-temperature (LT) borates formed in evaporitic rocks, whereas the other ten minerals are known only in high-temperature (HT) geological formations: granitic pegmatites, HT metamorphic or metasomatic rocks, or post-volcanic HT assemblages. These HT Al borates and oxoborates do not contain H_2O molecules and have compact crystal structures that cause high hardness and mechanical and chemical stability in the majority of them. Data on the Mohs hardness of peprossiite-(Ce) and pseudosinhalite are absent in literature, aluminomagnesiohulsite has the Mohs hardness value of 6, and jeremejevite, johachidolite, londonite, mengxianminite, painite, rhodizite, and sinhalite demonstrate the Mohs hardness values between 7 to 8 [24]. All ten minerals have crystal structures with only one type of B-centered polyhedra, BO_3 triangles [aluminomagnesiohulsite, jeremejevite, mengxianminite, and painite], or BO_4 tetrahedra [johachidolite, londonite, peprossiite-(Ce), pseudosinhalite, rhodizite, and sinhalite], without OH groups coordinating B [25]. In LT formations, the crystal chemistry and properties of borate minerals with species-defining Al change drastically. Satimolite and krasnoshteinite are highly hydrated chloroborates which have low Mohs hardness values, 2 (satimolite) or 3 (krasnoshteinite), and dissolve even in diluted HCl. They contain complex borate polyanions composed of both B-centered triangles and tetrahedra with OH groups which participate in the tetrahedra. The structures of both satimolite and krasnoshteinite are microporous and zeolite-like. Thus, under LT conditions, aluminum octahedra in borates became a building unit of open-work aluminoborate structure motifs.

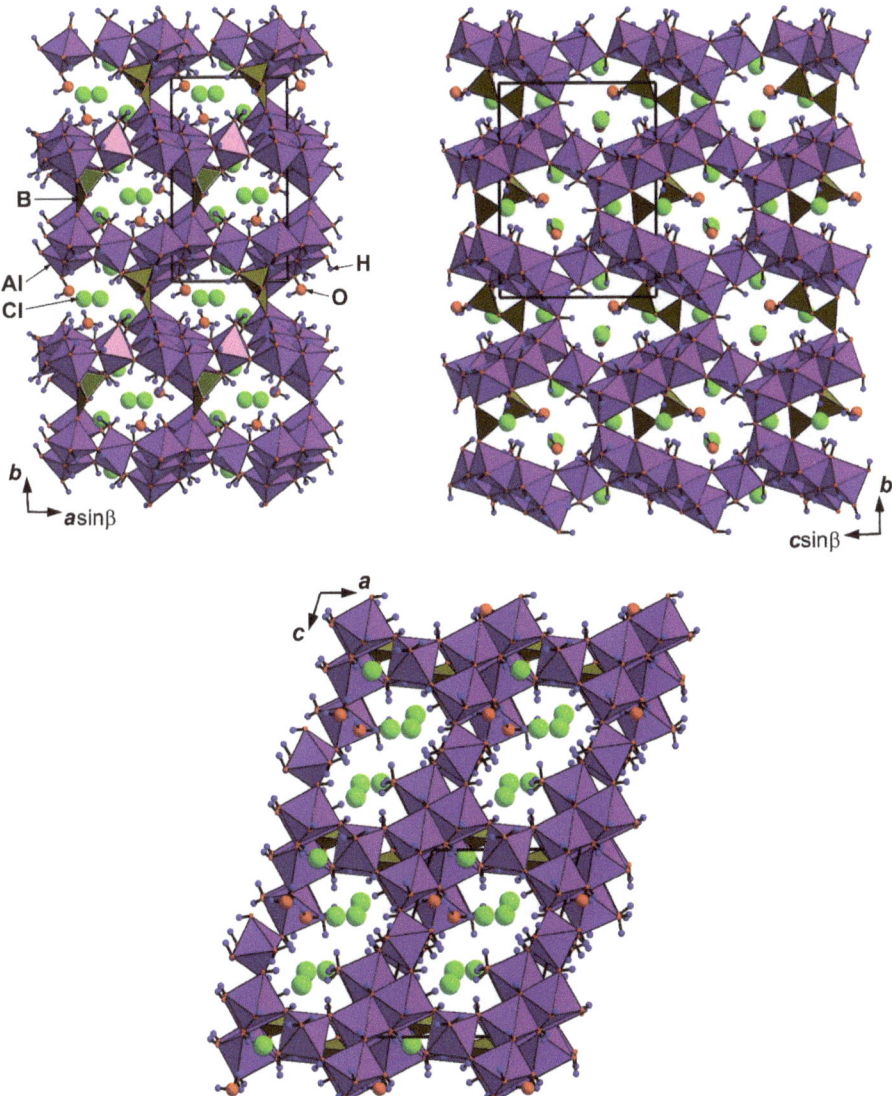

Figure 2. The crystal structure of krasnoshteinite in three projections. The unit cell is outlined.

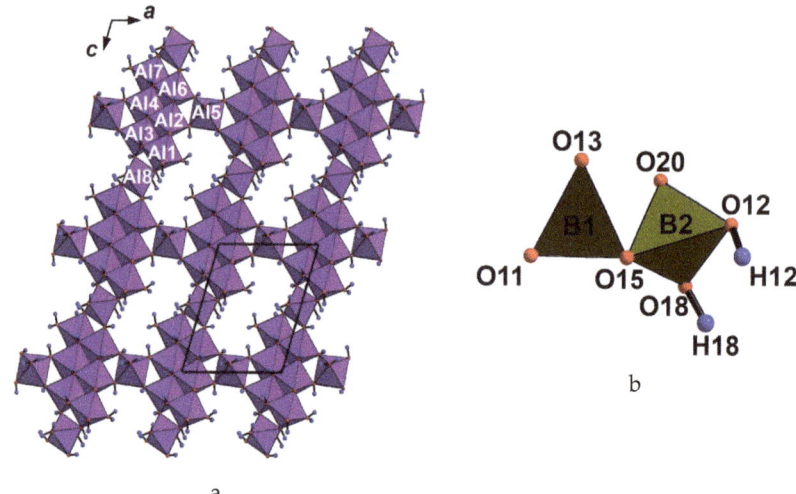

Figure 3. Octahedral layer (**a**) and insular $[B_2O_4(OH)_2]^{4-}$ group (**b**) in the structure of krasnoshteinite. For legend, see Table 3.

5. Conclusions

This paper is devoted to the new mineral species, krasnoshteinite. No mineral or synthetic compound related to it, in terms of crystal structure, has been found in literature and databases. Krasnoshteinite contains an earlier unknown borate polyanion, the insular $[B_2O_4(OH)_2]^{4-}$ group consisting of one BO_3 triangle and one $BO_2(OH)_2$ tetrahedron sharing corner. It was a surprise that such a simple anionic complex turned out novel for borates, both natural and synthetic, which is one of the most structurally diverse and best-studied classes of chemical compounds.

Krasnoshteinite ($Al_8[B_2O_4(OH)_2](OH)_{16}Cl_4 \cdot 7H_2O$) is the second, after jeremejevite ($Al_6(BO_3)_5$ $(F,OH)_3$), natural borate with only Al as a metal cation; and the second, after satimolite ($KNa_2(Al_5Mg_2)$ $[B_{12}O_{18}(OH)_{12}](OH)_6Cl_4 \cdot 4H_2O$), mineral with a zeolite-like aluminoborate framework motif in the structure. Due to the presence of a three-dimensional system of wide channels containing Cl^- anions and weakly bonded H_2O molecules, krasnoshteinite is of interest as a potential prototype of a novel family of microporous crystalline materials without large cations.

Borate minerals with species-defining Al formed in high-temperature and low-temperature geological formations are strongly different in crystal chemistry and physical and chemical properties. The high-temperature Al borates and oxoborates do not contain H_2O molecules, have compact crystal structures, and are typically characterized by high hardness and mechanical and chemical stability. Their crystal structures contain only one type of B-centered polyhedral, BO_3 triangles, or BO_4 tetrahedra. Unlike them, the low-temperature Al borates are highly hydrated, have low hardness, and are chemically unstable. They contain complex borate polyanions composed of both triangular and tetrahedral borate polyhedra with OH groups which participate in boron tetrahedra. Their structures are zeolite-like, being based upon open-work aluminoborate motifs.

Supplementary Materials: The following are available online at http://www.mdpi.com/2073-4352/10/4/301/s1.

Author Contributions: Conceptualization, I.V.P., N.V.Z., I.I.C., and D.Y.P.; Methodology, I.V.P., N.V.Z., and S.N.B.; Investigation, I.V.P., N.V.Z., I.I.C., E.P.C., D.I.B., V.O.Y., Y.V.B., and I.L.; Writing—Original Draft Preparation, I.V.P., N.V.Z., I.I.C., and D.Y.P.; Writing—Review and Editing, I.V.P. and N.V.Z.; Visualization, N.V.Z. All authors have read and agreed to the published version of the manuscript.

Funding: This work was supported by the Russian Foundation for Basic Research, grants 18-29-12007-mk (I.V.P., N.V.Z., and V.O.Y. for electron microprobe, XRD, and crystal structure studies) and 18-05-00046 (I.I.C. and E.P.C. for fieldwork and SEM studies).

Acknowledgments: The research has been carried out using facilities at the XRD Research Center of St. Petersburg State University in part of powder XRD study.

Conflicts of Interest: The authors declare no conflict of interest.

References

1. Rudnick, R.L.; Gao, S. The Composition of the Continental Crust. In *Treatise on Geochemistry, 3, The Crust*; Holland, H.D., Turekian, K.K., Eds.; Elsevier-Pergamon: Oxford, UK, 2003.
2. The Official IMA-CNMNC List of Mineral Names. Updated List of IMA-Approved Minerals. Available online: http://cnmnc.main.jp (accessed on 5 March 2020).
3. Anovitz, L.M.; Grew, E.S. Mineralogy, petrology and geochemistry of boron: An introduction. In Boron: Mineralogy, Petrology and Geochemistry. *Rev. Mineral.* **1996**, *33*, 1–40.
4. Hawthorne, F.C.; Burns, P.C.; Grice, J.D. The crystal chemistry of boron. In: Boron: Mineralogy, Petrology and Geochemistry. *Rev. Mineral.* **1996**, *33*, 41–115.
5. Ivanov, A.A.; Voronova, M.L. *Verkhnekamskoe Potassium Salt Deposit*; Nedra Publishing: Leningrad, Russia, 1975; pp. 1–219. (In Russian)
6. Kudryashov, A.I. *Verkhnekamskoe Salt Deposit*, 2nd ed.; Epsilon Plus: Moscow, Russia, 2013; pp. 1–368. (In Russian)
7. Zhitova, E.S.; Pekov, I.V.; Chaikovskiy, I.I.; Chirkova, E.P.; Yapaskurt, V.O.; Bychkova, Y.V.; Belakovskiy, D.I.; Chukanov, N.V.; Zubkova, N.V.; Krivovichev, S.V.; et al. Dritsite, $Li_2Al_4(OH)_{12}Cl_2 \cdot 3H_2O$, a new gibbsite-based hydrotalcite-supergroup mineral. *Minerals* **2019**, *9*, 492. [CrossRef]
8. Mandarino, J.A. The Gladstone–Dale compatibility of minerals and its use in selecting mineral species for further study. *Can. Miner.* **2007**, *45*, 1307–1324. [CrossRef]
9. Agilent Technologies. *CrysAlisPro Software System*; Version 1.171.37.35; Agilent Technologies UK Ltd.: Oxford, UK, 2014.
10. Sheldrick, G.M. Crystal structure refinement with SHELXL. *Acta Cryst.* **2015**, *A71*, 3–8.
11. Britvin, S.N.; Dolivo-Dobrovolsky, D.V.; Krzhizhanovskaya, M.G. Software for processing of X-ray powder diffraction data obtained from the curved image plate detector of Rigaku RAXIS Rapid II diffractometer. *Zap. Rmo.* **2017**, *146*, 104–107. (In Russian with English Abstract)
12. Bruker-AXS. *Topas V4.2: General Profile and Structure Analysis Software for Powder Diffraction Data*; Bruker: Karlsruhe, Germany, 2009.
13. Nespolo, M.; Ferraris, G. Twinning by syngonic and metric merohedry. Analysis, classification and effects on the diffraction pattern. *Zeit. Krist.* **2000**, *215*, 77–81. [CrossRef]
14. Brese, N.E.; O'Keeffe, M. Bond-valence parameters for solids. *Acta Crystallogr.* **1991**, *B47*, 192–197. [CrossRef]
15. Ferraris, G.; Ivaldi, G. Bond Valence vs. Bond Length in $O \cdots O$ Hydrogen Bonds. *Acta Cryst.* **1988**, *B44*, 341–344. [CrossRef]
16. Malcherek, T.; Schlüter, J. $Cu_3MgCl_2(OH)_6$ and the bond-valence parameters of the OH-Cl bond. *Acta Cryst.* **2007**, *B63*, 157–160. [CrossRef] [PubMed]
17. Burns, P.C.; Grice, J.D.; Hawthorne, F.C. Borate minerals. I. Polyhedral clusters and fundamental building blocks. *Can. Miner.* **1995**, *33*, 1131–1151.
18. Grice, J.D.; Burns, P.C.; Hawthorne, F.C. Borate minerals. II. A hierarchy of structures based upon the borate fundamental building block. *Can. Miner.* **1999**, *37*, 731–762.
19. Gao, W.; Wang, Y.; Li, G.; Liao, F.; You, L.; Lin, J. Synthesis and Structure of an Aluminum Borate Chloride Consisting of 12-Membered Borate Rings and Aluminate Clusters. *Inorg. Chem.* **2008**, *47*, 7080–7082. [CrossRef] [PubMed]
20. Ju, J.; Yang, T.; Li, G.; Liao, F.; Wang, Y.; You, L.; Lin, J. PKU-5: An Aluminoborate with Novel Octahedral Framework Topology. *Chem. Eur. J.* **2004**, *10*, 3901–3906. [CrossRef] [PubMed]
21. Yang, T.; Ju, J.; Li, G.; Liao, F.; Zou, X.; Deng, F.; Chen, L.; Su, Y.; Wang, Y.; Lin, J. Square-Pyramidal/Triangular Framework Oxide: Synthesis and Structure of PKU-6. *Inorg. Chem.* **2007**, *46*, 4772–4774. [CrossRef] [PubMed]
22. Pekov, I.V.; Zubkova, N.V.; Ksenofontov, D.A.; Chukanov, N.V.; Yapaskurt, V.O.; Korotchenkova, O.V.; Chaikovskiy, I.I.; Bocharov, V.M.; Britvin, S.N.; Pushcharovsky, D.Y. Redefinition of satimolite. *Miner. Mag.* **2018**, *82*, 1033–1047. [CrossRef]

23. Chen, H.; Ju, J.; Meng, Q.; Su, J.; Lin, C.; Zhou, Z.; Li, G.; Wang, W.; Gao, W.; Zeng, C.; et al. PKU-3: An HCl-Inclusive Aluminoborate for Strecker Reaction Solved by Combining RED and PXRD. *J. Am. Chem. Soc.* **2015**, *137*, 7047–7050. [CrossRef] [PubMed]
24. Anthony, J.W.; Bideaux, R.A.; Bladh, K.W.; Nichols, M.C. *Handbook of Mineralogy. Vol. V. Borates, Carbonates, Sulfates*; Mineral Data Publishing: Tucson, AZ, USA, 2003; pp. 1–813.
25. Pekov, I.V.; Zubkova, N.V.; Yapaskurt, V.O.; Ksenofontov, D.A.; Chaikovskiy, I.I.; Korotchenkova, O.V.; Chirkova, E.P.; Pushcharovsky, D.Y. Towards structural mineralogy and genetic crystal chemistry of boron: Novel crystal structures of borate and borosilicate minerals from different geological formations. In Proceedings of the XIX International Meeting on Crystal Chemistry, X-ray Diffraction and Spectroscopy of Minerals, Apatity, Russia, 2–6 July 2019; p. 98.

© 2020 by the authors. Licensee MDPI, Basel, Switzerland. This article is an open access article distributed under the terms and conditions of the Creative Commons Attribution (CC BY) license (http://creativecommons.org/licenses/by/4.0/).

Article

Crystal Chemistry of Stanfieldite, $Ca_7M_2Mg_9(PO_4)_{12}$ (M = Ca, Mg, Fe^{2+}), a Structural Base of $Ca_3Mg_3(PO_4)_4$ Phosphors

Sergey N. Britvin [1,2,*], Maria G. Krzhizhanovskaya [1], Vladimir N. Bocharov [3] and Edita V. Obolonskaya [4]

1. Department of Crystallography, Institute of Earth Sciences, St. Petersburg State University, Universitetskaya Nab. 7/9, 199034 St. Petersburg, Russia; mariya.krzhizhanovskaya@spbu.ru
2. Nanomaterials Research Center, Kola Science Center of Russian Academy of Sciences, Fersman Str. 14, 184209 Apatity, Russia
3. Centre for Geo-Environmental Research and Modelling, Saint-Petersburg State University, Ulyanovskaya ul. 1, 198504 St. Petersburg, Russia; bocharov@molsp.phys.spbu.ru
4. The Mining Museum, Saint Petersburg Mining University, 2, 21st Line, 199106 St. Petersburg, Russia; musmet11@yandex.ru
* Correspondence: sergei.britvin@spbu.ru

Received: 1 May 2020; Accepted: 25 May 2020; Published: 1 June 2020

Abstract: Stanfieldite, natural Ca-Mg-phosphate, is a typical constituent of phosphate-phosphide assemblages in pallasite and mesosiderite meteorites. The synthetic analogue of stanfieldite is used as a crystal matrix of luminophores and frequently encountered in phosphate bioceramics. However, the crystal structure of natural stanfieldite has never been reported in detail, and the data available so far relate to its synthetic counterpart. We herein provide the results of a study of stanfieldite from the Brahin meteorite (main group pallasite). The empirical formula of the mineral is $Ca_{8.04}Mg_{9.25}Fe_{0.72}Mn_{0.07}P_{11.97}O_{48}$. Its crystal structure has been solved and refined to $R_1 = 0.034$. Stanfieldite from Brahin is monoclinic, $C2/c$, a 22.7973(4), b 9.9833(2), c 17.0522(3) Å, β 99.954(2)°, V 3822.5(1)Å^3. The general formula of the mineral can be expressed as $Ca_7M_2Mg_7(PO_4)_{12}$ ($Z = 4$), where the M = Ca, Mg, Fe^{2+}. Stanfieldite from Brahin and a majority of other meteorites correspond to a composition with an intermediate Ca≈Mg occupancy of the M5A site, leading to the overall formula ~$Ca_7(CaMg)Mg_9(PO_4)_{12} \equiv Ca_4Mg_5(PO_4)_6$. The mineral from the Lunar sample "rusty rock" 66095 approaches the M = Mg end member, $Ca_7Mg_2Mg_9(PO_4)_{12}$. In lieu of any supporting analytical data, there is no evidence that the phosphor base with the formula $Ca_3Mg_3(PO_4)_4$ does exist.

Keywords: stanfieldite; phosphate; crystal structure; merrillite; meteorite; pallasite; mesosiderite; luminophore; bioceramics; powder diffraction; Raman spectroscopy

1. Introduction

It is known that the speciation of chemical elements in meteoritic substance significantly differs from their speciation in contemporary terrestrial lithosphere [1]. Concerning phosphorus, the main geochemical factors governing the diversity of terrestrial phosphorus-bearing minerals are (1) highly oxidative conditions typical of the present Earth and (2) the aquatic environment, which dramatically multiplies the number of possible pathways for phosphate geosynthesis. Contrary to Earth, the reductive and (in general) water-free conditions that accompanied the formation and early evolution of celestial bodies determined the limited number of meteoritic phosphorus-bearing minerals [2].

The most common meteoritic phosphates are the minerals related to the join merrillite–ferromerrillite, $Ca_9NaMg(PO_4)_7$–$Ca_9NaFe^{2+}(PO_4)_7$ [3,4]. They are the typical accessories of ordinary chondrites, lunar rocks, martian meteorites, iron and stony-iron meteorites [3–6]. Chlorapatite, $Ca_5(PO_4)_3Cl$, is the second abundant phosphate in meteorites [2]. A series of Mg-rich phosphates are characteristic of stony-iron meteorites—pallasites and mesosiderites [6,7]. These minerals usually occur in association with schreibersite, $(Fe,Ni)_3P$, and because of that, they are used for the assessment of phosphide–phosphate redox equilibria [8]. Stanfieldite, $\sim Ca_4Mg_5(PO_4)_6$, is the most common phosphate in the given assemblages [6,7]. The mineral was discovered in the Estherville mesosiderite [9] and is recognized in all well-studied pallasites [6,7,9–13], several mesosiderites [14,15] and even in the Lunar samples [16]. Being one of a very few Ca-rich phases occurring in pallasites, stanfieldite acts as a carrier of rare-earth elements substituting for Ca, and thereby is used in the studies of REE distribution among these meteorites [10,17]. Based on the overall observations and experimental data, stanfieldite can be regarded a late-stage cumulate of silicate-phosphate melts [18]. It is noteworthy that, in spite of an ordinary, "rock-forming" set of elements in the chemical composition, stanfieldite has never been encountered in terrestrial rocks. However, the phosphate identifiable as stanfieldite was reported as a constituent of prehistoric slags in Tyrol, Austria [19], and phosphorus-doped basaltic melts [20]. Stanfieldite of technogenic origin was detected as a component of bone-repairing bioceramics [21–28] and incinerated phosphate-based fertilizers [29–32].

In view of the notable role of stanfieldite in the mineralogy of pallasites, it looks unusual that the data on its chemical composition are rather ambiguous. The formula was first reported as $Ca_4Mg_5(PO_4)_6$ [9], but later on, the variations in the Ca/Mg ratio were shown to exist [12], and in many cases, stanfieldite formula is oversimplified as $Ca_3Mg_3(PO_4)_4$, e.g., Reference [33]. Synthetic $Ca_3Mg_3(PO_4)_4$ was reported as a phase in the $Ca_3(PO_4)_2$–$Mg_3(PO_4)_2$ system [34], and nowadays, the compounds having the same inferred formula are widely explored as luminophores (e.g., References [35–39]). However, the powder XRD (X-ray diffraction) data given in these works [35–39] refer not to $Ca_3Mg_3(PO_4)_4$ but to the compound $Ca_7Mg_9(Ca,Mg)_2(PO_4)_{12}$ [40], which was erroneously mislabeled as "$Ca_3Mg_3(PO_4)_4$" both in ICSD and ICDD databases. Synthetic $Ca_7Mg_9(Ca,Mg)_2(PO_4)_{12}$ was shown to have the same cell metrics as natural stanfieldite but it crystallizes in a different space group [9,40]. The latter discrepancy was discussed by Steele and Olsen in the abstract devoted to a crystal structure of natural stanfieldite [41]. However, no further structural data were provided by these authors, and as a consequence, no crystal structure of natural stanfieldite is available so far. In the course of a research of phosphate–phosphide assemblages of iron and stony-iron meteorites, we have found well-crystallized stanfieldite in the Brahin meteorite and carried out the detailed study of this mineral. We herein present the results and try to resolve some ambiguities related to a crystal chemistry of natural stanfieldite and its synthetic analogues.

2. Stanfieldite in the Brahin Pallasite

Brahin is a meteorite related to the main-group pallasites. It was first found in 1810 as two fragments (masses) of total weight ~80 kg at the Kaporenki village, Brahin district, Belarus. Since then, a few larger masses were recovered in the same district in 1968 and 2002. Nowadays, the total known weight of Brahin exceeds 800 kg [42]. Like other main-group pallasites, Brahin consists of round, nut-like and fragmented olivine crystals embedded into the Fe-Ni metal matrix. The less-common minerals are represented by schreibersite-nickelphosphide (Fe_3P-Ni_3P), chromite, troilite and daubreelite, $FeCr_2O_4$. The specific feature of oxygen-bearing minerals of Brahin is their depletion in Mn [12]. Phosphates are comprised by merrillite, $Ca_9NaMg(PO_4)_7$, and stanfieldite. The crystal structure of iron-free merrillite from this meteorite has been recently reported [4]. Stanfieldite in Brahin was studied with respect to the occurrence of fission tracks [43]. A 1 × 2 mm grain of colorless stanfieldite was found in the centimeter-sized Brahin fragment kindly provided for the study by the Mining Museum, Saint Petersburg Mining University (specimen M65/2, which originates from the first find in 1810). Stanfieldite and merrillite fill up the pocket bound by the fragmented olivine grains, schreibersite and (Fe, Ni) metal (Figure 1).

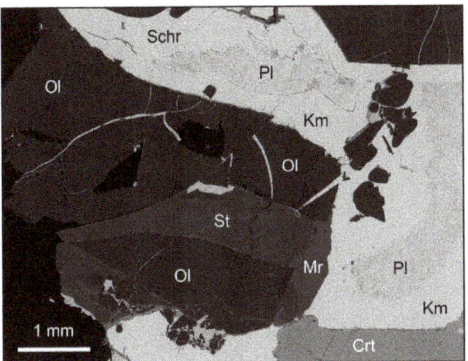

Figure 1. Stanfieldite (St) in the Brahin pallasite. Associated minerals: Ol, olivine; Mr, merrillite; Crt, chromite; Schr, schreibersite; Km, kamacite, α-(Fe,Ni); Pl, plessite (fine-grained aggregate of α- and γ-(Fe, Ni)). Polished section. SEM image of backscattered electrons. Image made by S.N.B.

3. Materials and Methods

A piece of the Brahin pallasite was polished and coated with a carbon film for electron microprobe study. SEM imaging (Figure 1) and microprobe analysis for the main elements were conducted by means of a CamScan 4 scanning electron microscope (SEM) (Cambridge, UK) equipped with a LINK AN1000 energy-dispersive analyzer (LINK Analytical, CA, USA). The following standards were used: chlorapatite (Ca-K, P-K), enstatite (Mg-K), hematite (Fe-K). The analysis was carried out at 20 kV acceleration voltage, 0.8 nA beam current, 1 µm estimated beam diameter and 60 s live acquisition time per spot. The check-up for minor constituents was performed with a Microspec WDX-2 wavelength-dispersive X-ray spectrometer (Microspec Corporation, CA, USA) attached to the same SEM. The Mn content was determined using Mn-$K\alpha$ line ($MnCO_3$ standard) at 20 kV and 15 nA, whereas the contents of Ni, Co, Na, K and Si were found to lie below the detection limit (less than 0.05 wt.%).

For the purposes of the X-ray structural study, the grain of stanfieldite was extracted from the section and crushed into a few fragments, which were examined under a polarizing microscope in the immersion oil. Several optically homogeneous grains were checked using a Rigaku Oxford diffraction Xcalibur single-crystal diffractometer equipped with a fine-focus sealed tube and graphite monochromator (Mo$K\alpha$, 50 kV, 40 mA). It was found that all checked fragments are optically irresolvable intergrowths, each of them being composed of two or more domains misoriented within 5–10°. The best selected two-domain grain (0.15 × 0.10 × 0.10 mm) was glued onto a plastic loop and subjected to further data collection. A hemisphere of reciprocal space was collected up to 70° at room temperature, and the details are provided in Table 1. Subsequent data processing routines (integration, scaling and *SHELX* files setup) were performed by means of a CrysAlisPro software (Rigaku Oxford diffraction) [44]. The crystal structure has been solved using an intrinsic phasing approach and refined by means of a *SHELX*-2018 set of programs [45] incorporated into the Olex2 operation environment [46]. The details of structure refinement are given in Table 1 and in the crystallographic information file (CIF) attached to the Supplementary Materials (S1).

Table 1. Crystal data, single-crystal and Rietveld refinement details for stanfieldite from Brahin.

Crystal Data:	Single Crystal	Rietveld Method
Crystal system, space group	Monoclinic, $C2/c$	Monoclinic, $C2/c$
Crystal size (mm)	0.15 × 0.10 × 0.10	ball Ø 0.15
a (Å)	22.7973(4)	22.8036(2)
b (Å)	9.9833(2)	9.9832(1)
c (Å)	17.0522(3)	17.0558(2)
β (°)	99.954(2)	99.964(1)
V (Å^3)	3822.5(1)	3824.2(1)
Z	4	4
D_x (g cm^{-1})	2.990	2.988
Data collection and refinement: Single Crystal		
Diffractometer	\multicolumn{2}{l	}{Rigaku Oxford Diffraction Xcalibur EoS}
Radiation		Mo$K\alpha$
μ (mm^{-1})		2.12
No. of meas., independent and obs. [$I > 2\sigma(I)$] reflections		18,163, 5420, 3949
h, k, l range		$-27\rightarrow32; -14\rightarrow13; -23\rightarrow23$
R_{int}, R_σ		0.044, 0.048
R_1 (|F_o|≥4σ_F), wR_2, GoF		0.034, 0.082, 0.92
$\Delta\sigma_{min}, \Delta\sigma_{max}$ (e Å^{-3})		$-0.56, 0.86$
Data collection and refinement: Rietveld method		
Diffractometer		Rigaku RAXIS Rapid II (imaging plate)
Radiation		Co$K\alpha_1$/Co$K\alpha_2$
μ (mm^{-1})		26.11
Exposure time (s)		1800
Calculation step (°)		0.02
2Θ range (°)		6–132
Peak shape description		Modified Pseudo-Voigt
Background subtraction		28-coefficient Chebyshev polynomial
R_p, R_{wp}, R_B (%), GOF		0.39, 0.69, 0.41, 1.79

The X-ray powder diffraction pattern (Table 2) was obtained with a Rigaku R-AXIS Rapid II difractometer (Rigaku Corporation, Tokyo, Japan) equipped with a curved (semi-cylindrical) imaging plate. A ~150 µm ball was prepared from the stanfieldite powder mixed with an epoxy resin and was picked onto a glass fiber. The image acquisition conditions were: Co$K\alpha$-radiation, rotating anode with microfocus optics, 40 kV, 15 mA, Debye-Scherrer geometry, r = 127.4 mm, exposure 30 min. The imaging plate was calibrated against Si standard. The image-to-profile data conversion was performed with an osc2xrd program [47]. The unit-cell parameters and occupancies of Mg1–Mg5 sites were refined by the Rietveld method (Table 1, Figure 2) using Bruker TOPAS v. 5.0 software (Bruker Inc., Wisconsin). The occupancies at the M5A site were fixed at the values determined by single-crystal refinement. The atomic coordinates were not refined but were fixed according to single-crystal data. The XPRD pattern (Table 2) was indexed on the basis of theoretical values calculated with STOE WinXPOW v. 1.28 software (Stoe & Cie GmbH, Darmstadt, Germany).

The micro-Raman spectrum was recorded from a random powder sample using a Horiba Jobin-Yvon LabRam 800 instrument (HORIBA Jobin Yvon GmbH, Bensheim, Germany), equipped with a 50× confocal objective. The instrument was operated with a 514 nm Ar$^+$ laser at a 1 nm lateral resolution and 2 cm^{-1} spectral resolution. The optics were preliminarily calibrated using a Si reflection standard.

Table 2. X-ray powder diffraction data (d in Å) for stanfieldite from the Brahin meteorite.

I_{meas}	d_{meas}	I_{calc} [1]	d_{calc}	hkl	I_{meas}	d_{meas}	I_{calc} [1]	d_{calc}	hkl
1	11.31	2	11.23	200	48	2.500	30	2.500	623
1	9.13	<1	9.12	110			20	2.496	040
23	8.32	21	8.26	−111	4	2.423	4	2.420	910
<1	6.26	1	6.23	202	6	2.401	6	2.398	241
23	6.00	21	5.97	112	2	2.379	2	2.377	−823
<1	5.625	<1	5.614	400	4	2.343	3	2.341	911
3	5.424	2	5.409	311	3	2.319	1, 1	2.322	135, 533
3	5.247	2	5.230	−312	5	2.295	5	2.293	624
4	5.108	3	5.091	−402	6	2.283	1, 3	2.281	−441, 440
16	5.006	13	4.992	020	8	2.259	6	2.257	−716
2	4.800	1	4.785	021	2	2.240	1, 1	2.240	−442, 441
1	4.662	1	4.629	113	5	2.205	3	2.202	−626
4	4.604	3	4.585	312	9	2.161	3	2.163	027
2	4.406	0.5	4.405	−313			2	2.160	−136
3	4.334	2	4.326	221			2	2.159	732
5	4.186	4	4.177	−204	4	2.138	1, 1	2.138	−734, −825
3	4.104	2	4.096	510	9	2.127	5, 4	2.126	823, 426
42	3.849	36	3.842	511	3	2.077	2	2.076	640
			3.742	600	2	2.066	1, 1	2.066	−10.2.2, 227
			3.733	−421	2	2.028	1	2.029	−10.2.3
100	3.738	100	3.730	420	7	2.006	2	2.008	−245
			3.727	204			2	2.003	045
			3.726	023			2	2.001	−518
10	3.688	8	3.681	−404	4	1.960	5	1.960	914
3	3.592	3	3.586	−513	2	1.901	1, 1	1.901	−736, 318
2	3.282	2	3.272	422	3	1.884	2	1.883	−153
9	3.248	7	3.246	−131	15	1.867	4	1.871	12.0.0
		2	3.234	314			8	1.867	−842
3	3.217	1	3.213	024			4	1.864	408
2	3.118	1	3.114	404			4	1.863	046
10	3.087	7,2	3.083	115, 513	7	1.840	7	1.841	−808
18	3.042	15	3.039	132	3	1.827	1, 1	1.828	841, 915
7	2.955	6	2.952	423	4	1.801	4	1.800	551
2	2.873	2	2.870	−133	2	1.789	2	1.787	−338
10	2.834	7	2.834	−206	3	1.753	1, 2	1.751	−429, 138
		12	2.825	−515	2	1.742	1, 1	1.742	−247, −538
83	2.807	26	2.812	−802	3	1.697	2	1.697	11.1.4
		27	2.807	800	2	1.652	2	1.653	−555
		24	2.800	−225	3	1.621	2, 2	1.622	−356, 10.0.6
12	2.731	3,5	2.730	−116, 712	3	1.605	2	1.605	−829
6	2.706	5	2.705	622	6	1.597	6	1.597	429
13	2.685	11	2.682	−316	3	1.543	3	1.543	339
2	2.639	2	2.642	424	3	1.516	2	1.515	−13.1.7
2	2.621	2	2.627	−532	3	1.481	2	1.481	−848
3	2.601	4	2.599	531					

[1] Intensities were normalized to $\Sigma I[(600) + (-421) + (420) + (204) + (023)] = 100$. Reflections having relative intensity less than 2 at $d < 4.00$ Å have been omitted.

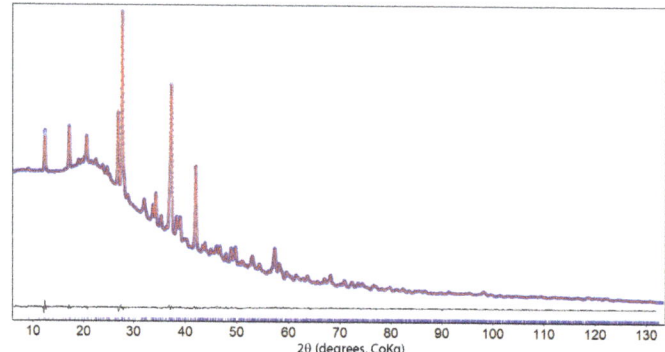

Figure 2. Rietveld refinement plot of stanfieldite from the Brahin meteorite.

4. Results and Discussion

4.1. Stanfieldite: A Complete Structure-Composition Dataset

The X-ray examination of Brahin stanfieldite was carried out by two different methods. Both single-crystal and Rietveld refinements of the unit cell had led to almost identical parameters (deviation between the unit-cell volumes is 0.04%, Table 1). The refined Mg/Fe occupancies were also well converged (Table 3). The chemical composition of studied stanfieldite (wt.%, average of 3 points): CaO 26.21, MgO 21.66, FeO 3.02, MnO 0.27, P_2O_5 49.38, total 100.54 can be recalculated to the formula $Ca_{8.04}Mg_{9.25}Fe_{0.72}Mn_{0.07}P_{11.97}O_{48}$. The latter is close to the composition determined by the structure refinement, $Ca_{7.97}Mg_{9.47}Fe_{0.56}P_{12}O_{48}$. Based on these results, one can state that the bond lengths and derivative bond-valence sums (Table 3) are herein calculated with a good confidence.

Fuchs [9], in 1969, could reliably determine the unit-cell metrics, but misrecognized the space group of the mineral (Table 4), perhaps due to the same pseudo-twinning of the crystals [40,41] which we observed on our studied stanfieldite.

Table 3. Selected structural parameters of cation sites in natural and synthetic stanfieldite [1].

Site	CN [2]	Length [3]	BVS [4]	Mg [5]	Length [3]	BVS [4]	Length [3]	Mg [5]
		Brahin (Present Work)			$Ca_7(Ca,Mg)Mg_9(PO_4)_{12}$ Synthetic [40]		Imilac [41]	
Ca1 (4*e*)	8	2.543	1.86		2.547	1.85	2.542	
Ca2 (8*f*)	7	2.445	2.09		2.444	2.10	2.445	
Ca3 (8*f*)	8	2.566	1.69		2.569	1.68	2.573	
Ca4 (8*f*)	8	2.496	2.02		2.498	2.01	2.498	
Mg1 (4*e*)	4	1.995	1.79	0.934(5)/0.91	1.999	1.75		0.928(4)
Mg2 (8*f*)	6	2.096	2.08	0.945(3)/0.94	2.095	2.07	2.097	0.942(3)
Mg3 (8*f*)	5	2.068	1.83	0.985(4)/0.98	2.071	1.81		0.969(3)
Mg4 (8*f*)	5	2.049	1.92	0.983(3)/1.00	2.054	1.89		0.980(3)
Mg5 (8*f*)	6	2.132	1.89	0.919(3)/0.90	2.130	1.88		0.903(3)
M5A (8*f*)	6	2.257	1.89	See Table 5	2.270	1.92	2.282	
P1 (8*f*)	4	1.535	4.83		1.540	4.76		
P2 (8*f*)	4	1.535	4.83		1.538	4.79		
P3 (8*f*)	4	1.537	4.81		1.541	4.75		
P4 (8*f*)	4	1.532	4.85		1.538	4.78		
P5 (8*f*)	4	1.534	4.83		1.536	4.81		
P6 (8*f*)	4	1.530	4.89		1.532	4.86		

[1] Complete listings of atomic coordinates, thermal displacement parameters and bond lengths are given in Supplementary Tables S1–S3. [2] CN, coordination number. [3] Average cation-oxygen bond lengths (Å). [4] BVS, bond-valence sums (in valence units), calculated by summation of individual element contributions based on parameters reported by Brese and O'Keeffe [48]. [5] Refined Mg occupancies for Mg1–Mg5 sites. Data for Brahin include both single-crystal and Rietveld refinement results separated by slash.

Table 4. Unit cell parameters of natural and synthetic stanfieldite refined from single-crystal data.

Source	Brahin	Synthetic	Imilac	Estherville
Space group	C2/c	C2/c	C2/c	P2/c or Pc [1]
$a(Å)$	22.7973(4)	22.841(3)	22.81	17.16(3)
$b(Å)$	9.9833(2)	9.994(1)	9.993	10.00(2)
$c(Å)$	17.0522(3)	17.088(5)	17.09	22.88(4)
$β(°)$	99.954(2)	99.63(3)	99.96	100.3(2)
$V(Å^3)$	3822.5	3845.8	3836.8	3862.9
Reference	This work	[40]	[41]	[9]

[1] Space group assignment and cell axes permutation are discussed in References [40,41].

Dickens and Brown [40] have synthesized the synthetic, Fe-free analogue of stanfieldite and thoroughly described its crystal structure. However, the latter authors did not perform independent determination of the chemical composition of synthesized material—as one will see, this is an essential requirement in view of the widely varying composition of at least one structural site of stanfieldite. Steele and Olsen [41] have reported the preliminary results of structural examination of natural stanfieldite from the Imilac pallasite. They gave the analytical chemical formula of the mineral, but did not provide full structural data, confining the results to unit cell metrics, average bond lengths and selected site occupancies (Tables 3 and 4). As a consequence, no complete structure-composition dataset for stanfieldite is available so far, and the data provided herein are the first report of that type.

4.2. General Features of Stanfieldite Structure and Its Formula

The crystal structure of stanfieldite is a complex framework composed of 10 metal sites and 6 phosphate groups (Table 3, Figure 3, Supplementary Table S1). Dickens and Brown [40] gave the detailed description of each site in the structure of synthetic analogue of the mineral, and the present paragraph aims to overview stanfieldite structure and highlight its features. The most interesting one is a pseudo-hexagonal character of the framework which can be best viewed via the arrangement of [PO$_4$] tetrahedra along the [10–2] axis (Figure 3A). In principle, stanfieldite, being presented by the oversimplified formula $M_3(PO_4)_2$ (M = Mg, Ca; Z = 24), can be regarded as a derivative of the well-known glaserite structure type, $K_3Na(SO_4)_2$ [49,50]. Dickens and Brown [40] discuss the relationships between stanfieldite and glaserite-related phosphates belonging to α- and β-$Ca_3(PO_4)_2$ structural types. The latter is known as a basement of whitlockite-group mineral structures [51], two of which, merrillite and ferromerrillite, are of fundamental importance in the mineralogy of meteorites [3,4]. In view of the common and intimate association of stanfieldite and merrillite in pallasite meteorites (Figure 1), these relationships could be of particular interest. However, contrary to Dickens and Brown [40], we would not overestimate the similarity of stanfieldite and merrillite structures. The unusual face-sharing of adjacent [MO$_6$] octahedron and [PO$_4$] tetrahedron characteristic of merrillite [3] does not occur in stanfieldite structure.

The refinement of occupancies of four Ca-sites in the stanfieldite structure showed no evidence for either Mg or Fe substitution. However, refinement of four Mg sites using both single-crystal and Rietveld methods concordantly leads to a partial substitution of Mg for Fe, with iron being preferentially concentrated in Mg1 (tetrahedral) and Mg5 (octahedral) positions (Table 3). The tetrahedral coordination of Mg1 is highly unusual; however, it is sometimes encountered in mineral structures such as åkermanite, $Ca_2MgSi_2O_7$ (melilite structure type), and spinel. The M5A site allows mixed occupancy by Ca, Mg and Fe, and thus will be discussed in the next section. Based on the structural data, the overall formula of stanfieldite can be written as $Ca_7M_2Mg_9(PO_4)_7$, where M = Ca, Mg or Fe^{2+}.

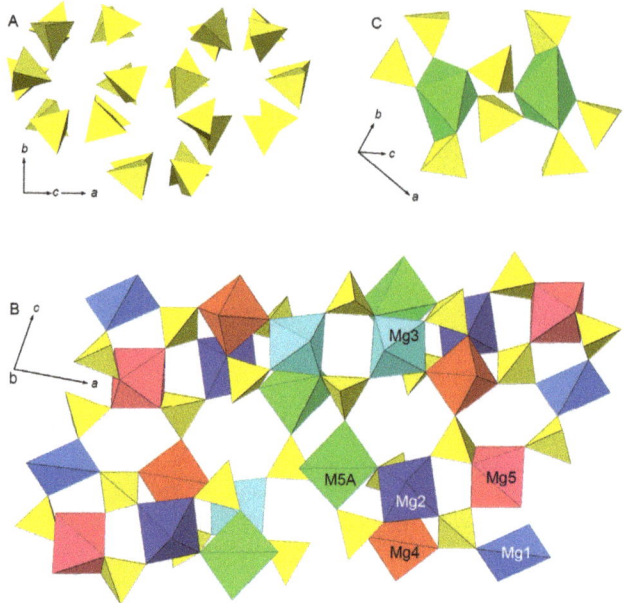

Figure 3. Crystal structure of stanfieldite. (**A**) Pseudo-hexagonal stacking of [PO$_4$] tetrahedra (yellow), projection along the [10–2] axis. The cations were hidden for clarity. (**B**) A slice of the structure on {010}. Arrangement of Mg-polyhedra and [PO$_4$] tetrahedra. Ca atoms reside in the spaces and have been hidden for clarity. (**C**) A view of a pair of distorted [M5AO6] octahedra corner-linked by [PO$_4$] tetrahedra.

4.3. The M5A Site, A Key to A Flexibility of Stanfieldite Composition

This cation site (which was previously referred to as Ca5 [40]) deserves a special discussion as it determines the variability of stanfieldite composition which, in turn, has led to misinterpretations of the chemical formula of the mineral and its synthetic analogues. The central atom resides in the general 8f position and coordinates to six oxygen atoms to form a highly distorted octahedron (Figure 3C), with the bond distances varying from 2.1 to almost 2.5 Å (Table 5).

Table 5. Bond lengths (Å), bond-valence sums (BVS, valence units) and occupancy factors for the M5A site of stanfieldite and its synthetic analogue [1].

	Brahin	Synthetic	Imilac
M5A—O4	2.194(2)	2.212(3)	
M5A—O5	2.291(2)	2.302(3)	
M5A—O18	2.494(2)	2.493(3)	
M5A—O19	2.200(2)	2.223(4)	
M5A—O22	2.099(3)	2.107(4)	
M5A—O23	2.262(3)	2.282(4)	
Mean M5A—O	2.257	2.270	2.282
BVS	1.89	1.92	
Ca	0.48(4)	0.51	0.55
Mg	0.43(2)	0.49	0.353(3)
Fe^{2+}	0.08(2)		0.097(3)
Reference	This work	[40]	[41]

[1] Bond-valence sums were calculated by summation of individual contributions for each element, based on parameters reported by Brese and O'Keeffe [48].

The M5A octahedra form paired clusters in the structure via corner-linking by phosphate tetrahedra (Figure 3C). Based on previous reports [40,41,52] and our data (Table 5), M5A may accommodate Ca, Mg, Fe^{2+} and Ni in different proportions, with the total occupancy equal to unity. Natural stanfieldite is a Mg-dominant mineral, and the refinement of M5A occupancy leads to a dominance of Mg over Fe^{2+} as well (Table 5). The latter is supported by calculation of the bond-valence sum, which is almost identical to that of synthetic analogue of stanfieldite (Table 3). The variability of M5A occupancy substantiates the existence of solid solution between hypothetical Mg and Ca end members.

The former would have the composition corresponding to $Ca_7Mg_2Mg_9(PO_4)_{12}$. The latter member would correspond to $Ca_7Ca_2Mg_9(PO_4)_{12}$, that is equal to $Ca_3Mg_3(PO_4)_4$. The intermediate composition having Ca = Mg in M5A results in a formula $Ca_7(CaMg)Mg_9(PO_4)_{12}$, or, in a simplified form, $Ca_4Mg_5(PO_4)_4$. One can see that the latter perfectly fulfils the ideal composition of stanfieldite proposed by Fuchs [9]. It is noteworthy that stanfieldite from Brahin described herein, the previously reported mineral from Imilac [41] and the synthetic analogue of stanfieldite [40] have M5A occupancies almost equally shared between Ca and (Mg + Fe) (Table 5). This could lead to the assumption that the ordering between Ca and Σ(Mg, Fe) might exist in the M5A site. However, neither our observations nor previously reported data reveal superstructure reflections which would evidence the Ca/Mg ordering. In this respect, an overview of reported compositions of stanfieldite-like minerals and compounds would be of special interest. We have collected the chemically relevant data which are gathered in Table 6 and plotted in Figure 4. It can be seen that the overwhelming majority of stanfieldite compositions fall within the range corresponding to Ca ≈ (Mg + Fe) in the M5A site. Therefore, the above assumption on the possible Ca/Mg ordering, albeit speculative, has a statistically substantiated basis.

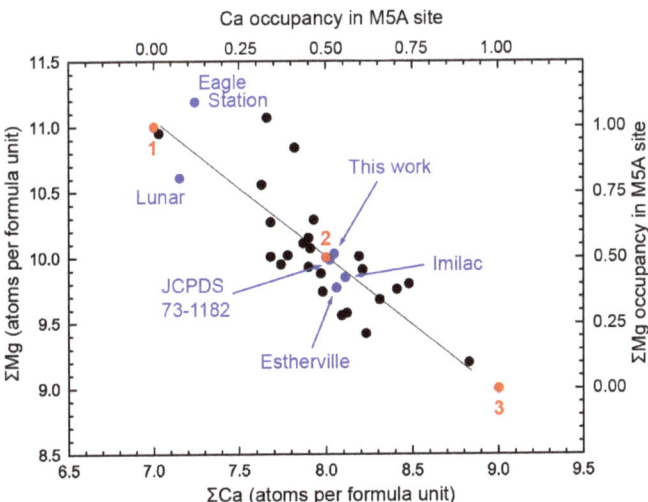

Figure 4. Plot of ΣCa versus ΣMg group element contents in natural, technogenic and synthetic stanfieldite. Left and bottom scale: formula amounts recalculated on the basis of 48 oxygen atoms per formula unit. Right and upper scale: expected occupancy factors for the M5A site. The grey straight line shows the linear fit for the depicted analytical data. The red dots denote theoretical (calculated) compositions corresponding to (1) $Ca_7Mg_2Mg_9(PO_4)_{12}$, (2) $Ca_7(CaMg)Mg_9(PO_4)_{12} \equiv Ca_4Mg_5(PO_4)_6$, (3) $Ca_7Ca_2Mg_9(PO_4)_{12} \equiv Ca_3Mg_3(PO_4)_4$. The blue dots and labels mark the compositions of particular interest which are discussed in the paper. References and source data are given in Table 6.

Table 6. Formula amounts of cations in stanfieldite and its analogues grouped by elements [1].

Source [2]	ΣCa [3]	ΣMg [4]	P + Si	Reference
Albin	8.23	9.42	12.15	[7]
Antofagasta	8.48	9.80	11.89	[7]
Antofagasta	8.41	9.76	11.93	[7]
Brahin	7.68	10.01	12.12	[43]
Brahin [5]	8.04	10.03	11.97	
Eagle Station	7.63	10.56	11.93	[7]
Eagle Station	7.24	11.19	11.86	[7]
Eagle Station	7.78	10.02	12.08	[10]
Estherville	8.06	9.77	12.07	[9]
Imilac	7.91	10.07	12.04	[7]
Imilac	8.11	9.85	12.02	[41]
Lunar	7.15	10.61	12.12	[16]
Mt. Vernon	8.21	9.91	11.95	[7]
Mt. Vernon	8.09	9.56	12.14	[7]
Mt. Vernon	8.31	9.68	12.02	[7]
Ollague	8.19	10.01	11.92	[7]
Rawlinna	7.66	11.07	11.79	[7]
Rawlinna	7.82	10.84	11.73	[7]
Santa Rosalia	7.90	9.93	12.08	[7]
Santa Rosalia	8.12	9.58	12.13	[7]
Santa Rosalia	7.98	9.74	12.11	[7]
Springwater	7.97	9.88	12.08	[7]
Springwater	7.74	9.95	12.14	[7]
Springwater	7.90	10.15	12.00	[7]
Springwater	7.87	10.11	12.01	[10]
Vaca Muerta	7.93	10.29	11.97	[15]
Slags	8.83	9.20	12.01	[19]
Slags	7.03	10.95	12.04	[19]
Slags	7.68	10.27	12.05	[19]
Synthetic	8.02	9.98	12.00	[40]

[1] Atoms per formula unit calculated on the basis of 48 oxygen atoms. [2] Meteorite names. Non-meteoritic sources are shown in italic type. [3] ΣCa includes (Ca, Na, K). [4] ΣMg includes (Mg, Fe, Mn, Al, Cr, Ti). [5] Present work.

The next interesting point is a significant departure of total cationic sums of many analyses from the ideal value requiring 18 cations per formula unit. These departures are readily revealed by the shifts of corresponding analytical points from the linear fit in Figure 4. At present, we have no explanation for the observed departures. They could imply the existence of analytical errors in the reported microprobe data. On the other hand, these shifts might mean the occurrence of vacancies in cationic sites of stanfieldite structure, and then they deserve a special investigation.

Although the majority of reported data fall within the central area of the plot in Figure 4, there are a few points showing significant prevalence of (Mg + Fe) sum over total Ca. These include one analysis from the Eagle Station pallasite [7] and the mineral found in the Lunar sample 66095 returned by the Apollo 16 mission [16]. These two analyses approach the $Ca_7Mg_2Mg_9(PO_4)_{12}$ end-member of the $M5A$ solid solution. At the opposite extreme of the plot, there is a single point approaching hypothetical $Ca_3M_3(PO_4)_4$ composition. This analysis, along with two more listed in Table 6, relate to a stanfieldite-like phosphate described from the ancient slags found in Tyrol [19]. The main feature of this compound is wide variations both in Ca/(Fe + Mg) and Fe/Mg ratios, up to nearly Fe-dominant compositions. Schneider, with co-authors [19], has provided Raman spectrum for this phosphate, but in the absence of Raman spectra for genuine stanfieldite, the comparison was not possible. We herein provide the Raman spectrum of stanfieldite from the Brahin meteorite (Figure 5). A comparison of this spectrum with that reported by Schneider with co-authors [19] shows that the latter can represent a poorly crystallized Fe-dominant analogue of stanfieldite.

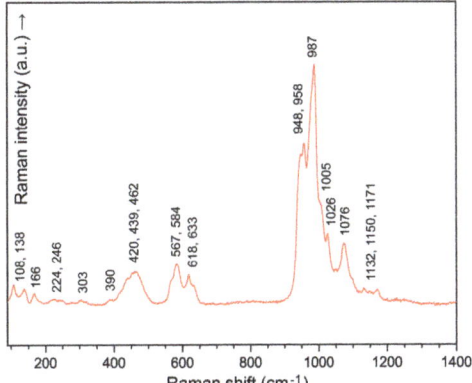

Figure 5. The Raman spectrum of stanfieldite from the Brahin meteorite. The bands between 900 and 1150 cm^{-1} correspond to stretching vibrations in [PO$_4$] tetrahedra. Bands in the range 400–650 cm^{-1} relate to bending modes of phosphate anion.

5. Ca$_3$Mg$_3$(PO$_4$)$_4$ Phosphors: Do They Exist?

In this section, we would like to clarify the mistake caused by the incorrect database assertion of primary structural data on the synthetic analogue of stanfieldite reported by Dickens and Brown [40]. In the title of their article, Dickens and Brown report the formula Ca$_7$Mg$_9$(Ca,Mg)$_2$(PO$_4$)$_{12}$, with Ca = Mg in the Ca5 site, leading to a bulk one Ca$_4$Mg$_5$(PO$_4$)$_6$. It looks obvious that the mistake was introduced in the stage of structural data transfer from the article tables to the ICSD database. The Ca5 site, equally occupied by Ca and Mg [40], was erroneously assigned to be fully occupied by Ca. The latter had led to a wrong formula, Ca$_3$Mg$_3$(PO$_4$)$_4$, which still appears in the ICSD database [53] (ICSD code 23642). Moreover, the calculated X-ray powder diffraction pattern has been further included into the ICDD (JCPDS) database under the reference number JCPDS-ICDD 73-1182 (Figure 4). There is a substantial interest to the family of luminophores (phosphors) based on the stanfieldite structure [35–39]. It is erroneous that the mistake in the chemical formula caused by the incorrect primary data transfer has passed first to the ICDD database and then to the papers devoted to a study of these phosphor materials [35–39]. Unfortunately, neither of the published articles does contain quantitative chemical data on synthesized phosphors. Thus, in lieu of any evidence supporting the existence of Ca$_3$Mg$_3$(PO$_4$)$_4$, one can state that these phosphors are in fact stanfieldite-based, Ca$_4$Mg$_5$(PO$_4$)$_4$ compounds.

Supplementary Materials: The following are available online at http://www.mdpi.com/2073-4352/10/6/464/s1: Supplementary crystallographic data for stanfieldite from the Brahin meteorite in Crystallographic Information File (CIF) format. Alternatively, CCDC reference number 1998335 contains the same data, and it can be obtained free of charge from the Cambridge Crystallographic Data Centre at www.ccdc.cam.ac.uk. S2, supplementary crystallographic Tables S1–S3.

Author Contributions: Conceptualization, writing—original draft preparation, S.N.B.; investigation, S.N.B., M.G.K. and V.N.B.; data curation, E.V.O. All authors have read and agreed to the published version of the manuscript.

Funding: This research was financially supported by the Russian Science Foundation, grant number 18-17-00079.

Acknowledgments: The authors gratefully acknowledge the curators of the Mining Museum, Saint Petersburg Mining University, for providing the sample of the Brahin meteorite used in this study. We are thankful to anonymous reviewers for their comments and suggestions. This research was supported by the resource Centre for X-ray diffraction studies and the Geomodel resource centre of Saint Petersburg State University.

Conflicts of Interest: The authors declare no conflict of interest.

References

1. Hystad, G.; Downs, R.T.; Grew, E.S.; Hazen, R.M. Statistical analysis of mineral diversity and distribution: Earth's mineralogy is unique. *Earth Planet Sci. Lett.* **2015**, *426*, 154–157. [CrossRef]
2. Rubin, A.E.; Ma, C. Meteoritic minerals and their origins. *Chem. Erde—Geochem.* **2017**, *77*, 325–385. [CrossRef]
3. Jolliff, B.L.; Hughes, J.M.; Freeman, J.J.; Zeigler, R.A. Crystal chemistry of lunar merrillite and comparison to other meteoritic and planetary suites of whitlockite and merrillite. *Am. Mineral.* **2006**, *91*, 1583–1595. [CrossRef]
4. Britvin, S.N.; Krivovichev, S.V.; Armbruster, T. Ferromerrillite, $Ca_9NaFe^{2+}(PO_4)_7$, a new mineral from the Martian meteorites, and some insights into merrillite-tuite transformation in shergottites. *Eur. J. Mineral.* **2016**, *28*, 125–136. [CrossRef]
5. Lewis, J.A.; Jones, R.H. Phosphate and feldspar mineralogy of equilibrated L chondrites: The record of metasomatism during metamorphism in ordinary chondrite parent bodies. *Meteor. Planet. Sci.* **2016**, *51*, 1886–1913. [CrossRef]
6. Buseck, P.R. Pallasite meteorites—Mineralogy, petrology and geochemistry. *Geochim. Cosmochim. Acta* **1977**, *41*, 711–740. [CrossRef]
7. Buseck, P.B.; Holdsworth, E.F. Phosphate minerals in pallasite meteorites. *Mineral. Mag.* **1977**, *41*, 91–102. [CrossRef]
8. Righter, K.; Arculus, R.J.; Paslick, C.; Delano, J.W. Electrochemical measurements and thermodynamic calculations of redox equilibria in pallasite meteorites—Implications for the eucrite parent body. *Geochim. Cosmochim. Acta* **1990**, *54*, 1803–1815. [CrossRef]
9. Fuchs, L.H. Stanfieldite: A new phosphate mineral from stony-iron meteorites. *Science* **1967**, *158*, 910–911. [CrossRef]
10. Davis, A.M.; Olsen, E.J. Phosphates in pallasite meteorites as probes of mantle processes in small planetary bodies. *Nature* **1991**, *353*, 637–640. [CrossRef]
11. Sharygin, V.V.; Kovyazin, S.V.; Podgornykh, N.M. Mineralogy of olivine-hosted inclusions from the Omolon pallasite. In Proceedings of the 37th Annual Lunar and Planetary Science Conference, League City, TX, USA, 13–17 March 2006. abstract no. 1235.
12. Boesenberg, J.S.; Delaney, J.S.; Hewins, R.H. A petrological and chemical reexamination of Main Group pallasite formation. *Geochim. Cosmochim. Acta* **2012**, *89*, 134–158. [CrossRef]
13. McKibbin, S.J.; Pittarello, L.; Makarona, C.; Hamann, C.; Hecht, L.; Chernonozhkin, S.M.; Goderis, S.; Claeys, P. Petrogenesis of main group pallasite meteorites based on relationships among texture, mineralogy, and geochemistry. *Meteor. Planet. Sci.* **2019**, *54*, 2814–2844. [CrossRef]
14. Kong, P.; Su, W.; Li, X.; Spettel, B.; Palme, H.; Tao, K. Geochemistry and origin of metal, olivine clasts, and matrix in the Dong Ujimqin Qi mesosiderite. *Meteor. Planet. Sci.* **2008**, *43*, 451–460. [CrossRef]
15. Greenwood, R.C.; Barrat, J.-A.; Scott, E.R.D.; Haack, H.; Buchanan, P.C.; Franchi, I.A.; Yamaguchi, A.; Johnson, D.; Bevan, A.W.R.; Burbine, T.H. Geochemistry and oxygen isotope composition of main-group pallasites and olivine-rich clasts in mesosiderites: Implications for the "Great Dunite Shortage" and HED-mesosiderite connection. *Geochim. Cosmochim. Acta* **2015**, *169*, 115–136. [CrossRef]
16. Shearer, C.K.; Sharp, Z.D.; Burger, P.V.; McCubbin, F.M.; Provencio, P.P.; Brearley, A.J.; Steele, A. Chlorine distribution and its isotopic composition in "rusty rock" 66095. Implications for volatile element enrichments of "rusty rock" and lunar soils, origin of "rusty" alteration, and volatile element behavior on the Moon. *Geochim. Cosmochim. Acta* **2014**, *139*, 411–433. [CrossRef]
17. Hsu, W. Minor element zoning and trace element geochemistry of pallasites. *Meteor. Planet. Sci.* **2003**, *38*, 217–1241. [CrossRef]
18. Tollari, N.; Toplis, M.J.; Barnes, S.-J. Predicting phosphate saturation in silicate magmas: An experimental study of the effects of melt composition and temperature. *Geochim. Cosmochim. Acta* **2006**, *70*, 1518–1536. [CrossRef]
19. Schneider, P.; Tropper, P.; Kaindl, R. The formation of phosphoran olivine and stanfieldite from the pyrometamorphic breakdown of apatite in slags from a prehistoric ritual immolation site (Goldbichl, Igls, Tyrol, Austria). *Mineral. Petrol.* **2013**, *107*, 327–340. [CrossRef]
20. Tarrago, M.; Garcia-Valles, M.; Martinez, S.; Pradell, T.; Bruna, P. Fe in P-doped basaltic melts: A Mössbauer spectroscopy study. *Mater. Lett.* **2018**, *228*, 57–60. [CrossRef]

21. LeGeros, R.Z.; Mijares, D.; Yao, F.; Tannous, S.; Catig, G.; Xi, Q.; Dias, R.; LeGeros, J.P. Synthetic bone mineral (SBM) for osteoporosis therapy: Part 1 - prevention of bone loss from mineral deficiency. *Key Eng. Mater.* **2008**, *361–363 Pt 1*, 43–46. [CrossRef]
22. Vorndran, E.; Ewald, A.; Müller, F.A.; Zorn, K.; Kufner, A.; Gbureck, U. Formation and properties of magnesium-ammonium-phosphate hexahydrate biocements in the Ca-Mg-PO$_4$ system. *J. Mater. Sci. Mater. Med.* **2011**, *22*, 429–436. [CrossRef] [PubMed]
23. Alkhraisat, M.H.; Cabrejos-Azama, J.; Rodriguez, C.R.; Jerez, L.B.; Cabarcos, E.L. Magnesium substitution in brushite cements. *Mater. Sci. Eng. C* **2013**, *33*, 475–481. [CrossRef] [PubMed]
24. Christel, T.; Geffers, M.; Klammert, U.; Nies, B.; Höß, A.; Groll, J.; Kübler, A.C.; Gbureck, U. Fabrication and cytocompatibility of spherical magnesium ammonium phosphate granules. *Mater. Sci. Eng. C* **2014**, *42*, 130–136. [CrossRef] [PubMed]
25. Khan, N.I.; Ijaz, K.; Zahid, M.; Khan, A.S.; Abdul Kadir, M.R.; Hussain, R.; Anis-ur-Rehman; Darr, J.A.; Ihtesham-ur-Rehman; Chaudhry, A.A. Microwave assisted synthesis and characterization of magnesium substituted calcium phosphate bioceramics. *Mater. Sci. Eng. C* **2015**, *56*, 286–293. [CrossRef]
26. Singh, S.S.; Roy, A.; Lee, B.; Banerjee, I.; Kumta, P.N. Synthesis, characterization, and in-vitro cytocompatibility of amorphous β-tri-calcium magnesium phosphate ceramics. *Mater. Sci. Eng. C* **2016**, *67*, 636–645. [CrossRef]
27. Blum, C.; Brückner, T.; Ewald, A.; Ignatius, A.; Gbureck, U. Mg:Ca ratio as regulating factor for osteoclastic in vitro resorption of struvite biocements. *Mater. Sci. Eng. C* **2017**, *73*, 111–119. [CrossRef]
28. Ammar, H.; Nasr, S.; Ageorges, H.; Salem, E.B. Sintering and mechanical properties of magnesium containing hydroxyfluorapatite. *J. Aust. Ceram. Soc.* **2019**, in press. [CrossRef]
29. Hernandez, A.B.; Ferrasse, J.-H.; Chaurand, P.; Saveyn, H.; Borschneck, D.; Roche, N. Mineralogy and leachability of gasified sewage sludge solid residues. *J. Hazard. Mater.* **2011**, *191*, 219–227. [CrossRef]
30. Zhang, Q.; Liu, H.; Li, W.; Xu, J.; Liang, Q. Behavior of phosphorus during co-gasification of sewage sludge and coal. *Energy Fuels* **2012**, *26*, 2830–2836. [CrossRef]
31. Qian, T.-T.; Jiang, H. Migration of phosphorus in sewage sludge during different thermal treatment processes. *ACS Sustain. Chem. Eng.* **2014**, *2*, 1411–1419. [CrossRef]
32. Zhao, Y.; Ren, Q.; Na, Y. Potential utilization of phosphorus in fly ash from industrial sewage sludge incineration with biomass. *Fuel Process. Technol.* **2019**, *188*, 16–21. [CrossRef]
33. Huminicki, D.M.C.; Hawthorne, F.C. The crystal chemistry of the phosphate minerals. In *Phosphates—Geochemical, Geobiological, and Materials Importance*; Reviews in Mineralogy and Geochemistry; Kohn, M.L., Rakovan, J., Hughes, J.M., Eds.; Mineralogical Society of America: Chantilly, VA, USA, 2002; Volume 48, p. 148.
34. Jumpei, A. Phase diagrams of Ca$_3$(PO$_4$)$_2$-Mg$_3$(PO$_4$)$_2$ and Ca$_3$(PO$_4$)$_2$-CaNaPO$_4$ systems. *Bull. Chem. Soc. Jpn.* **1958**, *31*, 201–205. [CrossRef]
35. Wu, W.; Xia, Z. Synthesis and color-tunable luminescence properties of Eu^{2+} and Mn^{2+}-activated Ca$_3$Mg$_3$(PO$_4$)$_4$ phosphor for solid state lighting. *RSC Adv.* **2013**, *3*, 6051–6057. [CrossRef]
36. Ju, G.; Hu, Y.; Chen, L.; Wang, X.; Mu, Z. Blue persistent luminescence in Eu^{2+} doped Ca$_3$Mg$_3$(PO$_4$)$_4$. *Opt. Mater.* **2014**, *36*, 1183–1188. [CrossRef]
37. Nair, G.B.; Dhoble, S.J. White light emission through efficient energy transfer from Ce^{3+} to Dy^{3+} ions in Ca$_3$Mg$_3$(PO$_4$)$_4$ matrix aided by Li$^+$ charge compensator. *J. Lumin.* **2017**, *192*, 1157–1166. [CrossRef]
38. Nair, G.B.; Dhoble, S.J. Orange light-emitting Ca$_3$Mg$_3$(PO$_4$)$_4$:Sm^{3+} phosphors. *Luminescence* **2017**, *32*, 125–128. [CrossRef]
39. Li, H.; Wang, Y. Effect of oxygen vacancies on the reduction of Eu^{3+} in Mg$_3$Ca$_3$(PO$_4$)$_4$ in air atmosphere. *Inorg. Chem.* **2017**, *56*, 10396–10403. [CrossRef]
40. Dickens, B.; Brown, W.E. The crystal structure of Ca$_7$Mg$_9$(Ca,Mg)$_2$(PO$_4$)$_{12}$. *Tscherm. Mineral. Petrogr. Mitt.* **1971**, *16*, 79–104. [CrossRef]
41. Steele, I.M.; Olsen, E. Crystal structure of natural stanfieldite from the Imilac pallasite. In Proceedings of the Lunar and Planetary Science Conference, Houston, TX, USA, 16–20 March 1992; pp. 1355–1356.
42. Brahin. Available online: https://www.lpi.usra.edu/meteor/metbull.php?code=5130 (accessed on 27 April 2020).
43. Bondar, Y.V.; Perelygin, V.P. Fission track age of the Brahin pallasite. *Solar Syst. Res.* **2001**, *35*, 299–306. [CrossRef]

44. Rigaku Oxford Diffraction. *CrysAlisCCD, CrysAlisRED and CrysAlisPRO*; Oxford Diffraction Ltd.: Oxford, UK, 2017.
45. Sheldrick, G. *SHELXT*—Integrated space-group and crystal-structure determination. *Acta Crystallogr. A* **2015**, *71*, 3–8. [CrossRef]
46. Dolomanov, O.V.; Bourhis, L.J.; Gildea, R.J.; Howard, J.A.K.; Puschmann, H. Olex2: A complete structure solution, refinement and analysis program. *J. Appl. Crystallogr.* **2009**, *42*, 339–341. [CrossRef]
47. Britvin, S.N.; Dolivo-Dobrovolsky, D.V.; Krzhizhanovskaya, M.G. Software for processing the X-ray powder diffraction data obtained from the curved image plate detector of Rigaku RAXIS Rapid II diffractometer. *Zapiski Russ. Mineral. Soc.* **2017**, *146*, 104–107. (In Russian)
48. Brese, N.E.; O'Keeffe, M. Bond-valence parameters for solids. *Acta Crystallogr.* **1991**, *47*, 192–197. [CrossRef]
49. Moore, P.B. Bracelets and pinwheels: A topological-geometrical approach to the calcium orthosilicate and alkali sulfate structures. *Am. Mineral.* **1973**, *58*, 32–42.
50. Eysel, W. Crystal chemistry of the system Na_2SO_4–K_2SO_4–K_2CrO_4–Na_2CrO_4 and of the glaserite phase. *Am. Mineral.* **1973**, *58*, 736–747.
51. Gopal, R.; Calvo, C. Structural relationship of whitlockite and β-$Ca_3(PO_4)_2$. *Nat. Phys. Sci.* **1972**, *237*, 30–32. [CrossRef]
52. El Bali, B.; Boukhari, A.; Holt, E.M.; Aride, J. Calcium nickel orthophosphate: Crystal structure of $Ca_{8.5}Ni_{9.5}(PO_4)_{12}$. *Z. Kristallogr.* **1995**, *210*, 838–842.
53. Inorganic Crystal Structure Database ICSD. Available online: https://icsd.fizkarlsruhe. (accessed on 30 April 2020).

© 2020 by the authors. Licensee MDPI, Basel, Switzerland. This article is an open access article distributed under the terms and conditions of the Creative Commons Attribution (CC BY) license (http://creativecommons.org/licenses/by/4.0/).

Article

The High Pressure Behavior of Galenobismutite, PbBi$_2$S$_4$: A Synchrotron Single Crystal X-ray Diffraction Study

Paola Comodi [1,*], Azzurra Zucchini [1], Tonci Balić-Žunić [2], Michael Hanfland [3] and Ines Collings [3]

1. Dipartimento di Fisica e Geologia, Università di Perugia, 06123 Perugia, Italy; azzurra.zucchini@unipg.it
2. Department of Geosciences and Natural Resource Management, University of Copenhagen, 1017 Copenhagen, Denmark; toncib@ign.ku.dk
3. European Synchrotron Radiation Facility, 71 avenue des Martyrs, 38000 Grenoble, France; hanfland@esrf.fr (M.H.); ines.collings@esrf.fr (I.C.)
* Correspondence: paola.comodi@unipg.it; Tel.: +39-075-5852656

Received: 27 March 2019; Accepted: 16 April 2019; Published: 18 April 2019

Abstract: High-pressure single-crystal synchrotron X-ray diffraction data for galenobismutite, PbBi$_2$S$_4$ collected up to 20.9 GPa, were fitted by a third-order Birch-Murnaghan equation of state, as suggested by a F_E-f_E plot, yielding V_0 = 697.4(8) Å^3, K_0 = 51(1) GPa and K' = 5.0(2). The axial moduli were M_{0a} = 115(7) GPa and M_a' = 28(2) for the a axis, M_{0b} = 162(3) GPa and M_b' = 8(3) for the b axis, M_{0c} = 142(8) GPa and M_c' = 26(2) for the c axis, with refined values of a_0, b_0, c_0 equal to 11.791(7) Å, 14.540(6) Å 4.076(3) Å, respectively, and a ratio equal to M_{0a}:M_{0b}:M_{0c} = 1.55:1:1.79. The main structural changes on compression were the M2 and M3 (occupied by Bi, Pb) movements toward the centers of their respective trigonal prism bodies and M3 changes towards CN8. The M1 site, occupied solely by Bi, regularizes the octahedral form with CN6. The eccentricities of all cation sites decreased with compression testifying for a decrease in stereochemical expression of lone electron pairs. Galenobismutite is isostructural with calcium ferrite CaFe$_2$O$_4$, the suggested high pressure structure can host Na and Al in the lower mantle. The study indicates that pressure enables the incorporation of other elements in this structure, increasing its potential significance for mantle mineralogy.

Keywords: galenobismutite; high pressure; single-crystal X-ray synchrotron diffraction; equation of state; calcium ferrite structure type; lone electron pair

1. Introduction

Galenobismutite PbBi$_2$S$_4$ is a Bi-sulfosalt usually found in hydrothermal veins or associated with fumarolic deposits [1,2]. Like other Bi-minerals, it has an important role in the reconstruction of the formation of ore deposits as it is sensitive to physical–chemical fluctuations and can constrain the genesis of ore.

According to sulfosalt classification [3] galenobismutite is classified among commensurate composite derivatives of cannizzarite in a sub-group with angelaite, Cu$_2$AgPbBiS$_4$ [4], nuffieldite, Cu$_{1.4}$Pb$_{2.4}$Bi$_{2.4}$Sb$_{0.2}$S$_7$ [5] and weibullite, Ag$_{0.33}$Pb$_{5.33}$Bi$_{8.33}$(S,Se)$_{18}$ [6]. It has a distinctly different crystal structure from the chemically (stoichiometrically) similar berthierite FeSb$_2$S$_4$ [7], garavellite FeSbBiS$_4$ [8] and clerite MnSb$_2$S$_4$ [9] which form a berthierite isotypic series [3].

Chemical substitution of Sb for Bi and Fe for Pb are common in galenobismutite. It is illustrated by galenobismutite from Beiya porphyry- and skarn-type deposits that contain Sb up to 0.39% and Fe up to 0.42% [10]. Selenium can replace sulfur in galenobismutite. In galenobismutite from Vulcano Island (Italy) [1] heterogeneous distribution of selenium in the Sulphur sites was found, with a total amount of up to 0.13 atoms of Se per formula unit. Moreover, galenobismutite from Vulcano shows an unusual presence of Cl, according to the coupled heterovalent substitution scheme: $Pb^{2+} + Cl^- = Bi^{3+} + S^{2-}$ [2].

The crystal structure of galenobismutite is orthorhombic, space group *Pnam*, and was described firstly by Wickman [11], then by Iitaka and Nowacki [12], and later classified by Makovicky [13] as being representative of a specific subgroup of cannizzarite-type structures.

The crystal structure contains three cation positions. M1 has a slightly distorted octahedral coordination and forms fragments of galena-like structure two octahedra wide by sharing edges with a conjugated M1 octahedron. M2 is surrounded by seven S atoms forming a "lying" (prism axis is perpendicular to the *c* crystal axis) mono-capped trigonal prism. M3 polyhedron is a "standing" (prism axis parallel to the *c* crystal axis) bi-capped trigonal prism (CN8) with one of the capping ligands relatively distant, so the coordination can also be described as a 7+1 (Figure 1).

The determination of the distribution of Bi and Pb in the three sites by refinement of X-ray diffraction data is practically impossible due to the similar number of electrons (83 and 82, respectively). It is therefore based on the bond valence calculations. According to Pinto et al. [1], who used bond lengths, bond valences and geometrical characteristics of the coordination polyhedral to interpret the occupancy on the M1, M2 and M3 positions. M1 is considered fully occupied by Bi, while M2 and M3 are mixed sites dominated by Bi and Pb, respectively.

The configuration of this structure, isotypic with calcium ferrite (CF) $CaFe_2O_4$, is of a particular interest for high pressure mineral physics. Finger and Hazen [14] include among the seven structure-types with exclusively six-coordinated silicon $NaAlSiO_4$ [15,16], the analogue of CF. This structure bears a close relationship to hollandite, another ^{VI}Si-coordinated structure type. Both structures consist of double octahedral chains which are joined to form 'tunnels' parallel to *c* that accommodate the alkali or alkaline-earth cations. In hollandite four double chains form square tunnels, whereas in CF four double chains define triangular tunnels.

Akaogi et al. [17] were the first who synthesized $NaAlSiO_4$ with CF structure at pressures higher than 18–20 GPa. Tutti et al. [18] found that this phase is stable at pressure up to at least 70–75 GPa and temperatures 800–2200 °C indicating it as an important carrier of Na and Al in the lower Mantle.

The most regular CF-type structure known is that of $PbSc_2S_4$ ([19], Figure 1b). The reported crystal structure of $NaAlSiO_4$ ([16], Figure 1c), obtained from a powder sample, has a substantially more distorted octahedral coordination. Dubrovinsky et al. [20] reported the structure of $NaAlSiO_4$ at 35 GPa, likewise done on a powder sample (Figure 1d). The data suggest that the M1 coordination becomes more regular without significant contraction, whereas M2 and Na coordinations significantly contract keeping their general shapes. Compared to the other CF structures, galenobismutite differs in having CN7 coordination of the M2 site and a significantly distorted coordination of the M3 site. M1 coordination is eccentric, unlike in $PbSc_2S_4$ or $NaAlSiO_4$. The increased CN of M2 and distortions of the other two coordinations in galenobismutite are explained as a stereochemical effect of the lone electron pair of Bi^{3+} [21].

The first high-pressure study of galenobismutite was done by Olsen et al [21] at pressures up to 8.9 GPa with single crystal X-ray diffraction. They found a bulk modulus of K_0 equal to 43.9(7) GPa and a K' of 6.9 (3). No phase transition was observed in this pressure range and, interestingly, although the stereochemical activity of Lone Electron Pair (LEP)'s decreased with pressure, the structure did not approach the CF isotype but moved further away from its typical configuration, keeping its distinct character [21].

The present paper extends the study of the baric behavior of galenobismutite over a significantly larger pressure range, up to 20.9 GPa, by a synchrotron single crystal X-ray diffraction study in order to obtain a more complete picture of its behavior under high pressure. Really the relevance of high pressure single crystal X-ray diffraction data with respect those from high pressure powder diffraction was very recently highlight in several papers i.e. [22,23].

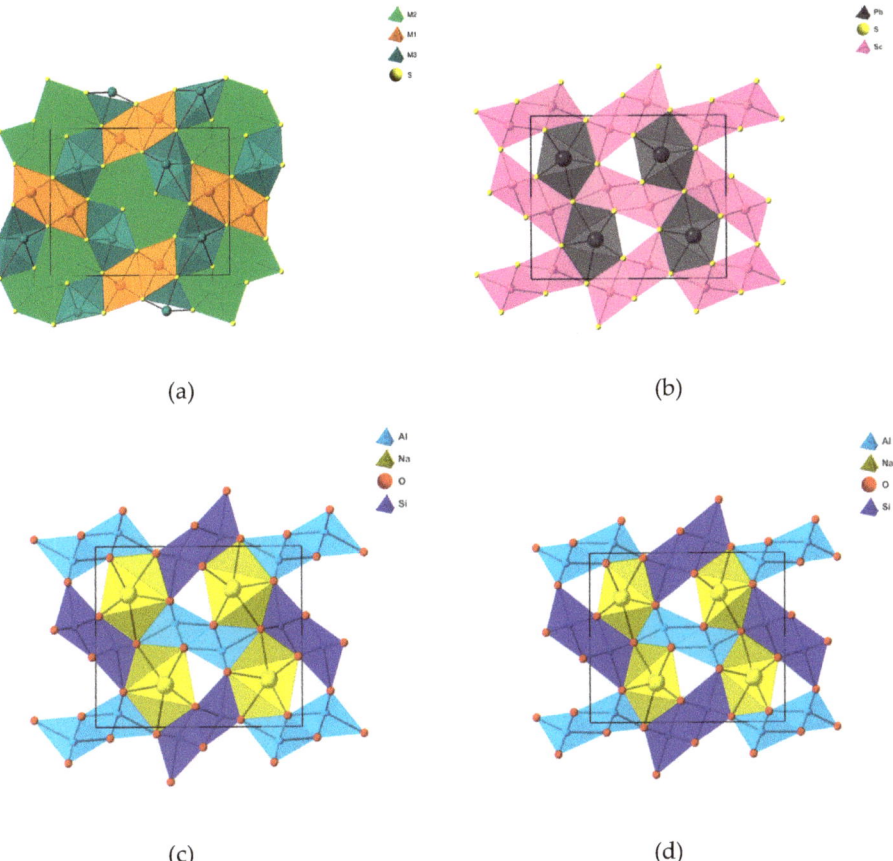

Figure 1. Crystal structures of: (**a**) galenobismutite, projected along [001] with x axis on the vertical and y axis on the horizontal line; (**b**) PbSc$_2$S$_4$ [19]; (**c**) NaAlSiO$_4$ at room pressure [16]; (**d**) NaAlSiO$_4$ at 35 GPa [20].

2. High Pressure Experiments

The HP synchrotron single-crystal X-ray diffraction experiments were carried out at ID-15B beamline at ESRF (Grenoble) dedicated to the determination of structural properties of solids at high pressure using angle-dispersive-diffraction with diamond anvil cells. A membrane-type Diamond Anvil Cell with an opening angle of +/−32 degrees, equipped with 600 μm diamond culets was used. Helium was used as a pressure transmitting medium. According to Singh [24] helium is superior in ensuring near to hydrostatic conditions at pressures of 20 GPa or over compared to argon.

Ruby sphere was loaded as a fluorescent P calibrant together with the galenobismutite sample (80 × 10 × 20 μm^3) in the 300 μm hole in the center of a pre-indented stainless steel gasket of 80 μm thickness. Pressure was measured before and after each data collection.

The sample-to-detector distance was 279.88 mm and calibrated, along with the wavelength, using Si standard and Fit2D software [25]. The same synthetic sample of galenobismutite used in Olsen et al 2007 was selected to collect the present set of measurements, to avoid differences in compressibility ascribable to different chemical compositions in the samples.

The X-ray beam was monochromatized to a wavelength of 0.41125 Å and focused down to 10 × 10 μm area. Data were collected with a DAC rotating 64° around the ω-axis (from −32 to +32°)

with angular step of 0.5° and counting time of 1s per step. The scattered radiation was collected by a MAR555 flat panel detector, with 430 × 350 mm (555mm diagonal) active area.

The extraction and correction of the intensity data, merging of reflections, and the refinements of the crystal lattice parameters were done by means of the CrysAlis software (Agilent technologies) [26]. Measurements were performed at different pressures from 0.5 to 20.9 GPa on increasing pressure, and at 16.43, 8.12, and 2.1 GPa on decreasing pressure to evaluate the reversibility and hysteresis phenomena of structural changes. The absorption correction was applied by means of ABSORB-7 software [27].

The structure refinements were carried out with ShelXle [28] on F^2. Scattering curves for neutral atoms were used. Table 1 summarizes details of data collection and structure refinements up to 20.9 GPa. Final atomic coordinates and isotropic displacement factors are listed in Table 2. Bond lengths, polyhedral volumes and polyhedral distortion parameters at different pressures, are reported in Table 3. Cif files with the hkl, i.e., the Miller indices of the collected reflections, of individual refinements are in Table S1 (deposited).

Table 1. Details of data refinements, lattice parameters, and density at different pressures.

P (GPa)	0.50	1.66	2.13 *	4.94	6.20	8.10 *	8.80	11.40	15.10	16.43 *	17.10	20.90
a (Å)	11.7338(7)	11.6511(4)	11.6167(7)	11.4505(5)	11.3956(6)	11.319(8)	11.2841(6)	11.2038(6)	11.0890(8)	11.0611(5)	11.0423(5)	10.9733(6)
b (Å)	14.5066(4)	14.3951(3)	14.3541(3)	14.1314(4)	14.0521(4)	13.9355(3)	13.8803(3)	13.7379(4)	13.5330(5)	13.4491(3)	13.4115(3)	13.2252(4)
c (Å)	4.05910(6)	4.03492(5)	4.02572(4)	3.97480(5)	3.95717(5)	3.93304(5)	3.92120(5)	3.89439(6)	3.86003(7)	3.84944(5)	3.84254(5)	3.81906(5)
V (Å^3)	690.93(5)	676.73(3)	671.28(4)	643.17(3)	633.67(4)	620.39(5)	614.16(4)	599.41(4)	579.26(5)	572.65(3)	569.06(3)	554.24(4)
ρ (g/cm^3)	7.243	7.395	7.455	7.780	7.897	8.066	8.148	8.348	8.639	8.739	8.794	9.029
Data collection												
# Meas. refl.	1004	1091	1132	1056	1052	1026	1015	1052	1022	1080	1033	1106
# Unique refl.	536	581	551	555	549	510	536	531	501	500	504	511
# Obs. refl. **	506	548	516	514	517		518	520	487	485	488	491
R_{int}	0.0314	0.0314	0.0266	0.0284	0.0315	0.0265	0.0296	0.0321	0.0263	0.0247	0.0237	0.0222
$2\theta_{max}$ (°)	41.84	40.93	41.05	41.29	40.96	41.16	41.02	41.10	41.63	41.16	41.26	41.30
Range hkl	$-10 \leq h \leq 12$	$-9 \leq h \leq 12$	$-7 \leq h \leq 10$	$-12 \leq h \leq 9$	$-9 \leq h \leq 12$	$-9 \leq h \leq 7$	$-11 \leq h \leq 9$	$-11 \leq h \leq 9$	$-9 \leq h \leq 11$	$-11 \leq h \leq 8$	$-9 \leq h \leq 11$	$-9 \leq h \leq 11$
	$-17 \leq k \leq 19$	$-19 \leq k \leq 17$	$-19 \leq k \leq 20$	$-19 \leq k \leq 17$	$-16 \leq k \leq 19$	$-20 \leq k \leq 19$	$-18 \leq k \leq 17$	$-18 \leq k \leq 16$	$-17 \leq k \leq 18$	$-18 \leq k \leq 17$	$-17 \leq k \leq 18$	$-16 \leq k \leq 18$
	$-6 \leq l \leq 5$	$-5 \leq l \leq 5$	$-5 \leq l \leq 5$	$-5 \leq l \leq 5$	$-5 \leq l \leq 5$	$-5 \leq l \leq 5$	$-5 \leq l \leq 5$	$-5 \leq l \leq 5$	$-5 \leq l \leq 5$	$-5 \leq l \leq 5$	$-5 \leq l \leq 5$	$-5 \leq l \leq 5$
senθ/λ (Å^{-1})	0.87	0.85	0.85	0.86	0.85	0.85	0.85	0.85	0.86	0.85	0.86	0.86
2θ max	41.8	40.9	41.1	41.3	41.0	41.2	41.0	41.1	41.6	41.2	41.3	41.3
Completeness (%) ***	25.76	30.11	29.09	29.32	30.13	28.73	29.98	30.18	28.53	29.58	29.66	30.75
Refinement												
R1 (|F$_0$| > 4σ)	0.0246	0.0213	0.0199	0.0223	0.0213	0.0221	0.0210	0.0221	0.0193	0.0163	0.0164	0.0158
wR2	0.0626	0.0537	0.0489	0.0538	0.0539	0.0586	0.0552	0.0553	0.0442	0.0372	0.0367	0.0360
Goodness of Fit (GooF)	1.071	1.023	1.044	1.016	1.032	1.027	1.084	1.078	1.006	1.019	1.016	1.037
# Parameters	44	44	44	44	44	44	44	44	44	44	44	44

* Data collected during decompression, ** Observed reflections |F0| > 4σ, *** Completeness is reported relative to 2θ max.

Table 2. Final atomic coordinates and isotropic displacement factors at different pressures.

	GPa	0.50	1.66	2.13*	4.94	6.20	8.10*	8.80	11.4	15.10	16.43*	17.10	20.90
Bi1	x	0.06702(6)	0.06582(4)	0.06536(7)	0.06325(6)	0.06267(6)	0.06199(7)	0.06172(6)	0.06115(6)	0.06082(6)	0.06074(6)	0.06076(6)	0.06071(6)
	y	0.39075(3)	0.39092(3)	0.39102(3)	0.39170(3)	0.39196(3)	0.39247(3)	0.39268(3)	0.39333(3)	0.39446(3)	0.39490(3)	0.39508(3)	0.39644(3)
	z	0.25	0.25	0.25	0.25	0.25	0.25	0.25	0.25	0.25	0.25	0.25	0.25
	Uiso	0.0150(2)	0.01358(19)	0.0122(2)	0.0112(2)	0.0106(2)	0.0108(3)	0.0103(2)	0.0102(2)	0.01034(18)	0.01120(17)	0.01005(17)	0.01091(16)
	U11	0.0144(7)	0.0138(6)	0.0117(8)	0.0109(6)	0.0090(6)	0.0110(9)	0.0095(6)	0.0093(6)	0.0108(6)	0.0123(5)	0.0113(5)	0.0125(5)
	U22	0.0168(4)	0.0149(3)	0.0135(3)	0.0128(4)	0.0136(4)	0.0123(3)	0.0123(4)	0.0127(4)	0.0117(3)	0.0120(3)	0.0106(3)	0.0115(3)
	U33	0.0139(2)	0.0121(2)	0.0115(2)	0.0097(2)	0.0093(2)	0.0091(2)	0.0091(2)	0.0085(2)	0.00849(18)	0.00937(17)	0.00829(17)	0.00868(16)
	U23	0	0	0	0	0	0	0	0	0	0	0	0
	U13	0	0	0	0	0	0	0	0	0	0	0	0
	U12	0.00162(16)	0.00131(13)	0.00121(18)	0.00082(16)	0.00074(14)	0.00024(18)	0.00052(14)	0.00024(14)	0.00030(14)	0.00024(13)	0.00023(13)	0.00013(13)
Bi2	x	0.10329(6)	0.10545(5)	0.10621(7)	0.10974(6)	0.11084(6)	0.11234(7)	0.11270(6)	0.11413(6)	0.11562(6)	0.11623(6)	0.11627(5)	0.11716(5)
	y	0.90590(3)	0.90648(3)	0.90666(3)	0.90753(4)	0.90789(3)	0.90843(3)	0.90869(3)	0.90935(3)	0.91036(3)	0.91074(3)	0.91097(3)	0.91199(3)
	z	0.25	0.25	0.25	0.25	0.25	0.25	0.25	0.25	0.25	0.25	0.25	0.25
	Uiso	0.0182(2)	0.0158(2)	0.0140(3)	0.0123(2)	0.0112(2)	0.0106(3)	0.0107(2)	0.0101(2)	0.00983(18)	0.01075(17)	0.00945(16)	0.01001(16)
	U11	0.0193(7)	0.0174(6)	0.0141(9)	0.0130(6)	0.0100(6)	0.0101(9)	0.0111(6)	0.0095(6)	0.0104(6)	0.0125(5)	0.0113(5)	0.0115(5)
	U22	0.0230(4)	0.0190(3)	0.0178(3)	0.0146(3)	0.0147(3)	0.0132(3)	0.0126(3)	0.0126(3)	0.0109(3)	0.0107(3)	0.0093(3)	0.0099(2)
	U33	0.0122(2)	0.0110(2)	0.0103(2)	0.0092(2)	0.0090(2)	0.0085(2)	0.0086(2)	0.0081(2)	0.00821(19)	0.00908(18)	0.00768(17)	0.00863(16)
	U23	0	0	0	0	0	0	0	0	0	0	0	0
	U13	0	0	0	0	0	0	0	0	0	0	0	0
	U12	0.00020(18)	0.00009(14)	0.00001(19)	0.00000(16)	0.00005(14)	−0.00003(17)	−0.00007(15)	−0.00010(15)	−0.00006(14)	−0.00009(13)	−0.00008(13)	−0.00030(13)
Pb3	x	0.24848(7)	0.25147(5)	0.25280(7)	0.25793(6)	0.25963(6)	0.26198(7)	0.26265(6)	0.26497(6)	0.26766(6)	0.26874(6)	0.26905(5)	0.27136(5)
	y	0.65233(4)	0.65209(3)	0.65203(3)	0.65201(4)	0.65203(3)	0.65214(3)	0.65219(4)	0.65241(3)	0.65264(3)	0.65274(3)	0.65279(3)	0.65308(3)
	z	0.25	0.25	0.25	0.25	0.25	0.25	0.25	0.25	0.25	0.25	0.25	0.25
	Uiso	0.0229(3)	0.0198(2)	0.0180(3)	0.0153(2)	0.0141(2)	0.0132(3)	0.0126(2)	0.0118(2)	0.01114(19)	0.01206(18)	0.01068(17)	0.01094(17)
	U11	0.0274(8)	0.0247(7)	0.0219(9)	0.0199(7)	0.0164(7)	0.0167(9)	0.0160(6)	0.0145(6)	0.0142(6)	0.0166(6)	0.0149(5)	0.0145(5)
	U22	0.0221(5)	0.0185(4)	0.0172(3)	0.0141(4)	0.0146(4)	0.0131(3)	0.0122(4)	0.0125(3)	0.0113(3)	0.0109(3)	0.0096(3)	0.0104(3)
	U33	0.0192(2)	0.0163(2)	0.0150(2)	0.0119(2)	0.0111(2)	0.0099(2)	0.0096(2)	0.0085(2)	0.00793(19)	0.00876(18)	0.00746(17)	0.00790(16)
	U23	0	0	0	0	0	0	0	0	0	0	0	0
	U13	0	0	0	0	0	0	0	0	0	0	0	0
	U12	0.0053(2)	0.00440(17)	0.0040(2)	0.00297(18)	0.00267(17)	0.00241(18)	0.00217(16)	0.00197(15)	0.00169(15)	0.00157(14)	0.00158(14)	0.00153(13)
S1	x	0.3343(4)	0.3330(3)	0.3335(5)	0.3313(4)	0.3306(4)	0.3309(5)	0.3303(4)	0.3295(4)	0.3285(4)	0.3280(4)	0.3281(4)	0.3267(4)
	y	0.0167(3)	0.0164(2)	0.0169(2)	0.0175(3)	0.0182(2)	0.0189(2)	0.0187(2)	0.0200(2)	0.0210(2)	0.0214(2)	0.0217(2)	0.0228(2)
	z	0.25	0.25	0.25	0.25	0.25	0.25	0.25	0.25	0.25	0.25	0.25	0.25
	Uiso	0.0161(9)	0.0148(8)	0.0150(16)	0.0124(10)	0.0127(9)	0.0108(16)	0.0122(10)	0.0113(10)	0.0108(9)	0.0117(8)	0.0119(9)	0.0110(8)
	U11	0.010(4)	0.010(4)	0.015(6)	0.011(4)	0.016(4)	0.006(6)	0.012(4)	0.010(4)	0.009(4)	0.010(4)	0.016(4)	0.008(3)
	U22	0.021(2)	0.0182(19)	0.0141(19)	0.012(2)	0.0079(19)	0.0134(16)	0.0132(2)	0.0115(19)	0.0102(2)	0.0106(18)	0.0066(18)	0.0117(17)
	U33	0.0178(10)	0.0162(10)	0.0163(11)	0.0137(11)	0.0147(10)	0.0133(10)	0.0122(10)	0.0122(10)	0.0134(10)	0.0145(10)	0.0128(10)	0.0131(10)
	U23	0	0	0	0	0	0	0	0	0	0	0	0
	U13	0	0	0	0	0	0	0	0	0	0	0	0
	U12	0.0001(11)	0.0025(10)	0.0014(14)	0.0008(11)	−0.0003(11)	0.0002(13)	0.017(12)	0.0014(11)	−0.0006(11)	−0.0002(10)	−0.0012(10)	0.0000(10)

Table 2. Cont.

GPa		0.50	1.66	2.13 *	4.94	6.20	8.10 *	8.80	11.4	15.10	16.43 *	17.10	20.90
S2	x	0.2595(4)	0.2591(3)	0.2601(5)	0.2602(4)	0.2597(4)	0.2602(5)	0.2605(4)	0.2614(4)	0.2626(4)	0.2618(4)	0.2620(4)	0.2627(4)
	y	0.2961(2)	0.29633(19)	0.2965(2)	0.2964(2)	0.2967(2)	0.2969(2)	0.2969(2)	0.2971(2)	0.2972(2)	0.2973(2)	0.2973(2)	0.2979(2)
	z	0.25	0.25	0.25	0.25	0.25	0.25	0.25	0.25	0.25	0.25	0.25	0.25
	U_{iso}	0.0135(10)	0.0123(8)	0.0118(14)	0.0091(9)	0.0104(9)	0.0105(14)	0.0098(10)	0.0085(8)	0.0097(9)	0.0098(8)	0.0096(8)	0.0098(8)
	U_{11}	0.011(4)	0.008(4)	0.010(5)	0.002(4)	0.008(4)	0.009(5)	0.007(4)	0.003(4)	0.007(4)	0.008(3)	0.010(3)	0.010(3)
	U_{22}	0.017(2)	0.0150(19)	0.0140(17)	0.014(2)	0.013(2)	0.0123(17)	0.013(2)	0.013(2)	0.013(2)	0.0107(18)	0.0115(19)	0.0103(18)
	U_{33}	0.0127(10)	0.0142(9)	0.0113(10)	0.0115(11)	0.0106(9)	0.0100(9)	0.0092(10)	0.0094(9)	0.0091(9)	0.0107(9)	0.0078(9)	0.0092(9)
	U_{23}	0	0	0	0	0	0	0	0	0	0	0	0
	U_{13}	0	0	0	0	0	0	0	0	0	0	0	0
	U_{12}	−0.0001(11)	0.0011(9)	−0.0008(12)	−0.0010(11)	−0.0011(10)	−0.0002(12)	−0.0005(10)	−0.0005(11)	−0.0004(11)	0.0002(9)	0.0000(10)	−0.0009(9)
S3	x	0.0547(4)	0.0543(3)	0.0543(5)	0.0535(5)	0.0534(4)	0.0530(5)	0.0525(4)	0.0518(4)	0.0510(4)	0.0510(4)	0.0504(4)	0.0499(4)
	y	0.0940(2)	0.0948(2)	0.0947(2)	0.0966(2)	0.0975(2)	0.0984(2)	0.0988(2)	0.1004(2)	0.1020(2)	0.1029(2)	0.1032(2)	0.1052(2)
	z	0.25	0.25	0.25	0.25	0.25	0.25	0.25	0.25	0.25	0.25	0.25	0.25
	U_{iso}	0.0149(10)	0.0139(9)	0.0116(14)	0.0112(10)	0.0109(9)	0.0121(17)	0.0101(9)	0.0096(9)	0.0096(9)	0.0116(9)	0.0087(8)	0.0099(8)
	U_{11}	0.011(4)	0.013(4)	0.007(5)	0.009(4)	0.008(4)	0.014(6)	0.007(4)	0.007(4)	0.009(4)	0.013(4)	0.008(3)	0.012(3)
	U_{22}	0.017(2)	0.0139(17)	0.0148(17)	0.013(2)	0.015(2)	0.0107(18)	0.012(2)	0.012(2)	0.0096(19)	0.0122(19)	0.0094(18)	0.0088(17)
	U_{33}	0.0174(11)	0.0152(9)	0.0132(10)	0.0116(11)	0.0104(10)	0.0117(9)	0.0110(9)	0.0091(9)	0.0097(10)	0.0099(9)	0.0084(9)	0.0093(9)
	U_{23}	0	0	0	0	0	0	0	0	0	0	0	0
	U_{13}	0	0	0	0	0	0	0	0	0	0	0	0
	U_{12}	0.0007(10)	−0.0013(9)	−0.0011(12)	0.0003(11)	−0.0003(10)	−0.0028(12)	0.0000(10)	0.0001(10)	0.0002(9)	−0.0011(10)	−0.0007(9)	−0.0005(9)
S4	x	0.0197(4)	0.0197(4)	0.0191(5)	0.0194(4)	0.0190(4)	0.0194(5)	0.0194(4)	0.0185(4)	0.0183(4)	0.0188(4)	0.0184(4)	0.0185(4)
	y	0.7124(2)	0.71258(19)	0.7129(2)	0.7142(2)	0.7146(2)	0.7151(2)	0.7154(2)	0.7156(2)	0.7157(2)	0.7154(2)	0.7157(2)	0.7151(2)
	z	0.25	0.25	0.25	0.25	0.25	0.25	0.25	0.25	0.25	0.25	0.25	0.25
	U_{iso}	0.0143(9)	0.0130(7)	0.0109(13)	0.0108(10)	0.0106(9)	0.0103(15)	0.0110(9)	0.0108(9)	0.0098(9)	0.0113(9)	0.0105(8)	0.0101(8)
	U_{11}	0.018(4)	0.015(3)	0.011(5)	0.011(4)	0.010(4)	0.009(5)	0.010(4)	0.010(4)	0.007(4)	0.013(3)	0.012(3)	0.010(3)
	U_{22}	0.0123(19)	0.0118(17)	0.0100(16)	0.013(2)	0.0131(19)	0.0118(16)	0.014(2)	0.0136(19)	0.0135(19)	0.0101(17)	0.0101(17)	0.0102(16)
	U_{33}	0.0127(9)	0.0122(8)	0.0115(9)	0.0081(10)	0.0091(9)	0.0101(9)	0.0085(9)	0.0091(9)	0.0093(9)	0.0108(9)	0.0092(9)	0.0101(9)
	U_{23}	0	0	0	0	0	0	0	0	0	0	0	0
	U_{13}	0	0	0	0	0	0	0	0	0	0	0	0
	U_{12}	−0.0031(11)	−0.0034(9)	−0.0023(12)	−0.0036(11)	−0.0040(10)	−0.0035(12)	−0.0040(11)	−0.0051(11)	−0.0037(11)	−0.0039(10)	−0.0040(10)	−0.0019(9)

* Data collected during decompression.

Table 3. Polyhedral evolution with Pressure. (a) M1, (b) M2, (c) M3 polyhedra. For each polyhedra, single bond lengths (in A), average bond lengths (in Å) and polyhedral volumes (in Å^3) are reported. The standard errors are on the last digit. Measurements at 0.0001 GPa come from Olsen et al. [21].

P (GPa)	M1-S2	M1-S4$_{(X2)}$'	M1-S1$_{(X2)}$	M1-S1'	Vol	Mean M1-S
0.0001	2.6890	2.1788	2.9646	3.0623	30.44	2.843
0.50	2.643	2.719	2.966	3.043	30.43	2.843
1.66	2.632	2.699	2.953	3.023	29.90	2.826
4.93	2.624	2.661	2.931	2.953	28.84	2.783
6.20	2.614	2.650	2.922	2.931	28.47	2.782
8.80	2.608	2.633	2.893	2.887	27.76	2.759
11.40	2.604	2.613	2.885	2.855	27.24	2.742
15.10	2.596	2.592	2.857	2.819	26.50	2.719
17.10	2.580	2.581	2.843	2.801	26.07	2.705
20.90	2.571	2.565	2.822	2.781	25.52	2.688
16.43	2.582	2.584	2.848	2.810	26.11	2.710
8.10	2.609	2.639	2.905	2.893	27.95	2.765
2.13	2.638	2.690	2.949	3.000	29.68	2.819

P (GPa)	M2-S3$_{(x2)}$	M2-S3'	M2-S4	M2-S2$_{(x2)}$	M2-S1	Vol	Mean M2-S
0.0001	2.7394	2.7962	2.9913	3.0559	3.1960	36.195	2.939
0.50	2.749	2.788	2.973	3.041	3.151	35.68	2.927
1.66	2.745	2.776	2.965	3.012	3.087	34.85	2.906
4.93	2.7319	2.747	2.924	2.935	2.974	32.89	2.854
6.20	2.725	2.744	2.911	2.921	2.945	32.40	2.842
8.80	2.707	2.725	2.882	2.881	2.891	31.25	2.811
11.40	2.695	2.716	2.869	2.849	2.852	30.38	2.789
15.10	2.677	2.691	2.847	2.810	2.795	29.26	2.758
17.10.	2.667	2.679	2.833	2.797	2.771	28.76	2.744
20.90	2.656	2.660	2.820	2.768	2.727	27.96	2.722
16.43	2.676	2.683	2.840	2.802	2.775	28.98	2.751
8.10.	2.716	2.731	2.892	2.892	2.914	31.67	2.822
2.13	2.744	2.766	2.960	2.994	3.078	34.52	2.897

P (GPa)	M3-S4	M3-S2$_{(x2)}$	M2-S1$_{(x2)}$	M3-S3$_{(x2)}$	M2-S4'	Vol	Mean M3-S
0.0001	2.8400	2.9320	3.0083	3.2375	3.7572	38.91	3.028
0.50	2.822	2.912	2.989	3.189	3.739	37.78	3.000
1.66	2.837	2.898	2.976	3.142	3.683	37.01	2.981
4.93	2.869	2.860	2.931	3.037	3.543	35.08	2.933
6.20	2.880	2.845	2.917	3.007	3.500	34.47	2.917
8.80	2.882	2.819	2.894	2.957	3.431	33.39	2.889
11.40	2.895	2.798	2.867	2.918	3.370	32.50	2.866
15.10	2.894	2.768	2.835	2.870	3.301	31.30	2.834
17.10	2.893	2.751	2.817	2.848	3.270	30.70	2.818
20.90	2.893	2.730	2.788	2.810	3.224	29.78	2.793
16.43	2.890	2.757	2.823	2.851	3.286	30.86	2.822
8.10	2.883	2.828	2.902	2.969	3.451	33.70	2.897
2.13	2.852	2.894	2.970	3.122	3.600	36.72	2.975

3. Results

3.1. Compressibility

The evolution of the unit-cell of galenobismutite with pressure is reported in Figure 2 and in Table 1. The behavior of the cell parameters shows no discontinuities in the investigated pressure range and indicates that no phase transition occurs in galenobismutite structure up to 20.9 GPa. The volume-pressure data were fitted with a third-order Birch-Murnaghan equations-of-state, using the EOSFIT7-GUI software [29], as suggested by f_E-F_E, namely the "Eulerian finite strain" versus "normalized stress" plot [30], (Figure 3). The third order Birch-Murnaghan Equation of State (EoS) fit yields $V_0 = 697.4(8)$ Å^3, $K_0 = 51(1)$ GPa and $K' = 5.0(2)$. The bulk modulus and the first derivative values were in good agreement with the values obtained from the f_E-F_E plot [30]. The intercept value and the slope obtained by a linear regression give F_{E0} and K' values equal to 51(1) GPa and 4.8(8), respectively.

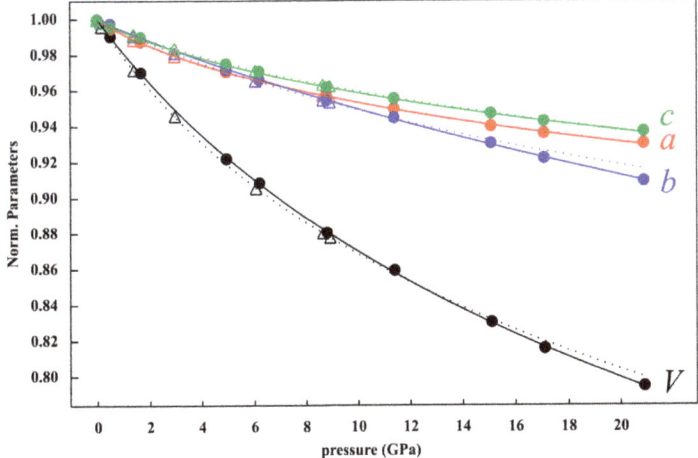

Figure 2. Evolution of the unit cell volume and a, b, c lattice parameters normalized to the values at room conditions as a function of pressure (GPa), fitted by a third-order Birch-Murnaghan EoS. Olsen et al. [21] data are shown by stippled lines and triangles for comparison.

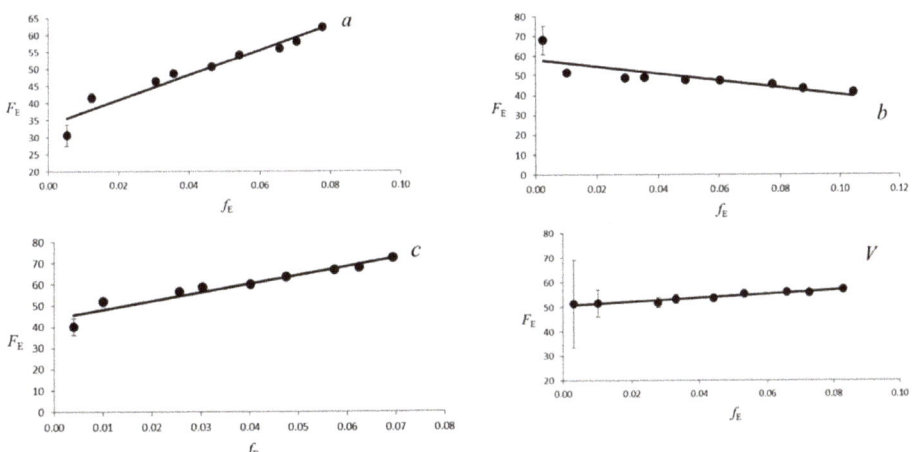

Figure 3. Evolution of the Eulerian finite strain f_E versus the "normalized stress" F_E. The solid line is the weighted linear fit of the data for V, a, b and c lattice parameters.

The lattice parameter moduli, calculated using a third-order Birch-Murnaghan equation of state, were for the axis M_{0a} = 115(7) GPa and M_a' = 28(2), for the b axis M_{0b} = 162(3) GPa and M_b' = 8(3), for the c axis M_{0c} = 142(8) GPa and M_c' = 26(2), with refined values of a_0 11.791(7) Å, b_0 = 14.540(6) Å, c_0 = 4.076(3) Å, respectively. Since the results gave large differences in M' parameters, the lattice parameter moduli were calculated using the second order Birch-Murnaghan equation of state, fixing M' to 12 in order to evaluate the anisotropic behavior. The results of this fitting give M_{0a} 191 (9) GPa with a_0 equal to 11.74 (2) Å, M_{0b} = 123(5) GPa with b_0 equal to 14.96 (2) Å and M_{0c} = 226 (10) GPa with c_0 equal to 4.058(6) Å. The compressional anisotropy of crystallographic axes, showed that b and c were the most and the least compressible lattice parameters, respectively, with the anisotropic ratio $M_{0a}:M_{0b}:M_{0c} = 1.55:1:1.84$.

Density of galenobismutite changed from 7.243 g/cm^3 at 0.5 GPa to 9.029 g/cm^3 at 20.9 GPa, with an increase of about 22% in the investigated pressure range.

To compare the present data with those of other sulfides of metalloids from literature (galena [31], bismuthinite [32], stibnite [33], chalcostibite [34], lillianite [35], heyrovskyite [36], berthierite [37]) a K' vs K$_0$ plot was elaborated (Figure 4). In the plot, the confidence ellipses at 90 and 68 % of confidence level for the present data and those reported by Olsen et al. [21] are shown. In order to allow a more direct comparison of K$_0$ and K' calculated with the two data sets and to evaluate if the observed differences were due to the different pressure range, we also calculated K$_0$ and K' restricting our data to the same pressure range investigated by Olsen et al [21]. We observed a strong negative correlation between K' and K$_0$ in agreement with the data presented by Olsen et al. [21]. However, the ellipsoides for the two data sets did not overlap, even if they are quite close. The reason might be that the results of Olsen et al. [21] were biased by an unequal distribution of pressures at which the data were measured.

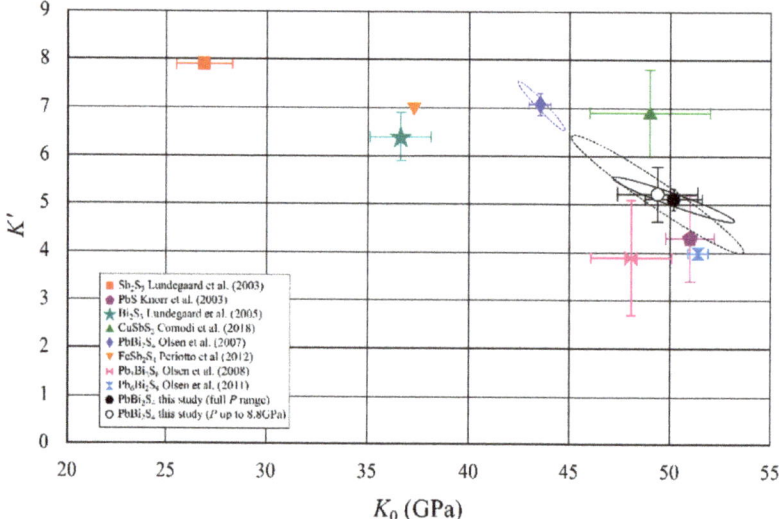

Figure 4. Bulk Moulus (K$_0$) vs its pressure derivative (K') for different sulfides. Confidence ellipses at 90% of confidence level are reported for K$_0$ and K' calculated with the present data collected up to 20.9 GPa (solid black line) as well as with data limited at 8.8 GPa (stippled black line). Confidence ellipse at 90% for Olsen et al. [21] data is also shown (stippled blue line).

K$_0$ and K' values for galena (Figure 4), PbS, before the phase transition, were very close to those observed for galenobismutite. On the other hand, K$_0$ for bismuthinite, Bi$_2$S$_3$, were significantly lower (Figure 4). Olsen et al. [21] suggested an empirical relation between the bulk modulus of galenobismutite and those of PbS and Bi$_2$S$_3$ corresponding to the proportion of Bi and Pb in galenobismutite: K$_{PbBi2S4}$ = (2KBi$_2$S$_3$ + KPbS)/3. Although this relation holds approximately for the data from the previous study, the present corrected data for galenobismutite does not support this observation. We can conclude that a simple relation between a bulk modulus for a complex composition cannot be derived straightforwardly from the bulk moduli of its simpler constituents [38] even if they contain the same general structural modules (like in sulfosalts). Obviously, a more complex cooperative mechanism between the structural modules should be involved [39]. For sulfosalts it is important to take into account that they contain cations with active lone electron pairs (LEPs), which can strongly affect the polyhedral distortion, and the overall structural compressibility to different extents. Sb^{3+}

LEP's stereochemical activity is generally higher than that of Bi^{3+}, evaluated from the measurements of the eccentricity of Sb and Bi polyhedra at room pressure conditions, which show larger difference in interatomic Sb-S distances compared to Bi-S ones. Under high pressure, the polyhedra become more regular and the eccentricity reduces more rapidly for Sb^{3+} polyhedra with respect to those of Bi^{3+}, because the longest interatomic contacts in atomic coordinations generally compress faster than the shortest ones. As a consequence, Sb sulfosalts have bulk moduli lower than the corresponding Bi sulfosalts, as illustrated by the isomorphic chalcostibite-emplectite series [34].

Pb^{2+} also contains a LEP, but it is generally less expressed than that of Bi^{3+}. LEP of Pb^{2+} is even fully suppressed in several structures, like in galena or the earlier mentioned $PbSc_2S_4$. To the best of our knowledge the only observed regular coordination of Bi^{3+} is the octahedral coordination in the mineral kupcikite, $Cu_4Bi_5S_{10}$ [40,41]. It is interesting that the pressure can force coordinations with suppressed LEP to a structure with highly expressed stereochemical activity through phase transition [35,36,42].

Very few theoretical calculations provide an analysis of the relation between electronic structure, lone electron pairs and structural geometrical parameters. Olsen et al 2011 [43], by using SIESTA DFT code considered the effect of pressure in Bi_2S_3 and compared the theoretical with experimental data. Their data on the effective Bi s-p hydridization support the origin of the stereochemically active lone pair and its evolution with pressure increases.

A comparison of the bulk modulus of galenobismutite to those of CF type structures shows much larger differences. Dubrovinsky et al. [20] reported the CF type $NaAlSiO_4$ bulk modulus measured up 40 GPa. Their data gave a very high bulk modulus of 220 GPa and its pressure derivative was equal to 4.1(1), similar to the values measured for other compounds with a calcium ferrite structure. For example, the value of K_0, with K' fixed to 4, of $MgAl_2O_4$ measured by Yutani et al. [44] was 241(1) GPa, whereas K_0 measured for Fe_3O_4 by Haavik et al. [45] was 202(7) GPa, with K' equal to 4. The general rule, suggested by Anderson et al. [46], KV = constant, where V represents the molar volume (36.58 cm^3/mol for $NaAlSiO_4$, 36.13 cm^3/mol for $MgAl_2O_4$ and 41.89 cm^3/mol for Fe_3O_4), seems to be followed by this group of calcium ferrite structures [20]. In comparison, galenobismutite has a higher molar volume (105.0 cm^3/mol) but, at the same time, a much lower bulk modulus, resulting in a violation of the Anderson's relation. This is most probably due to a large difference in chemistry, influenced by both cation and anion electronic configurations and especially by the presence of cation LEPs in galenobismutite.

3.2. Structural Evolution with Pressure

The M1, M2 and M3 polyhedral evolution with pressure was analyzed through changes in bond lengths and polyhedral volumes reported in Table 3.

Figures 5 and 6 show the changes of bond distances and volumes with pressure. The bulk moduli of M1, M2 and M3 polyhedra, calculated as the reciprocals of linear compressibilities are 114 (3) GPa, 86(2) GPa and 84(2) GPa, respectively. The values agree with the general relationship suggested by Finger and Hazen [14], which relates the polyhedral bulk moduli to inverse of the mean cation-anion distances for several oxides, silicates as well as sulfides and selenides and several other types of compounds.

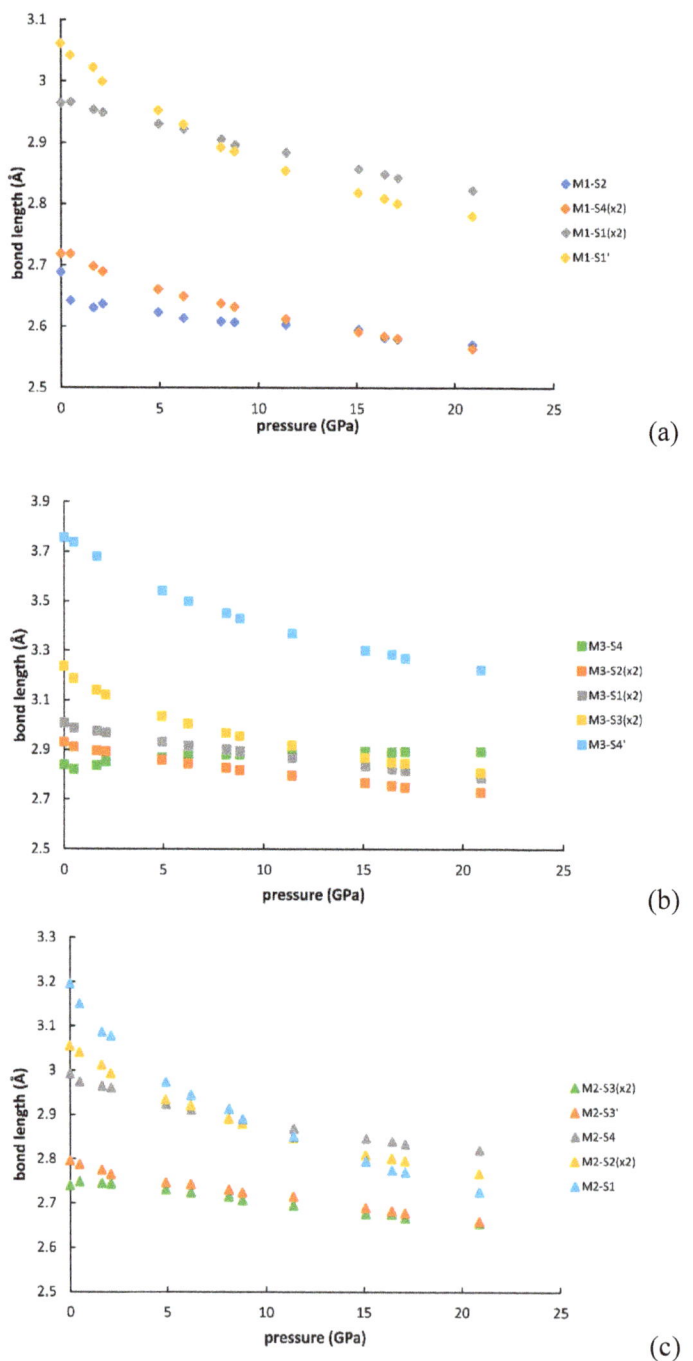

Figure 5. Evolution of the bond distances with pressure for M1 (**a**), M2 (**b**), M3(**c**) polyhedra.

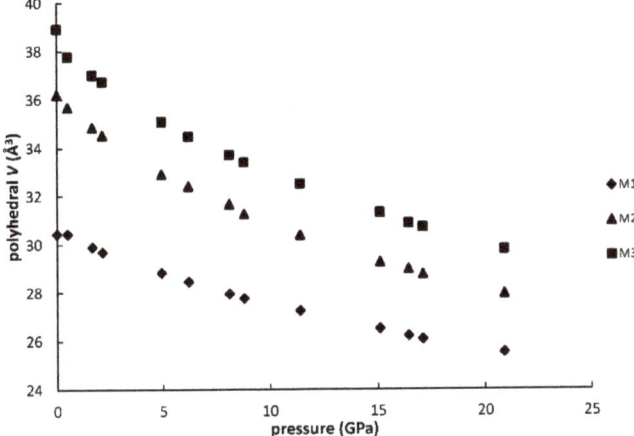

Figure 6. Variation of polyhedral volume for M1 (a), M2 (b), M3 (c) polyhedra with pressure.

The distortion parameters of the coordination polyhedra can give an additional insight in the compressibility behavior of atomic coordinations. Figure 7 shows the development of the eccentricities, asphericities and shape distortions (or volume distortions [47]). For M3 we calculated the parameters for both CN7 and CN8, because of its specific character. The eccentricities of all coordinations decreased continuously with pressure but much faster for M2 and M3 than for M1. After 4 GPa M1 reached the most eccentric coordination in spite of its smallest CN. It is interesting that the eccentricity of M3 related to only the closest seven S atoms levels off after 12 GPa and does not show further changes with pressure. However, for CN8 it continued to decrease, due to a continuous approach of the eight S atom. The asphericities showed much smaller changes with pressure. Note that M1 from the start had negligible asphericity, meaning that all six S atoms fit practically perfectly to a common sphere. It is interesting that the asphericity of the M3 coordination for CN7 actually increased with pressure, in spite of a constant decrease in asphericity calculated for CN8. It must, however, be noted that the asphericity for CN8 was significantly higher. The shape distortion, which shows the departure of the arrangement of ligands compared to an ideal polyhedron, shows an increase with pressure for all coordination polyhedra. The parameters are in all cases calculated compared to the ideal polyhedron which shows the smallest V_S/V_P ratio for a given CN, where V_S and V_P are the volumes of the circumscribed sphere and the polyhedron, respectively. For CN6 this is the regular octahedron, for CN7 the regular pentagonal bipyramid and for CN8 the "maximum volume" bisdisphenoid. Compared to the latter two, an ideal monocapped trigonal prism would have a "shape distortion" of 0.159, and an ideal bicapped trigonal prism would have a "shape distortion" of 0.073. In this respect, the values calculated for M2 and M3 (both for CN7 and for CN8) are actually a sign of approaching the shapes closer to ideal monocapped, respectively bicapped trigonal prism. M1, however, departed more from an ideal octahedron shape with increasing pressure.

The orientation and expression of a LEP can be calculated from the relative positions of the central atom in a coordination and the centroid of the ligand arrangement [48]. The black spheres in Figure 8 have their centers in centroids of coordinations, thus, they illustrate the orientations and the expressions of the LEPs of cations.

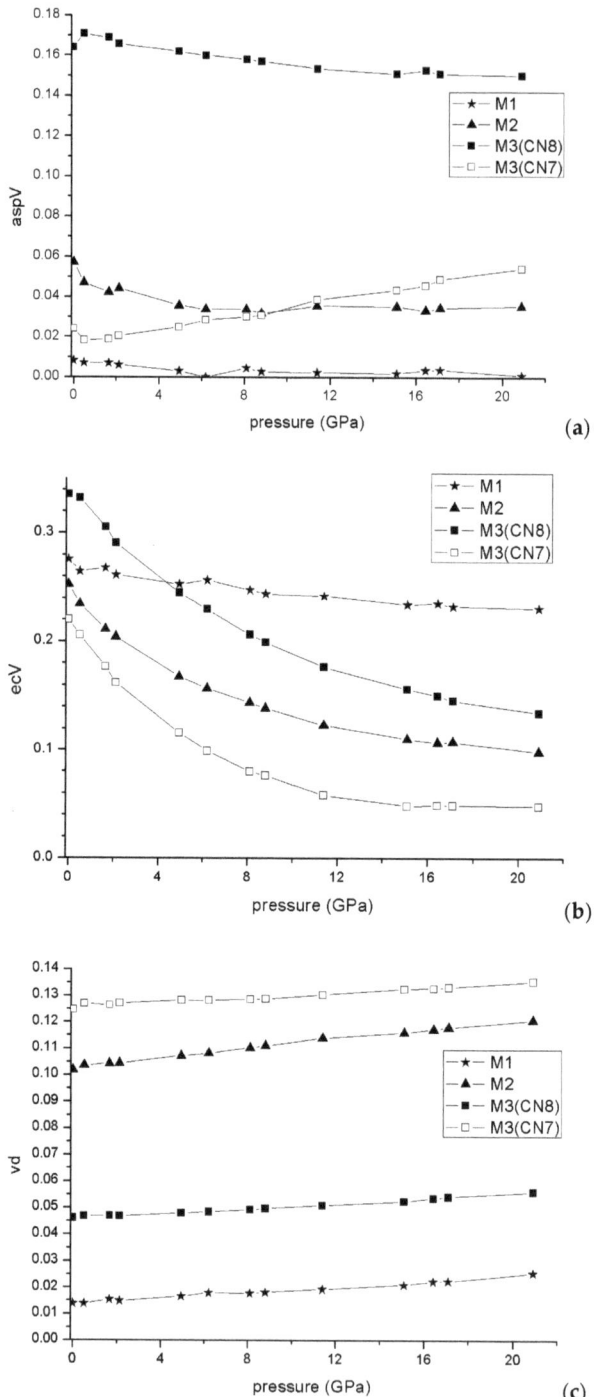

Figure 7. Asphericity (**a**), Eccentricity (**b**) and shape distortion (**c**) evolution with pressure for M1, M2, and M3 polyhedra.

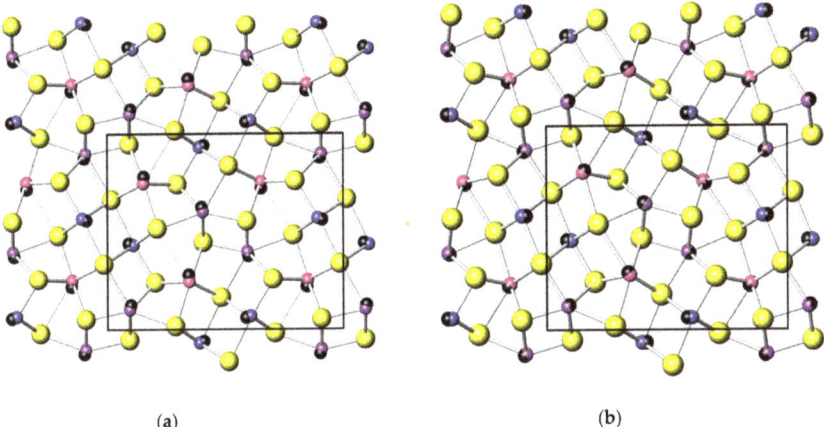

Figure 8. Galenobismutite at room pressure (**a**) and at 20.9 GPa (**b**). Black spheres are centered on centroids of coordinations and indicate the orientations of M1, M2 and M3 lone electron pairs.

Taking into account the changes in bond distances and distortion parameters plus the global aspects of the crystal structure, the changes that occur in galenobismutite under increasing pressure can be summarized as follows: The main change is that both M2 and M3 atoms move towards the centers of the bodies of respective trigonal prisms. It can be visually verified by comparing the crystal structures at 1 bar and 20.9 GPa, as represented in Figure 8, and by checking the development of the bond lengths, as in Figure 5. Here, the atoms making the body of the trigonal prism were two S3 atoms, two S2 atoms plus S3 and S1 for M2. Note that bond distances to these six S atoms showed a merging tendency with increasing pressure. The distance to the capping S4 atom decreased with a much lower gradient than the ones to two S2 plus S1 atom (that are longer at 1 bar) and actually became the longest one from 12 GPa on. In the case of M3, two S2, two S1 and two S3 atoms formed the prism body and one can observe the same tendency of merging the bond distances up to approximately 10 GPa; above this pressure they became the shortest bond distances in the coordination polyhedron. It is true that the longest distance to one of the capping S4 atoms had a significant decrease during the whole measurement range, but with a gradient that was similar to the one of the two S3 atoms belonging to prism body. On the contrary, the distance to the other S4 capping atom actually slightly increased under compression. This all testifies also in this case that M3 moves inside the body of the trigonal prism with a consequence that it also moves away from the closest capping S4 atom. As the distance to the other one largely decreases due to its approach to the prism body, the two distances to the capping S atoms show a merging tendency and we can assume that the coordination's character changes from the 7+1 type towards the real CN8, becoming a more regular bicapped trigonal prism (also confirmed by the values of the shape distortion in Figure 7c).

The changes in the M1 coordination were very small compared to M2 and M3. The eccentricity of this site changed very little (Figure 7a) as the difference between the three shortest and three longest bonds remained almost the same (Figure 5a). There was actually a slight but constant increase in the distortion of the octahedral shape (Figure 7c). The main change in this coordination is due to the polyhedral accommodation to the contraction of the b axis that had the largest compressibility (Figures 2 and 8a). The expression of the LEP of M1 slightly changed, but its orientation, seen from the atomic nucleus, changed more significantly from the diagonal one, oriented towards the space between the two neighboring M1 coordinations, to a direction along the b axis (Figure 8b). The changes in the expression of LEPs of M2 and M3 were more significant and their orientations changed to directions closer to the M2-capping S and M3-most distant capping S, in accordance to the movement of M2 and M3 towards the centers of their respective trigonal prisms.

4. Discussion and Conclusion

The comparison of data collected at different pressures on galenobismutite allows the following conclusions:

(a) The structural evolution is completely reversible with pressure increase up to 20.9 GPa. The same values were measured increasing and decreasing the pressure and the same equation of state is measured by using values collected increasing or decreasing pressure. No evidence of hysteresis in the changes were observed, meaning that the changes are completely elastic.

(b) The change in atomic coordinations bring the M3 coordination polyhedron closer to the shape observed in other members of the CF structural family (from CN7+1 to CN8). However, unlike other CF crystal structures, M2 keeps and even equalizes its seven-fold coordination with increasing pressure. This emphasizes the specific character of galenobismutite in this structural family. We suggest that the main reason is a comparatively large size of the M2 cation, comparable to that of the M3, unlike the other examples of CF structures, where M2 is significantly smaller than M3.

(c) The structure remains stable at very high pressures (up to 20 GPa) notwithstanding the moderate bulk modulus, at least under the structural point of view, since there are no incompatible distances up to 20.9 GPa. All sulfur-sulfur distances, which could indicate instability of the structure, remained quite large with the shortest S3-S4 distance equal to 3.140 Å.

(d) Calcium ferrite structure type reveals enough flexibility in incorporating various element combinations through the example of galenobismutite. Thus, not only Al and Na, incompatible in the periclase or perovskite crystal structures under the lower mantle conditions, can be considered to prefer this structure type, but it might incorporate also some other important or less abundant elements or combinations of elements.

Supplementary Materials: The following is available online at http://www.mdpi.com/2073-4352/9/4/210/s1, Table S1: HKL at different pressure of galenobismutite.

Author Contributions: Conceptualization, P.C. and T.B.-Z.; methodology, M.H. and I.C.; software, M.H., I.C. and A.Z.; formal analysis, A.Z., data curation, A.Z.; writing of original draft preparation, P.C. and T.B.-Z.; writing—review and editing, P.C., T.B.-Z., A.Z.

Funding: This research received no external funding.

Acknowledgments: The European Synchrotron Facility is acknowledged for allocating beam-time for the experiment ES-723 (main proposer P.C.).

Conflicts of Interest: The authors declare no conflict of interest.

References

1. Pinto, D.; Balic-Zunic, T.; Garavelli, A.; Makovicky, E.; Vurro, F. Comparative crystal-structure of Ag-free lillianite and galnbismutite from Vulcano, Aeolian Island, Italy. *Can. Mineral.* **2006**, *44*, 159–175. [CrossRef]
2. Pinto, D.; Balic-Zunic, T.; Bonaccorsi, E.; Bordaev, Y.S.; Garavelli, A.; Garbarino, C.; Makovicky, E.; Mozgova, N.; Vurro, F. Rare Sulfosalts from Vulcano, Aeolian Island, Italy. VII. Cl-bearing galenobismutite. *Can. Mineral.* **2006**, *44*, 443–457. [CrossRef]
3. Moëlo, Y.; Makovicky, E.; Mozgova, N.N.; Jambor, J.L.; Cook, N.; Pring, A.; Paar, W.; Nickel, E.H.; Graeser, S.; Karup-Møller, S.; et al. Sulfosalt systematics: A review. Report of the sulpfosalt sub-committee of the IMA Commission on Ore Mineralogy. *Eur. J. Mineral.* **2008**, *20*, 7–46.
4. Brodtkorb, M.K.; Paar, W. Angelaíta en la paragénesis del distrito Los Manantiales, provincia del Chubut: Una nueva especie mineral. *Rev. Assoc. Geol. Argent.* **2004**, *59*, 787–789.
5. Moëlo, Y.; Meerschaut, A.; Makovicky, E. Refinement of the crystal structure of nuffieldite, $Pb_2Cu_{1.4}(Pb_{0.4}Bi_{0.2}Sb_{0.2})Bi_2S_7$: Structural relationships and genesis of complex lead sulfosalt structures. *Can. Mineral.* **1997**, *35*, 1497–1508.
6. Mumme, W.G. Seleniferous lead–bismuth sulphosalts from Falun, Sweden: Weibullite, wittite, and nordströmite. *Am. Mineral.* **1980**, *65*, 789–796.

7. Buerger, M.J. The crystal structure of berthierite. *Am. Mineral.* **1936**, *21*, 205–206.
8. Bindi, L.; Menchetti, S. Garavellite, FeSbBiS$_4$, from the Caspari mine, North Rhine-Westphalia, Germany: Composition, physical properties and determination of the crystal structure, Locality: Caspari mine, North Rhine-Westphalia, Germany. *Mineral. Petrol.* **2005**, *85*, 131–139. [CrossRef]
9. Bente, K.; Edenharter, A. Rontgenographische strukturanalyse von MnSb$_2$S$_4$ und strukturverfeinerung, von berthierit, FeSb$_2$S$_4$. *Z. Krist.* **1989**, *185*, 31–33.
10. Zhou, H.; Sun, X.; Fu, Y.; Lin, H.; Jiang, L. Mineralogy and mineral chemistry: Constrains of ore genesis of the Beiya giant porphyry-skarn gold deposit, southwestern China. *Ore Geol. Rev.* **2016**, *79*, 408–424. [CrossRef]
11. Wickman, F.E. The crystal structure of galenobismutite PbBi$_2$S$_4$. *Ark. Miner.* **1951**, *1*, 219.
12. Iitaka, Y.; Nowacki, W. A re-determination of the crystal structure of galenobismutite PbBi$_2$S$_4$. *Acta Crystallogr.* **1962**, *15*, 691–698. [CrossRef]
13. Makovicky, E. The building principles and classification of bismith-lead sulfosalts and related compounds. *Fortschr. Miner.* **1981**, *59*, 137–190.
14. Finger, L.W.; Hazen, R.M. Systematic of high-pressure silicate structures. *Rev. Mineral. Geochem.* **2000**, *41*, 123–155. [CrossRef]
15. Liu, L.G. High pressure NaAlSiO$_4$: The first silicate calcium ferrite isotype. *Geophys. Res. Lett.* **1977**, *4*, 183–186. [CrossRef]
16. Yamada, H.; Matsui, Y.; Ito, E. Crystal-chemical characterization of NaAlSiO$_4$ with the CaFe$_2$O$_4$ structure. *Mineral. Mag.* **1983**, *47*, 177–181. [CrossRef]
17. Akaogi, M.; Tanaka, A.; Kobayashi, M.; Fikushima, N.; Suzuki, T. High pressure transformation in NaAlSiO$_4$ and thermodynamic properties of jadeite, nepheline and calcium ferrite-type phase. *Phys. Earth Plant. Inter.* **2002**, *130*, 49–58. [CrossRef]
18. Tutti, F.; Dubrovisky, L.; Saxena, S.K. High pressure phase transformation of jadeite and stability of NaAlSiO$_4$ with calcium-ferrite type structure in the lower mantle conditions. *Geophs. Res. Lett.* **2000**, *27*, 2025–2028. [CrossRef]
19. Shemet, V.; Gulay, L.; Stepien-Damm, J.; Pietraszko, A.; Olekseyuk, I. Crystal structure of the Sc$_2$PbX$_4$ (X = S and Se) compounds. *J. Alloy. Compd.* **2006**, *407*, 94–97. [CrossRef]
20. Dubrovinsk, L.S.; Dubrovinskaia, N.A.; Prokopenko, V.B.; Le Bihan, T. Equation of state and crystal structure of NaAlSiO$_4$ with calcium-ferrite type structure in the conditions of the lower mantle. *High Press. Res.* **2002**, *22*, 495–499. [CrossRef]
21. Olsen, L.A.; Balic-Zunic, T.; Makovicky, E.; Ullrich, A.; Miletich, R. Hydrostatic compression of galenobismutite (PbBi$_2$S$_4$): Elastic properties and high-pressure crystal chemistry. *Phys. Chem. Miner.* **2007**, *34*, 467–475. [CrossRef]
22. Ruiz-Fuertes, J.; Friedrich, A.; Errandonea, D.; Segura, A.; Morgenroth, W.; Rodríguez-Hernández, P.; Muñoz, A.; Meng, Y. Optical and structural study of the pressure-induced phase transition of CdWO$_4$. *Phys. Rev. B* **2017**, *95*, 174105. [CrossRef]
23. Comboni, D.; Lotti, P.; Gatta, G.D.; Lacalamita, M.; Mesto, E.; Merlini, M.; Hanfland, M. Amstrongite at non-ambient conditions: An in-situ high pressure single-crystal X-ray diffraction study. *Microporous Mesoporous Mater.* **2019**, *274*, 171–175. [CrossRef]
24. Singh, A.K. Strength of solid helium under high pressure. *J. Phys. Conf. Ser.* **2012**, *377*, 012007. [CrossRef]
25. Hammersley, A.P.; Svensson, S.O.; Hanfland, M.; Fitch, A.N.; Hausermann, D. Two-dimensional detector software: From real detector to idealized image or two-theta scan. *High Press. Res.* **1996**, *14*, 235–245. [CrossRef]
26. Oxford Diffraction. *CrysAlis(Pro)*; Oxford Diffraction Ltd.: Abingdon, UK, 2006.
27. Angel, R.; Gonzalez-Platas, J. ABSORB-7 and ABSORB-GUI for single crystal absorption correction. *J. Appl. Crystall.* **2013**, *46*, 252–254. [CrossRef]
28. Hubschle, C.B.; Sheldrick, G.M.; Dittrich, B. ShelXle: A Qt graphical user interface for29 SHELXL. *J. Appl. Cryst.* **2011**, *44*, 1281–1284. [CrossRef] [PubMed]
29. Gonzalez-Platas, J.; Alvaro, M.; Nestola, F.; Angel, R. EOSFIT7-GUI: A new graphycal user interface for equation of state calculation, analyses and teaching. *J. Appl. Cryst.* **2016**, *49*, 1377–1382. [CrossRef]
30. Angel, R.J. Equation of state. In *High Temperature and High Pressure Crystal Chemistry*; Hazen, R.M., Downs, R.T., Eds.; Reviews in Mineralogy and Geochemistry; Mineralogical Society of America: Chantilly, VA, USA, 2000; Volume 41, pp. 117–211.

31. Knorr, K.; Ehm, L.; Hytha, M.; Winkler, B.; Depmeier, W. The high-pressure α/β phase transition in lead sulphide (PbS)—X-ray powder diffraction and quantum mechanical calculations. *Eur. Phys. J.* **2003**, *31*, 297–303. [CrossRef]
32. Lundegaard, L.; Makovicky, E.; Boffa Ballaran, T.; Balic-Zunic, T. Crystal structure and cation lone electron pair activity of Bi_2S_3 between 0 and 10 GPa. *Phys. Chem. Miner.* **2005**, *32*, 578–584. [CrossRef]
33. Lundegaard, L.F.; Miletich, R.; Balic-Zunic, T.; Makovicky, E. Equation of state and crystal structure of Sb_2S_3 between 0 and 10 GPa. *Phys. Chem. Miner.* **2003**, *30*, 463–468. [CrossRef]
34. Comodi, P.; Guidoni, F.; Nazzareni, S.; Balic-Zunic, T.; Zucchini, A.; Makoviscky, E.; Prakapenka, V. A high-pressure phase transition in chalcostibite, $CuSbS_2$. *Eur. J. Miner.* **2018**, *30*, 491–505. [CrossRef]
35. Olsen, L.A.; Balić-Žunić, T.; Makovicky, E. High-pressure anisotropic distortion of $Pb_3Bi_2S_6$: A pressure-induced, reversible phase transition with migration of chemical bonds. *Inorg. Chem.* **2008**, *47*, 6756–6762. [CrossRef]
36. Olsen, L.A.; Friese, K.; Makovicky, E.; Balić-Žunić, T.; Morgenroth, W.; Grzechnik, A. Pressure induced phase transition in $Pb_6Bi_2S_9$. *Phys. Chem. Miner.* **2011**, *38*, 1–10. [CrossRef]
37. Periotto, B.; Balic-Zunic, T.; Nestola, F. The role of the Sb^{3+} lone electron pairs and Fe^{2+} coordination in the high-pressure behavior of berthierite. *Can. Mineral.* **2012**, *50*, 201–218. [CrossRef]
38. Comodi, P.; Mellini, M.; Zanazzi, P.F. Scapolites: Variation of structure with pressure and possible role in the storage of fluids. *Eur. J. Mineral.* **1990**, *2*, 195–202. [CrossRef]
39. Gatta, G.D.; Comodi, P.; Zanazzi, P.F.; Boffa Ballaran, T. Anomalous elastic behaviour and high-pressure structural evolution of zeolite levyne. *Am. Mineral.* **2005**, *90*, 645–652. [CrossRef]
40. Mariolacos, K.; Kupčik, V.; Ohmasa, M.; Miehe, G. The Crystal Structure of $Cu_4Bi_5S_{10}$ and its Relation to the Structures of Hodrushite and Cuprobismutite. *Acta Cryst.* **1975**, *31*, 703–708. [CrossRef]
41. Topa, D.; Makovicky, E.; Balić-Žunić, T.; Paar, W.H. Kupčikite, $Cu_{3.4}Fe_{0.6}Bi_5S_{10}$, a new Cu-Bi sulfosalt from Felbertal, Austria, and its crystal structure. *Can. Mineral.* **2003**, *41*, 1155–1166. [CrossRef]
42. Grzechnik, A.; Friese, K. Pressure-induced orthorhombic structure of PbS. *J. Phys. Condens. Matter* **2010**, *22*, 095402. [CrossRef]
43. Olsen, L.A.; Lopez-Solano, J.; Garcia, A.; Balic-Zinic, T.; Makovicky, E. Dependence of lone pair of bismuth on coordination environment and pressure: Ab Initio study on $Cu_4Bi_5S_{10}$ and Bi_2S_3. *J. Solid State Chem.* **2010**, *18*, 2133–2143. [CrossRef]
44. Yutani, M.; Yagi, T.; Yusa, H.; Irifune, T. Compressibility of calcium ferrite-type $MgAl_2O_4$. *Phys. Chem. Miner.* **1997**, *24*, 340–344. [CrossRef]
45. Haavik, C.; Stolen, S.; Fjellvag, H.; Hanfland, M.; Häusermann, D. Equation of state of magnetite and its high-pressure modification: Thermodynamics of the Fe-O system at high pressure. *Am. Mineral.* **2000**, *85*, 514–523. [CrossRef]
46. Anderson, O.L. *Equation of State of Solids for Geophysics and Ceramic Science*; Oxford University Press: New York, NY, USA, 1994; p. 495.
47. Makovicky, E.; Balić-Žunić, T. New measure of distortion for coordination polyhedra. *Acta Cryst.* **1998**, *54*, 766–773. [CrossRef]
48. Balić Žunić, T.; Makovicky, E. Determination of the Centroid or "the Best Centre" of a Coordination Polyhedron. *Acta Cryst.* **1996**, *52*, 78–81. [CrossRef]

 © 2019 by the authors. Licensee MDPI, Basel, Switzerland. This article is an open access article distributed under the terms and conditions of the Creative Commons Attribution (CC BY) license (http://creativecommons.org/licenses/by/4.0/).

Article

Crystal Structures and High-Temperature Vibrational Spectra for Synthetic Boron and Aluminum Doped Hydrous Coesite

Yunfan Miao [1], Youwei Pang [1], Yu Ye [1,*], Joseph R. Smyth [2], Junfeng Zhang [1,3], Dan Liu [1], Xiang Wang [1] and Xi Zhu [1]

[1] State Key Laboratory of Geological Processes and Mineral Resources, China University of Geosciences, Wuhan 430074, China; miaoyfa@163.com (Y.M.); youweipang@126.com (Y.P.); jfzhang@cug.edu.cn (J.Z.); ldddcug@126.com (D.L.); xiangwang@cug.edu.cn (X.W.); zhuxi@cug.edu.cn (X.Z.)
[2] Department of Geological Sciences, University of Colorado, Boulder, CO 80309, USA; smyth@colorado.edu
[3] School of Earth Sciences, China University of Geosciences, Wuhan 430074, China
* Correspondence: yeyu@cug.edu.cn

Received: 5 November 2019; Accepted: 3 December 2019; Published: 5 December 2019

Abstract: Coesite, a high-pressure SiO_2 polymorph, has drawn extensive interest from the mineralogical community for a long time. In this study, we synthesized hydrous coesite samples with different B and Al concentrations at 5 and 7.5 GPa (1273 K). The B concentration could be more than 400 $B/10^6Si$ with about 300 ppmw H_2O, while the Al content can be as much as 1200 to 1300 $Al/10^6Si$ with C_{H2O} restrained to be less than 10 ppmw. Hence, B-substitution may prefer the mechanism of $Si^{4+} = B^{3+} + H^+$, whereas Al-substitution could be dominated by $2Si^{4+} = 2Al^{3+} + O_V$. The doped B^{3+} and Al^{3+} cations may be concentrated in the Si1 and Si2 tetrahedra, respectively, and make noticeable changes in the Si–O4 and Si–O5 bond lengths. In-situ high-temperature Raman and Fourier Transformation Infrared (FTIR) spectra were collected at ambient pressure. The single crystals of coesite were observed to be stable up to 1500 K. The isobaric Grüneisen parameters (γ_{iP}) of the external modes (<350 cm^{-1}) are systematically smaller in the Al-doped samples, as compared with those for the Al-free ones, while most of the OH-stretching bands shift to higher frequencies in the high temperature range up to ~1100 K

Keywords: coesite; high-temperature Raman; FTIR spectrum; single crystal structure; isobaric Grüneisen parameters; OH-stretching modes

1. Introduction

Coesite, a high-pressure polymorph of SiO_2, was firstly synthesized at 3.5 GPa (773–1073 K) in 1953 [1], and subsequently discovered in many locations, such as in the shocked sandstone ejecta samples from craters [2,3] as well as in eclogite [4–6]. Coesite is a very important index silica mineral for ultrahigh-pressure metamorphism [7,8], which provides key clue for the continental dynamics such as lithospheric subduction, exhumation, and reentry in extreme depths of more than 100 km. Furthermore, the physical and chemical properties of coesite at high-pressure and high-temperature conditions also attract a lot of interest from the community of mineral physics, like thermo-elasticity [9–12], phase transitions [13–15], and vibrational spectra under high-pressure (P) and high-temperature (T) conditions [16–20].

Water (OH^-) incorporation into coesite has a significant impact on the stability of coesite at high-P/T conditions [21,22], which is important for exploring preservation of coesite in the deep mantle. There could be up to 200–300 ppm ppmw H_2O in synthetic coesite samples [19,23–26] resulted from hydro-garnet substitution ($Si^{4+} + 4O^{2-} = V + 4OH^-$) as well as electrostatically coupled substitution with M^{3+} incorporation ($Si^{4+} = M^{3+} + H^+$; M = B, Al), although natural coesite has been found to be nearly dry so far [27,28].

In this study, we synthesized hydrous coesite samples with various compositions (Si-pure, B-doped, Al-doped, as well as B plus Al-doped), and explored the effects of B and Al on the hydration mechanism and internal structure of coesite. Taking advantage of in-situ high-temperature Raman and FTIR vibrational spectra, we have also studied thermal response of lattice vibration with the contributions from the trace elements of H, Al, and B, which may provide important constraints on thermodynamic properties of coesite (such as heat capacities and entropy) under deep mantle conditions.

2. Materials and Methods

2.1. Sample Synthesis and Characterization

A total of five coesite samples were synthesized at the *P-T* conditions of 5 and 7.5 GPa and 1273 K with heating durations of 9–12 hours (Table 1), using welded Pt capsules in sintered MgO octahedron assemblies in the 1000-ton multi-anvil press at China University of Geosciences (Wuhan). The corner truncation of the 25.4-mm tungsten carbide cubes was 12 mm for synthetic experiments at 5 GPa and 8 mm for the runs conducted at 7.5 GPa, respectively. Temperature was monitored with a W5Re95–W2Re74 (C-type) thermocouple, and a graphite furnace was used in our experiments. Analytical reagent SiO_2, $Al(OH)_3$, $B(OH)_3$ (purity of >99.99%) were adopted as the starting materials to synthesize hydrous coesite samples with different compositions: Si-pure (Run 503), B-doped (R663), Al-doped (R749), and B,Al-doped (R694 and R712). Excess liquid water (1 µL) was added in each capsule to guarantee the water fugacity. Single crystals up to 300 µm were recovered from these synthetic experiments, while no other crystallized phases were detected (by Raman spectra) in the run products.

In-situ analyses of the trace elements of B and Al in these synthetic coesite samples were conducted on an Agilent 7900 inductively coupled plasma mass spectrometry (ICP-MS) (Agilent Technology, Tokyo, Japan) combined with a Yb femtosecond laser ablation (fs-LA) system (GeoLas 2005, Lambda Physik, Göttingen, Germany), without applying an internal standard [29]. The ICP-MS works at a power of 1350 W with a plasma and an auxiliary gas flow rate of 15.0 and 1.0 L/min, respectively; while the fs-LA system (λ = 257 nm) is operated at a repetition rate of 8 Hz and a pulse length of 300 fs. The spot size is 24 µm with an energy density of 2.8 J/cm^2, and a mixture of He and Ar is used as the carrier gas. The element contents of B and Al were calibrated against multiple-reference materials (BCR-2G, BIR-1G, and BHVO-2G) using the 100% oxide normalization method [30] with the detection limits of 0.1 ppmw for B and 0.8 ppmw for Al, and the determined average B and Al concentrations in these reference materials show relative deviations of −5 to about −10% from the recommended values [31]. The derived B and Al concentrations are listed in Table 1.

Because the Al concentration in R749 is as high as about 400 ppmw as indicated by fs-LA–ICP-MS, we further checked the Al composition by a JEOL JXA-8100 electron probe micro analyzer (EPMA) (JEOL Ltd., Akishima, Japan), which is equipped with four wavelength-dispersive spectrometers (WDS). The EPMA system is operated at an accelerating voltage of 15 kV and a beam current of 5 nA, while the spot size is reduced to 10 nm to minimize the fluctuations of X-ray intensity as well as sample damage [32]. The certified mineral standards of pyrope garnet (for Al) and olivine (for Si) were adopted for quantification using ZAF wavelength-dispersive corrections. Totally, twelve points were selected for measurements on the sample of R749, and the derived Al_2O_3 content is 0.0742 ± 0.0096 wt.% with a detection limit of 100 ppmw (corresponding to 393 ± 51 ppmw for the Al element), which is consistent with the result from fs-LA–ICP-MS within the experimental uncertainties.

Table 1. Starting composition and synthetic condition for each run. The B and Al element concentrations are mainly measured by femtosecond laser ablation (fs-LA)–ICP-MS, while the H contents are estimated by FTIR.

Run No.	Starting Materials (wt.%)	Pressure (GPa)	Temperature (K)	Time (h)	B (ppmw)	B/10^6Si	Al (ppmw)	Al/10^6Si	H_2O (ppmw)	H/10^6Si
R503:	SiO_2 + 1 µL H_2O	7.5	1273	12	—	—	—	—	32.3 ± 14.2	215 ± 95
R663:	SiO_2 (96) + H_3BO_3 (4) + 1 µL H_2O	5	1273	10	74.6 ± 3.5	445 ± 21	—	—	51.0 ± 19.8	340 ± 132
R694:	SiO_2 (94) + H_3BO_3 (3) + Al(OH)$_3$ (3) + 1 µL H_2O	5	1273	10	73.4 ± 9.0	437 ± 54	115.6 ± 52.0	343 ± 154	22.8 ± 11.7	152 ± 78
R712:	SiO_2 (94) + H_3BO_3 (3) + Al(OH)$_3$ (3) + 1 µL H_2O	7.5	1273	9	39.7 ± 8.2	237 ± 49	139.8 ± 31.7	415 ± 94	24.1 ± 10.9	161 ± 73
R749:	SiO_2 (96) + Al(OH)$_3$ (4) + 1 µL H_2O	5	1273	9	—	—	445.8 ± 86.5 [a] 392.7 ± 50.8 [b]	1323 ± 257 [a] 1166 ± 151 [b]	7.2 ± 2.9	48 ± 19

[a]: measured by ICP-MS; [b]: determined by an electron probe micro analyzer (EPMA).

2.2. Single-Crystal X-ray Diffraction (XRD)

A single grain (with a diameter of 100–120 µm) from each synthetic sample was selected for XRD at ambient conditions. The unit-cell parameters (Table 2) were refined on a Rigaku XtalAB mini diffractometer (Rigaku, Akishima, Japan) with a 600-W rotating Mo-anode X-ray source, which is operated at 50 kV and 20 mA. A Saturn 724 HG CCD detector (with a resolution of 1024 × 1024) was mounted on this diffractometer. The average wavelength of Mo $K_{\alpha 1}$–$K_{\alpha 2}$ was calibrated to 0.71073 Å, and intensity data were collected in the 2θ scanning range of up to 52°. The refinements of atomic positions (Table 3) and anisotropic displacement parameters (Table S1) were conducted using the software package of CrysAlisPro/Olex2 [33]. The data collection parameters are also listed in Table 2, including the numbers of the measured equivalent and unique reflections, as well as the model fit values for Goof, R_1, and R_{int}. For all these synthetic single crystals, the Goof parameters remain below 1.1, while the R_1 and R_{int} values are lower than 2.9% and 1.5%, respectively. The Si^{4+} [34] and O^{2-} [35] ionic scattering factors were adopted, and the Si1 and Si2 occupancies were fixed at 1 (full) during the structural refinement procedures.

Table 2. Intensity data collection and unit-cell parameters for the synthetic coesite samples.

	R503	R663	R694	R712	R749
a (Å)	7.1458 (5)	7.1332 (9)	7.1355 (5)	7.1426 (13)	7.1437(7)
b (Å)	12.3922 (10)	12.3886 (5)	12.3678 (5)	12.3698 (8)	12.3964 (6)
c (Å)	7.1778 (8)	7.1828 (16)	7.1763 (8)	7.1788 (16)	7.1858 (12)
β (°)	120.293 (11)	120.31 (2)	120.358 (10)	120.37 (2)	120.292 (15)
V (Å^3)	548.82 (10)	548.01 (18)	546.47 (9)	547.24 (19)	549.46 (13)
No. total refl.	1028	777	2117	911	1346
No. unique total	366	436	523	467	487
No. unique $I > 4\sigma$	348	427	511	446	461
Goof	1.084	1.059	1.021	1.051	1.041
R_1 for all (%)	2.45	2.89	2.91	2.67	2.83
R_1 for $I > 4\sigma$ (%)	2.33	2.84	2.86	2.57	2.72
R_{int} (%)	1.33	0.62	1.04	1.23	1.47

The unit-cell angles: $\alpha = \gamma = 90°$.

2.3. Vibrational Spectra at Room and High Temperatures

Single grains of a diameter less than 150 µm were chosen for in-situ high-temperature Raman measurement, using a Horiba LabRAM HR Evolution system (HORIBA JobinYvon S.A.S., Paris, France) with a Ar$^+$ laser excitation source (λ = 532 nm) and a micro-confocal spectrometer. Each crystal piece was loaded on a sapphire plate in a Linkam TS 1500 heating stage (Linkam Scientific Instruments Ltd., Tadworth, Surrey, UK). High temperatures were generated by a resistance heater from 300 K up to 1500 K, with an increment of 50 K and a heating rate of 20 K/min. To further test the temperature dependence of these lattice vibrational modes, we also chose another grain from R503 for low-temperature Raman measurement. The sample piece was loaded on a sapphire window in a Linkam THMS 600 heating/cooling stage, and low temperatures were cooled down to 80 K by liquid nitrogen with a cooling rate of 15 K/min. The temperatures were automatically controlled with uncertainties less than 5 K. Each target temperature was maintained at least 5 minutes before measurement to guarantee thermal equilibrium.

To analyze the water contents in these synthetic coesite samples, 7–9 cleaned crystal pieces (in a diameter of 100–160 µm) were selected from each sample source for Mid-FTIR measurement at ambient condition. All these crystals were double-side polished to a thickness of 60–80 µm before measurements, and the water contents for each of these coesite samples are estimated as an average of these measured pieces in the following discussion. The IR spectra were collected using a Nicolet iS50 FTIR instrument (Thermofisher, Madison, WI, USA) coupled with a Continuum microscope, a KBr beam-splitter, and a MCT-A detector cooled by liquid N$_2$. For in-situ high-T FTIR measurement, four polished sample pieces (R503, R663, R694 and R749) was selected and loaded at the sapphire window of a custom HS1300G-MK2000 external heating stage (INSTC, Boulder, CO, USA). The FTIR spectra were obtained in the wavelength range above 3200 cm^{-1}, with a resolution of 4 cm^{-1} and an accumulation of 256 scans. Temperatures were measured from room temperature to about 1200 K with an interval of 50 K and a heating rate of 15 K/min. Background was also obtained after the measurement on the sample for each step.

Table 3. Refined atomic position coordinates.

		R503	R663	R694	R712	R749
Si1	x	0.14027 (12)	0.14048 (13)	0.14044 (9)	0.14049 (11)	0.14050 (11)
	y	0.10826 (6)	0.10833 (5)	0.10835 (4)	0.10832 (5)	0.10829 (5)
	z	0.07226 (10)	0.07240 (15)	0.07242 (9)	0.07232 (11)	0.07242 (11)
Si2	x	0.50653 (12)	0.50655 (13)	0.50668 (9)	0.50682 (11)	0.50675 (11)
	y	0.15808 (6)	0.15794 (5)	0.15796 (8)	0.15798 (5)	0.15795 (5)
	z	0.54061 (10)	0.54028 (15)	0.54069 (8)	0.54073 (11)	0.54079 (10)
O2	x	0.5	0.5	0.5	0.5	0.5
	y	0.1162 (2)	0.1164 (2)	0.1164 (15)	0.1163 (2)	0.11656 (17)
	z	0.75	0.75	0.75	0.75	0.75
O3	x	0.7340 (3)	0.7333 (3)	0.7336 (2)	0.7334 (3)	0.7335 (3)
	y	0.12289 (16)	0.12287 (15)	0.12290 (12)	0.12284 (15)	0.12280 (13)
	z	0.5603 (3)	0.5593 (4)	0.5594 (3)	0.5594 (3)	0.5595 (3)
O4	x	0.3115 (3)	0.3103 (4)	0.3109 (2)	0.3112 (3)	0.3107 (3)
	y	0.10385 (17)	0.10397 (15)	0.10386 (11)	0.10372 (15)	0.10377 (12)
	z	0.3282 (3)	0.3278 (4)	0.3278 (2)	0.3282 (3)	0.3276 (3)
O5	x	0.0172 (3)	0.0168 (3)	0.0174 (2)	0.0175 (3)	0.0175 (3)
	y	0.21182 (17)	0.21177 (15)	0.21174 (11)	0.21173 (15)	0.21168 (13)
	z	0.4784 (3)	0.4787 (4)	0.4786 (2)	0.4786 (3)	0.4786 (3)

The O1 site is at the inversion center with $x = y = z = 0$.

3. Results and Discussion

3.1. Hydration and B/Al Concentrations

The representative FTIR spectra for these synthetic coesite samples at ambient condition are shown in Figure 1. Four OH-stretching bands of v_1 (3572 cm^{-1}), $v_{2a,b}$ (3522 cm^{-1}), v_3 (3458 cm^{-1}), and v_4 (3300 cm^{-1}) are detected for all these synthetic samples, which are independent of presence of B or Al. The mode v_6 (3500 cm^{-1}) is clearly observed with most absorbance for the B-doped samples (R663, R694, R712), as compared to the B-free samples (R503 and R749), which is caused by B-substitution in coesite (Si^{4+} = B^{3+} + H$^+$) [25]. It should be noted that the B concentrations in this study are much higher than those synthesized in Koch-Müller et al. [25] (BR01-03), and the v_6 absorbance is consequently significantly stronger than that for $v_{2a,b}$.

The total H$_2$O content in coesite (C_{H2O}, wt%) can be calculated on the basis of Lambert–Beer law [24]:

$$C_{H2O} = \frac{1.8 \times A_i}{\rho \times \varepsilon_i \times d}, \quad (1)$$

where ρ is density (2.93 g/cm^3), d is the thickness of sample (cm^{-1}), while ε_i is the integrated molar absorption coefficient for H$_2$O, which was calibrated to be 190000 ± 30000 L·mol^{-1}·cm^{-2} for coesite [24]. The integrated absorbance A_i in the wavenumber range from v_1 to v_2 is expressed as

$$A_i = \int_{v_1}^{v_2} \log\left(\frac{I_0}{I}\right) \cdot dv \quad (2)$$

where I_0 and I are the intensities of incoming and transmitted radiation, respectively. For each coesite sample, several unoriented crystal pieces were selected and polished for FTIR measurement at room temperature, and similar I_{v3}/I_{v1} and I_{v2}/I_{v1} intensity ratios are observed among these FTIR spectra. The averaged hydration concentrations are listed in Table 1 with statistical uncertainties.

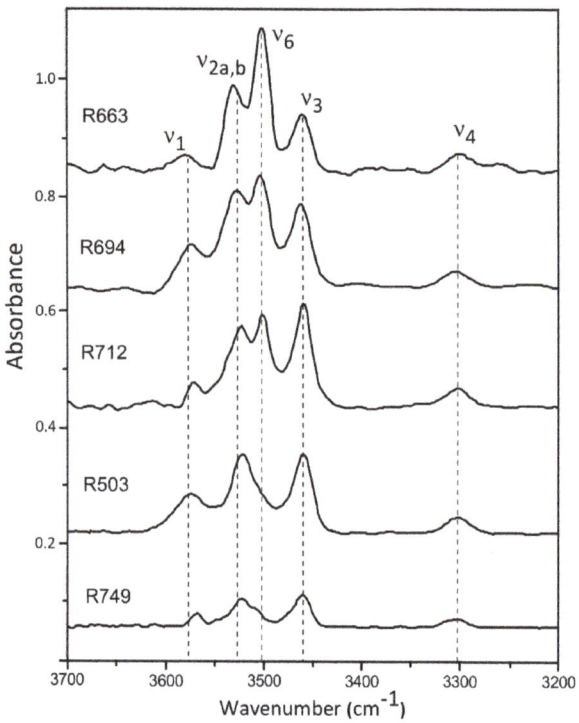

Figure 1. Representative FTIR spectra obtained at ambient conditions with the OH-stretching bands noted.

The hydration concentrations in these synthetic samples show a general trend: R663 (B-doped) > R503 (B,Al-free) > R694/R712 (B,Al-doped) > R749 (Al-doped). This observation can be satisfactorily interpreted as results of different incorporation mechanisms between B and Al in coesite. The predominant B-substitution mechanism in coesite should be an electrostatically coupled substitution $Si^{4+} = B^{3+} + H^+$ [25], which could increase hydration solubility, as compared with the Si-pure sample R503. In contrast, most of Al cations were incorporated into the internal structure of coesite by causing oxygen vacancies ($2Si^{4+} = 2Al^{3+} + O_V$), which is similar to the Al-substitution mechanism in stishovite [36,37]. Such Al-corporation may have an effect of reducing water solubility in coesite, according to the estimated water content in the sample R749. In the case of R749, the atomic concentration ratio of Al:H reaches more than 24:1, while in the B,Al-doped samples (R694 and R712), the sums of B and Al atomic concentrations are still four or five times of that for hydrogen. In addition, we also tried to collect Raman spectra on these samples in the similar frequency range of 3200–3700 cm^{-1}, but no OH-stretching modes were detected due to the low water concentrations (no more than 60 ppmw).

The magnitudes of the B/10^{6}Si and Al/10^{6}Si concentrations in this study are about one order of magnitude higher than those (BR01, BR02, BR03, and BRcal2) from Koch-Müller et al. [25], whereas the magnitudes of the measured water contents from both studies are in the same range (H/10^{6}Si in a few hundred atomic ppm). The synthetic conditions (including pressure, temperature, heating duration, as well as excessive B and Al in the starting materials) are similar or comparable for both studies, while the main difference is that the Ni:NiO buffer was adopted in Koch-Müller et al. [25] to control water (oxygen) fugacity. However, it should be also noted that Deon et al. [26] also synthesized a coesite sample with 1600 atomic ppm B (B/10^6Si) and 900 atomic ppm H (H/10^6Si), both of which are even higher than those in our sample R663, at a P-T condition of 9.1 GPa and 1673 K.

Hence, the B and Al solubilities in coesite at high P-T conditions still need to be carefully examined, and the effect of oxygen fugacity should also be taken into consideration. What is more important, Koch-Müller et al. [25] measured B and Al concentrations by ion microprobe [38], whereas we used fs-LA–ICP-MS in this study, as well as EPMA to cross check the Al content in the sample R749. Hence, discrepancies between different analytical methods in different laboratories should also be considered.

3.2. Crystal Structures

The space group of coesite is $C2/c$, and SiO_4 tetrahedra form an infinite three-dimensional framework of a (b-unique) monoclinic structure (Figure 2). There are a total of two Si sites (Si1 and Si2) and five O sites (special O1 and O2 sites, as well as general O3, O4, and O5 ones) in the lattice. The refined crystal structures in this study are consistent with the previous studies [10,39–43]. The measured unit-cell volumes of the B-doped (R663), B,Al-doped (R694 and R712), and Al-doped (R749) samples differ −0.15%, −0.3 to about −0.4%, and +0.1%, away from that for the Si-pure one (R503), respectively, while such differences are significantly larger than the experimental uncertainty from single-crystal XRD. Hence, even a few hundred ppm concentrations of B and Al trace elements could have noticeable impact on the volume of coesite, and a similar phenomenon was also noted by Koch-Müller et al. [25].

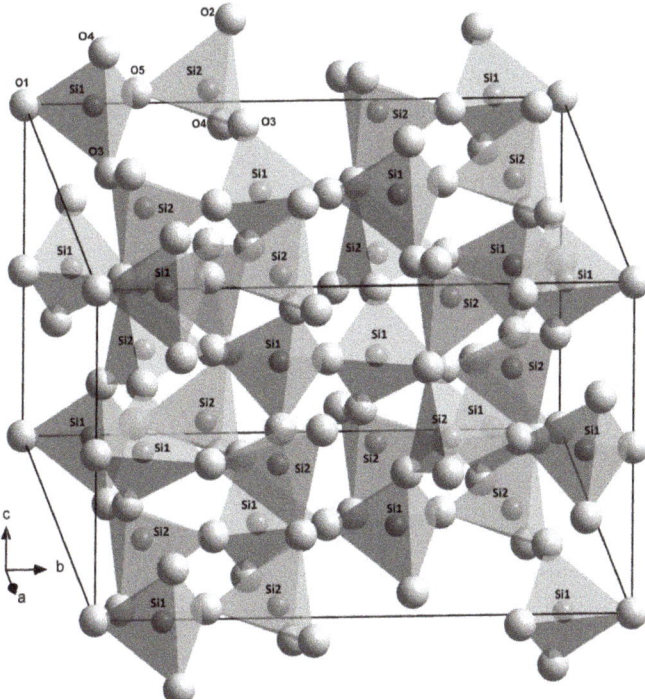

Figure 2. Crystal structure of coesite sketched on the basis of the structure refinement for the sample R503 in this study. The smaller (at centers of tetrahedra) and larger balls represent Si and O atoms, respectively.

To further investigate any B and Al effects on the internal structure of coesite, we conducted structure refinements on these five synthetic samples. The calculated bond lengths and angles are listed in Table 4, using the software package XTALDRAW [44]. As compared with the Si-pure sample (R503), the B-doped (R663) and B,Al-doped (R694 and R712) ones exhibit significantly shorter Si1–O4 and Si1–O5 bond lengths, while the Al-doped one (R749) shows noticeably longer Si2–O4 and Si2–O5 bond lengths, which are generally consistent with the order of cation sizes of $B^{3+} < Si^{4+} < Al^{3+}$. The Si2–O3 bond length for R503 is longer than those for other samples. In addition, there are no significant differences for the O–Si–O bond angles among these samples. Hence, we proposed that the B^{3+} cations may concentrate in the smaller Si1 tetrahedra, while the Al^{3+} cations would prefer the larger Si2 tetrahedra. The B,Al-coupled substitution seems to collapse the lattice structure even more, as compared with the case with only B substitution in coesite.

Table 4. Bond lengths (Å), bond angles (°), and polyhedral volumes (Å^3).

	R503	R663	R694	R712	R749
Si1–O1	1.5966 (8)	1.5968 (7)	1.5949 (5)	1.5953 (7)	1.5980 (6)
Si1–O3	1.6134 (16)	1.6130 (20)	1.6105 (15)	1.6120 (20)	1.6134 (17)
Si1–O4	1.6146 (19)	1.6090 (30)	1.6088 (16)	1.6120 (20)	1.6106 (19)
Si1–O5	1.6230 (20)	1.6184 (19)	1.6191 (14)	1.6205 (19)	1.6229 (16)
<Si1–O>	1.6119 (22)	1.6093 (19)	1.6083 (13)	1.6100 (16)	1.6112 (16)
Poly. Vol.	2.1485 (25)	2.1382 (30)	2.1344 (24)	2.1415 (21)	2.1457 (20)
Si2–O2	1.6130 (12)	1.6147 (12)	1.6109 (7)	1.6121 (11)	1.6121 (9)
Si2–O3	1.6194 (18)	1.6125 (19)	1.6150 (14)	1.6150 (20)	1.6160 (16)
Si2–O4	1.6050 (20)	1.6060 (30)	1.6049 (15)	1.6046 (19)	1.6111 (18)
Si2–O5	1.6200 (20)	1.6213 (19)	1.6191 (14)	1.6194 (19)	1.6283 (17)
<Si2–O>	1.6144 (23)	1.6136 (22)	1.6125 (14)	1.6128 (12)	1.6169 (6)
Poly. Vol.	2.1579 (24)	2.1563 (30)	2.1515 (18)	2.1522 (22)	2.1645 (20)
O1–Si1–O3	110.37 (9)	110.28 (9)	110.35 (6)	110.41 (8)	110.32 (7)
O1–Si1–O4	109.41 (8)	109.34 (8)	109.33 (5)	109.28 (8)	109.32 (6)
O1–Si1–O5	109.94 (8)	110.03 (16)	109.83 (6)	109.75 (8)	109.80 (7)
O4–Si1–O3	110.26 (11)	110.51 (12)	110.34 (9)	110.30 (11)	110.39 (11)
O4–Si1–O5	108.86 (10)	108.81 (12)	108.91 (8)	108.92 (10)	108.92 (9)
O3–Si1–O5	107.97 (10)	107.84 (11)	108.05 (7)	108.14 (10)	108.05 (9)
O2–Si2–O4	109.30 (11)	109.41 (10)	109.37 (7)	109.30 (10)	109.48 (8)
O2–Si2–O5	110.35 (11)	110.22 (12)	110.21 (8)	110.22 (12)	110.17 (10)
O3–Si2–O2	109.42 (8)	109.56 (10)	109.75 (7)	109.75 (9)	109.55 (8)
O3–Si2–O4	108.80 (10)	108.93 (12)	108.71 (8)	108.69 (11)	108.72 (10)
O3–Si2–O5	109.47 (10)	109.35 (10)	109.45 (7)	109.54 (10)	109.55 (8)
O5–Si2–O4	109.45 (11)	109.35 (11)	109.25 (7)	109.31 (10)	109.23 (9)

3.3. Lattice Vibrations and Grüneisen Parameters γ_{iP}

The Raman spectra measured at ambient condition (in the frequency range up to 1200 cm^{-1}) are shown in Figure 3 for these synthetic samples. The fitted peak positions are listed in Table S2, and the vibrational bands at 521 cm^{-1} are always detected with most intensity. The Raman spectra are essentially the same among these coesite samples, while the most noticeable difference is that for the Al-doped samples (R694, R712, and R749). The intensities of the Raman modes at 151 and 178 cm^{-1} are relatively stronger and even comparable to the one at 119 cm^{-1}, as compared with those for the Al-free samples (R503 and R663). There are a total of 33 Raman-active modes (16 A_g(R) + 17 B_g(R)) as well as 36 IR-active modes (18 A_u(IR) + 18 B_u(IR)) predicted for coesite [16,17], while fewer peaks are detected in this Raman measurement.

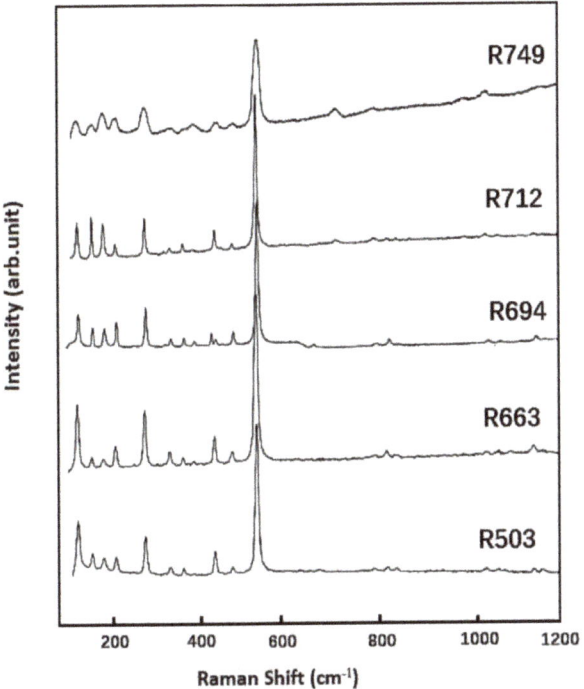

Figure 3. Raman spectra obtained at ambient condition for these synthetic coesite samples.

Next, we carried out in-situ high-temperature Raman experiments on the Si-pure (R503), B-doped (R663), Al-doped (R749), and B,Al-doped (R712) samples, as well as low-temperature measurement on R503. The representative Raman spectra for R503 at various temperatures are shown in Figure 4A as an example, while the high-T spectra for other coesite samples are deposited in the supplementary Figure S1 (See Supplementary Materials). The R503 sample was heated up to the temperature of 1500 K, and no phase transition was detected throughout the heating procedure. Although the signals got weaker and the background radiation became stronger especially at high temperatures above 1300 K, most of the Raman peaks could still be distinguished and fitted at the high temperatures. Another spectrum was recorded when the temperature was quenched to room temperature, and no clear shifts were observed among these Raman bands compared with those before heating (Figure 4B). Meanwhile, Bourova et al. [11] superheated a coesite sample to the temperature of 1776 K (at ambient pressure), which was 900 K higher than the predicted metastable melting point [45], and the coesite sample remained stable without any significant phase transition, melting, or amorphization. On the other hand, Liu et al. [19] reported amorphization of a hydrous coesite sample at a relatively low temperature of 1473 K. (See Supplementary Materials).

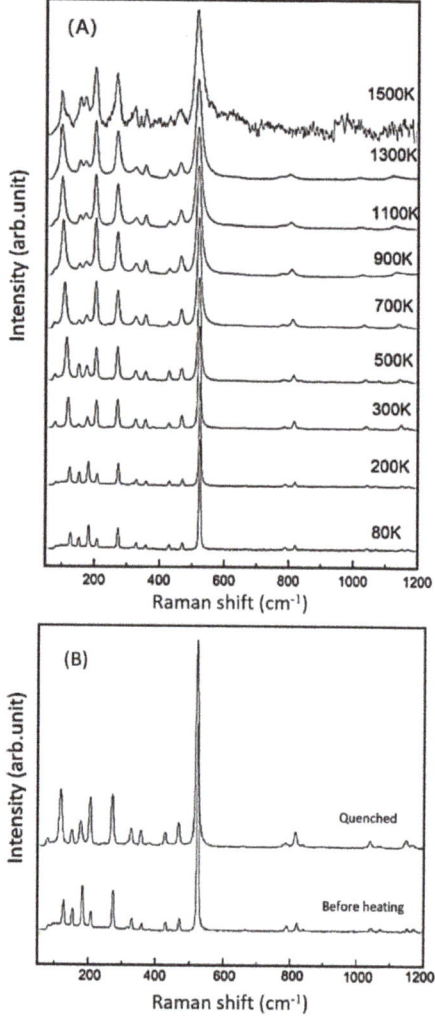

Figure 4. (**A**) Selected Raman spectra for the sample of R503 at various temperatures; (**B**) comparison of the Raman spectra taken before and after heating.

Variation of these Raman-active modes for R503 is plotted as a function of temperature in Figure 5A–C, and the data points at low temperatures are in consistence with those at high temperatures (Figures S2–S4 for R663, R712, and R749, individually). All these bands systematically shift to a lower frequency at elevated temperature, and linear regression was fitted to each mode with the negative slopes ($\delta v_i/\delta T$) (at $P = 0$ GPa) (Table S2). The values of $(\delta v_i/\delta T)_P$ are typically in the range of −0.01 to about −0.03 (cm^{-1}·K^{-1}) for the modes below 350 cm^{-1} or above 700 cm^{-1}, while −0.002 to about −0.007 (cm^{-1}·K^{-1}) for the ones in the range from 350 to 700 cm^{-1}. Our result is essentially in agreement with the previous high-temperature Raman studies on SiO$_2$-pure coesite [17,46].

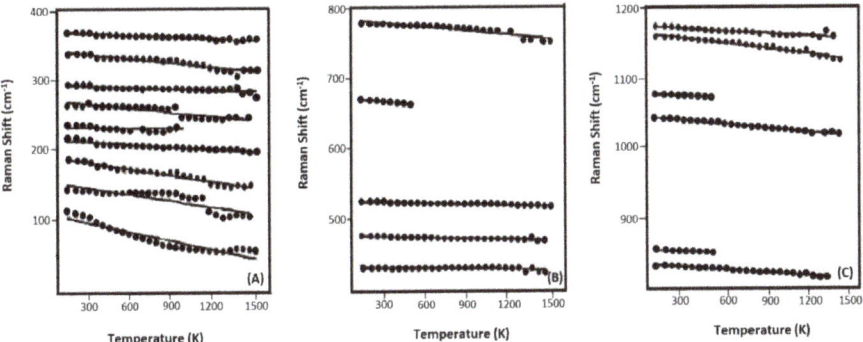

Figure 5. Variation of the frequencies for the Raman-active modes (R503) with temperature, in the frequency ranges of (**A**) 0–400 cm^{-1}, (**B**) 400–800 cm^{-1}, and (**C**) 800–1200 cm^{-1}. Linear regression is fitted for each dataset.

The isobaric mode Grüneisen parameter (γ_{iP}) is defined as

$$\gamma_{iP} = -\frac{1}{\alpha \cdot v_i} \cdot \left(\frac{\partial v_i}{\partial T}\right)_P, \tag{3}$$

where α is the averaged volumetric thermal expansion coefficient ($\alpha = 8.4 \times 10^{-6}$ K^{-1} for coesite [11]). The calculated γ_{iP} parameters are shown in Figure 6A for the samples of R503, R663, R712, and R749. The Raman-active modes above 400 cm^{-1} are mostly associated with the internal bending and stretching vibrations of SiO$_4$ tetrahedra linked in a three-dimensional framework for coesite [17,19,47]. The corresponding γ_{iP} parameters (1.4–3.2) are systematically larger than those internal modes (0–1.4) for isolated SiO$_4$ units as in forsterite (Mg-pure olivine) [48,49] and pyrope garnet [50], as well as a one-dimensional Si$_2$O$_6$ chain as in enstatite (MgSiO$_3$-orthopyroxene) [51], which are the most abundant minerals in the upper mantle above 410-km seismic discontinuity. Although the magnitudes of the ($\delta v_i/\delta T)_P$ slopes are similar among these studies, the thermal expansion coefficient for coesite [11] is much smaller as compared with these silicate minerals [52–54]. On the other hand, for the bands below 350 cm^{-1}, which are typically attributed to the external vibrations of SiO$_4$ tetrahedra in coesite, the values of γ_{iP} Grüneisen parameters are distributed in a much wider value range from −5 to 20.

Next, the differences of the γ_{iP} parameters among the samples of R663, R712, R749, and R503 (reference) are plotted in Figure 6B. The most significant difference is that in the frequency range below 350 cm^{-1}, the γ_{iP} parameters for the Al-doped samples (R712 and R749) are systematically lower than those for the Al-free ones (R503 and R663), while no such differences are observed above 400 cm^{-1}. When the Al^{3+} cations take the place of Si^{4+} in the tetrahedra, the thermal response of the enlarged tetrahedra units could get hindered to some extent at high temperature, while the smaller B^{3+} cations do not show such an effect on the external vibrations of the tetrahedra units in coesite.

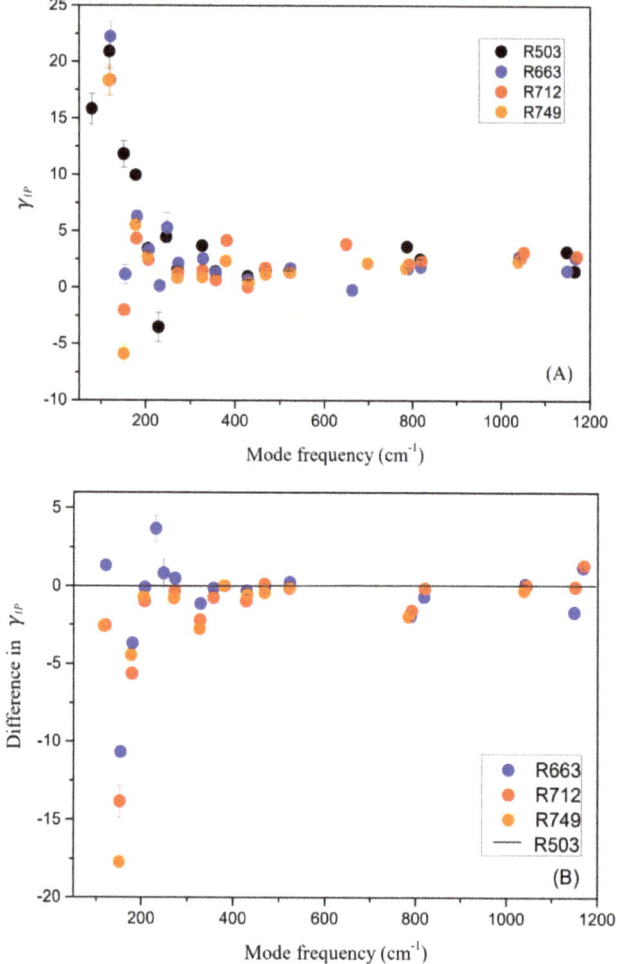

Figure 6. (**A**) The isobaric mode Grüneisen parameters (γ_{iP}) for the synthetic samples of R503, R663, R712, and R749; (**B**) comparison among the γ_{iP} parameters with the ones for the sample R503 set as reference.

3.4. OH-Stretching Modes at High Temperature

The selected high-temperature FTIR spectra for R503 are shown in Figure 7A in the temperature range up to 1150 K (Figure S5 for other coesite samples). The IR signal became weaker and broader at a higher temperature, which might be caused by rapid proton hopping between adjacent O atoms [55,56], as well as a black body radiation effect. Above 700 K, the broadened OH-stretching modes of v_1, $v_{2a,b}$, and v_3 (in the frequency range of 3450–3600 cm^{-1}) merge to be a broad hump and could not be distinguished from each other. The weak and discrete v_4 band (around 3300 cm^{-1}) vanishes very quickly and cannot be detected above 500 K. Another FTIR spectrum was collected when quenched to room temperature, and all these four OH-stretching bands could be clearly identified at the same positions as before heating (Figure 7B). The integrated absorbance for all these OH modes is about 80% of that before heating, and then 20% of the OH groups in the sample could be dehydrated during the heating procedure up to 1150 K. On the other hand, 30–40% dehydration was also observed for other samples at temperatures of up to 1000–1100 K, by comparing the integrated IR absorbance of the OH-stretching modes before and after heating. Meanwhile, Liu et al. [19] also conducted high-T FTIR measurement on hydrous coesite, and they observed noticeable dehydration above 870 K as well as completed dehydration at the temperature of 1473 K. In addition, we also conducted reflectance FTIR measurements [56,57] on these coesite samples. Nevertheless, due to the low water contents, the signals are significantly weaker as compared with those collected in the transmission method at ambient condition, and the OH bands could not be observed in the reflectance IR spectra above 500 K.

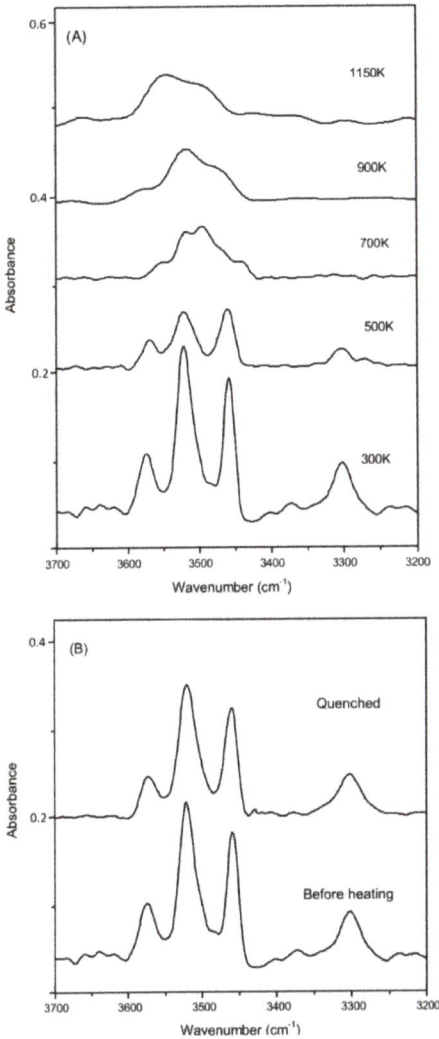

Figure 7. (**A**) Representative spectra for the sample of R503 at elevated temperatures; (**B**) comparison of the OH-stretching bands measured before and after heating.

Variations of the OH bands with temperature are plotted in Figure 8A–D for samples R503, R663, R694, and R749. Throughout the high-T measurments, the modes of $v_{2a,b}$, v_3, as well as v_6 (for B-doped samples of R663 and R712) were observed to show a 'blue-shift' with a slope $(\delta v_i/\delta T)_P$ of +0.01 to about +0.20 cm$^{-1}\cdot$K^{-1}, whereas a slight 'red shift' is detected for the v_1 mode (at a high frequency of around

3600 cm^{-1}) with a temperature derivative of −0.08 to about −0.20 cm^{-1}·K^{-1}. Above 600–700 K, these OH vibrations cannot be distinguished from each other, and the broad hump is observed to gradually move to a higher frequency at a higher temperature, with a temperature dependence of +0.05 to about +0.09 cm^{-1}·K^{-1}. In addition, the v_4 mode also shows a 'blue shift' with temperature increasing to below 500 K.

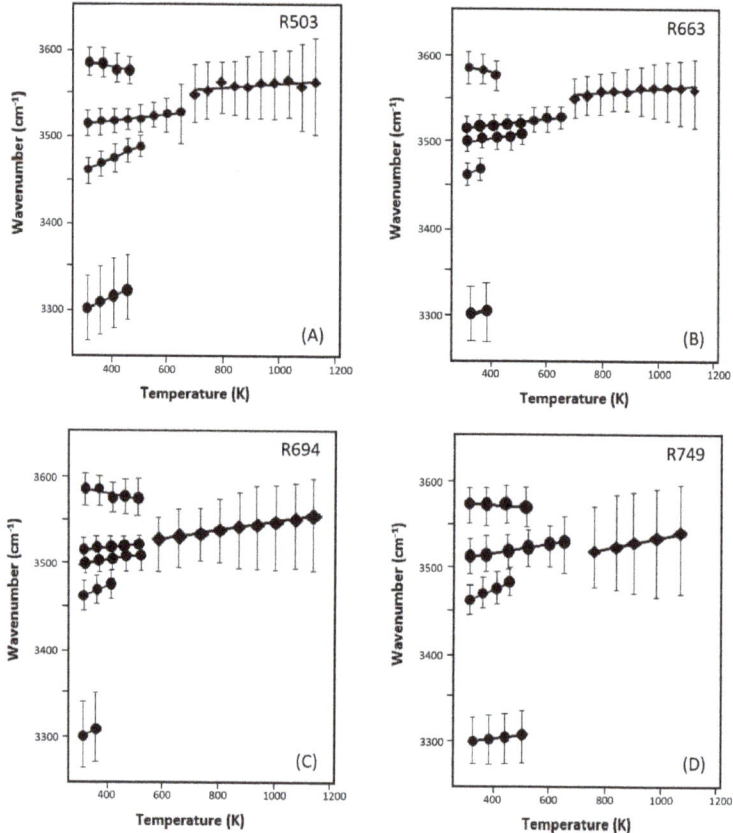

Figure 8. The frequencies of the OH-stretching modes as a function of temperature for the samples of (**A**) R503, (**B**) R663, (**C**) R694, and (**D**) R749. Linear regression lines are fitted (Table S3), and the vertical error bars represent the full-width of half maximum for each OH-stretching mode.

The OH-stretching modes observed in this study (in the frequency range above 3300 cm^{-1} with O...O distance of >2.74 cm^{-1}) should be attributed to protonation outside SiO$_4$ tetrahedra with the OH bonds pointing away from the centers of the tetrahedra [24–26]. The oxygen anion, that belong to different SiO$_4$ tetrahedra, may try to get away from each other during the thermal expansion and relaxation procedure at high temperature (i.e., the O...O distance between tetrahedra becomes larger). Consequently, we observe a 'blue shift' for most of the OH-stretching bands in the high-temperature FTIR measurements. On the other hand, Koch-Müller et al. [25] reported that some other OH-stretching modes (v_7, v_8, v_9, and v_{10} in the wavenumber range of 3370–3470 cm^{-1}) shift to higher frequencies at a high pressure of up to 10 GPa.

4. Conclusions

(1) We synthesized several hydrous coesite samples with different B and Al compositions at pressures of about 5–7.5 GPa (1273 K) in the multi-anvil press. The concentrations of the B/Al trace elements were measured by fs-LA–ICP-MS, while H_2O contents were estimated by FTIR. The B concentrations are more than 400 atomic ppm (B/10^6Si) with ~350 ppmw H_2O, while the Al^{3+} contents are about 1100–1300 atomic ppm, which were cross checked by both ICP-MS and EPMA. Al-substitution significantly reduces the hydrogen concentration in coesite. Hence, the mechanism controlled by oxygen vacancies ($2Si^{4+} = 2Al^{3+} + O_V$) may be dominant for the Al incorporation, which is similar to that in aluminous stishovite, while the B incorporation may prefer the electrostatically coupled substitution ($Si^{4+} = B^{3+} + H^+$);

(2) The doped B^{3+} and Al^{3+} cations would prefer the Si1 and Si2 tetrahedra, respectively, and the single-crystal structure refinements reveal that B^{3+} significantly shortens the Si1–O4 and Si1–O5 bond lengths, whereas Al^{3+} noticeably elongates the Si2–O4 and Si2–O5 distances;

(3) In-situ high-temperature Raman spectra were collected on these synthetic samples of up to 1500 K (at ambient condition), and no amorphization of phase transition was observed throughout the heating procedures. The derived isobaric mode Grüneisen parameters (γ_{iP}) for the external vibrations of SiO_4 units (below 350 cm^{-1}) are significantly reduced for the Al-doped samples, as compared with the Al-free ones. Hence, the relaxation of the SiO_4 units might be hindered to some extent due to the enlarged tetrahedra units by Al-substitution. On the other hand, the γ_{iP} parameters for the internal bending and stretching modes of SiO_4 tetrahedra in coesite (above 400 cm^{-1}) are significantly larger than those of most silicate minerals, due to the abnormally small thermal expansion coefficient for coesite;

(4) The OH-stretching modes v_1, $v_{2a,b}$, v_3, and v_4 are observed for all these hydrous samples with the various compositions, and another strong band v_6 is also observed for the B-doped ones. Most of these OH vibrational modes shift to higher frequencies at elevated temperatures (except the weak v_1 mode around 3600 cm^{-1}), implying that the O...O distances between different SiO_4 gets longer during the thermal relaxation of the lattice framework at a high temperature. On the other hand, about 20–40% dehydration of OH groups were observed for these hydrous coesite samples at high temperatures above 1000 K at ambient pressure.

Supplementary Materials: The following are available online at http://www.mdpi.com/2073-4352/9/12/642/s1, Figure S1. Selected high-temperature Raman spectra as well as the pattern taken when quenched to room temperature for the samples of (A) R663, (B) R712 and (C) R749. Figure S2. The frequencies for the Raman-active modes as a function of temperature for the sample of R712: (A) 0–400 cm^{-1}, (B) 400–800 cm^{-1} and (C) 800–1200 cm^{-1}. Linear regression is fitted for each dataset. Figure S3. The frequencies for the Raman-active modes as a function of temperature for the sample of R663: (A) 0–400 cm^{-1}, (B) 400–800 cm^{-1} and (C) 800–1200 cm^{-1}. Linear regression is fitted for each dataset. Figure S4. The frequencies for the Raman-active modes as a function of temperature for the sample of R749: (A) 0–400 cm^{-1}, (B) 400–800 cm^{-1} and (C) 800–1200 cm^{-1}. Linear regression is fitted for each dataset. Figure S5. Representative FTIR spectra obtained at high temperatures as well as when quenched to room temperature for the samples of (A) R663, (B) R694 and (C) R749. Table S1. Anisotropic displacement parameters (Å2) for the synthetic coesite samples in this study. Table S2. The frequencies of the Raman-active modes at ambient condition, as well as the temperature dependence and γ_{iP} parameters. Table S3. The frequencies of the OH bands by FTIR measurement at ambient temperature if not noted, as well as their temperature dependence. The cif files are for the single-crystal structure refinements for the five coesite samples (R503, R663, R694, R712 and R749).

Author Contributions: Conceptualization, Y.Y.; methodology and investigation, Y.M., Y.P., D.L., X.W., and X.Z.; writing—original draft preparation, Y.M.; writing—review and edition, Y.Y., J.R.S., and J.Z.; visualization, Y.M. and Y.Y.; supervision, Y.Y.; project administration, Y.Y.; funding acquisition, Y.Y. All authors discussed the results and commented on the manuscript.

Funding: This research was funded by the National Key Research and Development Program of China (Grant No. 2016YF0600204), the National Natural Science Foundation of China (Grant Nos. 41590621 and 41672041).

Acknowledgments: The multi-anvil press synthesis, fs-LA–ICP-MS, EPMA, and high-temperature Raman measurements were conducted at China University of Geosciences (CUG) at Wuhan; the single-crystal X-ray diffraction experiments were carried out at Huazhong University of Science and Technology (HUST); while the high-temperature FTIR spectra were collected at Zhejiang University. Many thanks to Tao Luo (CUG), Yan Qin (HUST), and Xiaoyan Gu (Zhejiang University) for their experimental assistance.

Conflicts of Interest: The authors declare no conflict of interest.

References

1. Coes, L. A new dense crystalline silica. *Science* **1953**, *118*, 131–132. [CrossRef] [PubMed]
2. Kieffer, S.W.; Phakey, P.P.; Christie, J.M. Shock processes in porous quartzite: Transmission electron microscope observations and theory. *Contrib. Mineral. Petrol.* **1976**, *59*, 41–93. [CrossRef]
3. Folco, L.; Mugnaioli, E.; Gemelli, M.; Masotta, M.; Campanale, F. Direct quartz-coesite transformation in shocked porous sandstone from Kamil Crater (Egypt). *Geology* **2018**, *9*, 739–742. [CrossRef]
4. Smyth, J.R.; Hatton, C.J. A coesite-sanidine grospydite from the Roberts-Victor Kimberlite. *Earth Planet. Sci. Lett.* **1977**, *34*, 284–290. [CrossRef]
5. Wang, X.M.; Liou, J.G.; Mao, H.K. Coesite-bearing eclogite from the Dabie mountains in central China. *Geology* **1989**, *17*, 1085–1088. [CrossRef]
6. Bi, H.; Song, S.; Dong, J.; Yang, L.; Qi, S.; Allen, M.B. First discovery of coesite in eclogite from East Kunlun, northwest China. *Sci. Bull.* **2018**, *63*, 1536–1538. [CrossRef]
7. Mosenfelder, J.L.; Bohlen, S.R. Kinetics of the coesite to quartz transformation. *Earth Planet. Sci. Lett.* **1997**, *153*, 133–147. [CrossRef]
8. Chopin, C. Coesite and pure pyrope in high-grade blueschists of the Western Alps: A first record and some consequences. *Contrib. Mineral. Petrol.* **1984**, *86*, 107–118. [CrossRef]
9. Angel, R.J.; Mosenfelder, J.L.; Shaw, C.S.J. Anomalous compression and equation of state of coesite. *Phys. Earth Planet. Inter.* **2001**, *124*, 71–79. [CrossRef]
10. Angel, R.J.; Shaw, C.S.J.; Gibbs, G.V. Compression mechanisms of coesite. *Phys. Chem. Miner.* **2003**, *30*, 167–176. [CrossRef]
11. Bourova, E.; Pichet, P.; Petitet, J.-P. Coesite (SiO_2) as an extreme case of superheated crystal: An X-ray diffraction study up to 1776 K. *Chem. Geol.* **2006**, *229*, 57–63. [CrossRef]
12. Chen, T.; Gwanmesia, G.D.; Wang, X.; Zou, Y.; Liebermann, R.C.; Michaut, C.; Li, B. Anomalous elastic properties of coesite at high pressure and implications for the upper mantle X-discontinuity. *Earth Planet. Sci. Lett.* **2015**, *412*, 42–51. [CrossRef]
13. Haines, J.; Leger, J.M.; Gorelli, F.; Hanfland, M. Crystalline post-quartz phase in silica at high pressure. *Phys. Rev. Lett.* **2001**, *87*, 155503. [CrossRef]
14. Černok, A.; Ballaran, T.B.; Caracas, R.; Miyajima, N.; Bykova, E.; Prakapenka, V.; Libermann, H.P.; Dubrovinsky, L. Pressure-induced phase transitions in coesite. *Am. Miner.* **2014**, *99*, 755–763. [CrossRef]
15. Hu, Q.Y.; Shu, J.-F.; Cadien, A.; Meng, Y.; Yang, W.G.; Sheng, H.W.; Mao, H.-K. Polymorphic phase transition mechanism of compressed coesite. *Nat. Commun.* **2015**, *6*, 6630. [CrossRef] [PubMed]
16. Hemley, R.J. Pressure dependence of Raman spectra of SiO2 polymorphs: α-quartz, coesite, and stishovite. In *High Pressure Research in Mineral Physics*; Manghnani, M.H., Syono, Y., Eds.; Geophysics Monograph Series 39; AGU: Washington, DC, USA, 1987; pp. 347–359.
17. Gillet, P.; Le, C.A.; Madon, M. High-temperature Raman spectroscopy of SiO_2 and GeO_2 polymorphs: Anharmonicity and thermodynamic properties at high-temperatures. *J. Geophys. Res.* **1990**, *95*, 21635–21655. [CrossRef]
18. Williams, Q.; Hemley, R.L.; Kruger, M.B.; Jeanloz, R. High pressure infrared spectra of α-quartz, coesite, stishovite and silica glass. *J. Geophys. Res.* **1993**, *98*, 157–170. [CrossRef]
19. Liu, X.; Ma, Y.; He, Q.; He, M. Some IR features of SiO_4 and OH in coesite, and its amorphization and dehydration at ambient pressure. *J. Asian Earth Sci.* **2017**, *148*, 315–323. [CrossRef]
20. He, M.; Yan, W.; Chang, Y.; Liu, K.; Liu, X. Fundamental infrared absorption of α-quartz: An unpolarzied single-crystal absorption infrared spectroscopic study. *Vib. Spectrosc.* **2019**, *101*, 52–63. [CrossRef]
21. Mosenfelder, J.L.; Schertl, H.-P.; Smyth, J.R.; Liou, J.G. Factors in the preservation of coesite: The importance of fluid infiltration. *Am. Miner.* **2005**, *90*, 779–789. [CrossRef]

22. Lathe, C.; Koch-Müller, M.; Wirth, R.; Van Westrenen, W.; Mueller, H.-J.; Schilling, F.; Lauterjung, J. The influence of OH in coesite on the kinetics of the coesite-quartz phase transition. *Am. Miner.* **2005**, *90*, 36–43. [CrossRef]
23. Mosenfelder, J.L. Pressure dependence of hydroxyl solubility in coesite. *Phys. Chem. Miner.* **2000**, *27*, 610–617. [CrossRef]
24. Koch-Müller, M.; Fei, Y.; Hauri, E. Location and quantitative analysis of OH in coesite. *Phys. Chem. Miner.* **2001**, *28*, 693–705. [CrossRef]
25. Koch-Müller, M.; Dera, P.; Fei, Y.; Reno, B.; Sobolev, N.; Hauri, E.; Wysoczanski, R. OH$^-$ in synthetic and natural coesite. *Am. Miner.* **2003**, *88*, 1436–1445. [CrossRef]
26. Deon, F.; Koch-Müller, M.; Hövelmann, J.; Rhede, D.; Thomas, S.-M. Coupled boron and hydrogen incorporation in coesite. *Eur. J. Mineral.* **2009**, *21*, 9–16. [CrossRef]
27. Rossman, G.R.; Smyth, J.R. Hydroxyl contents of accessory minerals in mantle eclogites and related rocks. *Am. Miner.* **1990**, *75*, 775–780.
28. Zhang, J.F.; Shi, F.; Xu, H.J.; Wang, L.; Feng, S.Y.; Liu, W.L.; Wang, Y.F.; Green, H.W. Petrofabric and strength of SiO_2 near the quartz-coesite phase boundary. *J. Metamorph. Geol.* **2013**, *31*, 83–92. [CrossRef]
29. Luo, T.; Ni, Q.; Hu, Z.; Zhang, W.; Shi, Q.; Günther, D.; Liu, Y.; Zong, K.; Hu, S. Comparison of signal intensities and elemental fractionation in 257 nm femtosecond LA-ICP-MS using He and Ar as carrier gases. *J. Anal. At. Spectrom.* **2017**, *32*, 2217–2225. [CrossRef]
30. Li, Z.; Hu, Z.; Liu, Y.; Gao, S.; Li, M.; Zong, K.; Chen, H.; Hu, S. Accurate determination of elements in silicate glass by nanosecond and femtosecond laser ablation ICP-MS at high spatial resolution. *Chem. Geol.* **2015**, *400*, 11–23. [CrossRef]
31. Jochum, K.P.; Willbold, M.; Raczek, I.; Stoll, B.; Herwig, K. Chemical characterization of the USGS reference glasses GSA-1G, GSC-1G, GSD-1G, GSE-1G, BCR-2G, BHVO-2G and BIR-1G using EPMA, ID-TIMS, ID-ICP-MS and LA-ICP-MS. *Geostand. Geoanal. Res.* **2005**, *29*, 285–302. [CrossRef]
32. Wang, X.; Xu, X.; Ye, Y.; Wang, C.; Liu, D.; Shi, X.; Wang, S.; Zhu, X. In-situ high-temperature XRD and FTIR for calcite, dolomite and magnesite: Anharmonic contribution to the thermodynamic properties. *J. Earth Sci.* **2019**, *30*, 964–976. [CrossRef]
33. Dolomanov, O.V.; Blake, A.J.; Champness, N.R.; Schroder, M. Olex: New software for visualization and analysis of extended crystal structures. *J. Appl. Crystallogr.* **2003**, *36*, 1283–1284. [CrossRef]
34. Cromer, D.T.; Mann, J. X-ray scattering factors computed from numerical Hartree-Fock wave functions. *Acta Crystallogr.* **1968**, *A24*, 321–325. [CrossRef]
35. Tokonami, M. Atomic scattering factor for O^{2-}. *Acta Crystallogr.* **1965**, *19*, 486. [CrossRef]
36. Pawley, A.R.; McMillan, P.F.; Holloway, J.R. Hydrogen in stishovite, with implications for mantle water content. *Science* **1993**, *261*, 1024–1026. [CrossRef]
37. Litasov, K.D.; Kagi, H.; Shatskiy, A.; Ohtani, E.; Lakshtanov, D.L.; Bass, J.D.; Ito, E. High hydrogen solubility in Al-rich stishovite and water transport in the lower mantle. *Earth Planet. Sci. Lett.* **2007**, *262*, 620–634. [CrossRef]
38. Shimizu, N.; Hart, S.R. Applications of the ion microprobe to geochemistry and cosmochemistry. *Annu. Rev. Earth Planet. Sci.* **1982**, *10*, 483–526. [CrossRef]
39. Araki, T.; Zoltai, T. Refinement of a coesite structure. *Z. Krist.* **1969**, *129*, 381–387. [CrossRef]
40. Levien, L.; Prewitt, C.T. High-pressure crystal structure and compressibility of coesite. *Am. Miner.* **1981**, *66*, 324–333.
41. Smyth, J.R.; Artioli, G.; Smith, J.V.; Kvick, A. Crystal structure of coesite, a high-pressure form of SiO_2, at 15 and 298 K from single-crystal neutron and X-ray diffraction data: Test of bonding models. *J. Phys. Chem.* **1987**, *91*, 988–992. [CrossRef]
42. Sasaki, S.; Chen, H.K.; Prewitt, C.T.; Nakajima, Y. Re-examination of "$P2_1/a$ coesite". *Z. Krist.* **1983**, *164*, 67–77. [CrossRef]
43. Ikuta, D.; Kawame, N.; Banno, S.; Hirajima, T.; Ito, K.; Rakovan, J.F.; Downs, R.T.; Tamada, O. First in situ X-ray diffraction identification of coesite and retrograde quartz on a glass thin section of an ultrahigh-pressure metamorphic rock and their crystal structure details. *Am. Miner.* **2007**, *92*, 57–63. [CrossRef]
44. Downs, R.T.; Bartelmehs, K.L.; Gibbs, G.V.; Boisen, M.B. Interactive software for calculating and displaying X-ray or neutron power diffractometer patterns of crystalline materials. *Am. Miner.* **1993**, *78*, 1104–1107.

45. Richet, P. Superheating, melting and vitrification through decompression of high-pressure minerals. *Nature.* **1988**, *331*, 56–58. [CrossRef]
46. Liu, L.G.; Mernagh, T.P.; Hibberson, W.O. Raman spectra of high-pressure polymorphs of SiO_2 at various temperatures. *Phys. Chem. Miner.* **1997**, *24*, 396–402. [CrossRef]
47. Kieffer, S.W. Thermodynamics and lattice vibrations of minerals: Lattice dynamics and an approximation for minerals with application to simple substances and framework silicates. *Rev. Geophys.* **1979**, *17*, 35–39. [CrossRef]
48. Gillet, P.; Daniel, I.; Guyot, F. Anharmonic properties of Mg_2SiO_4-forsterite measured from the volume dependence of the Raman spectrum. *Eur. J. Mineral.* **1997**, *9*, 255–262. [CrossRef]
49. Yang, Y.; Wang, Z.; Smyth, J.R.; Liu, J.; Xia, Q. Water effects on the anharmonic properties of forsterite. *Am. Miner.* **2015**, *100*, 2185–2190. [CrossRef]
50. Gillet, P.; Fiquet, G.; Malezieux, J.M.; Geiger, C.A. High-pressure and high-temperature Raman spectroscopy of end-member garnets: Pyrope, grossular and andradite. *Eur. J. Mineral.* **1992**, *4*, 651–664. [CrossRef]
51. Zucker, R.; Shim, S.-H. In situ Raman spectroscopy of $MgSiO_3$ enstatite up to 1550 K. *Am. Miner.* **2009**, *94*, 1638–1646. [CrossRef]
52. Kroll, H.; Kirfel, A.; Heinemann, R.; Barbier, B. Volume thermal expansion and related thermophysical parameters in the Mg, Fe olivine solid-solution series. *Eur. J. Mineral.* **2012**, *24*, 935–956. [CrossRef]
53. Du, W.; Clark, S.M.; Walker, D. Thermo-compression of pyrope-grossular garnet solid solutions: Non-linear compositional dependence. *Am. Miner.* **2015**, *100*, 215–222. [CrossRef]
54. Jackson, J.M.; Palko, J.W.; Andrault, D.; Sinogeikin, S.V.; Lakshtanov, D.L.; Wang, J.; Bass, J.D.; Zha, C.-S. Thermal expansion of natural orthoenstatite up to 1472 K. *Eur. J. Mineral.* **2003**, *97*, 6842–6866.
55. Keppler, H.; Bagdassarov, N.S. High-temperature FTIR spectra of H_2O in rhyolite melt to 1300 °C. *Am. Miner.* **1993**, *78*, 1324–1327.
56. Grzechnik, A.; McMillan, P.F. Temperature dependence of the OH^- absorption in SiO_2 glass and melt to 1975 K. *Am. Miner.* **1998**, *83*, 331–338. [CrossRef]
57. Moore, G.; Chizmshya, A.; McMillan, P.F. Calibration of a reflectance FTIR method for determination of dissolved CO_2 concentration in rhyolitic glasses. *Geochim. Gosmochim. Acta* **2000**, *64*, 3571–3579. [CrossRef]

© 2019 by the authors. Licensee MDPI, Basel, Switzerland. This article is an open access article distributed under the terms and conditions of the Creative Commons Attribution (CC BY) license (http://creativecommons.org/licenses/by/4.0/).

Article

Bacterial Effect on the Crystallization of Mineral Phases in a Solution Simulating Human Urine

Alina R. Izatulina [1,*], Anton M. Nikolaev [1,2], Mariya A. Kuz'mina [1], Olga V. Frank-Kamenetskaya [1] and Vladimir V. Malyshev [3]

1. Crystallography Department, Institute of Earth Sciences, St. Petersburg State University, St. Petersburg 199034, Russia; floijan@gmail.com (A.M.N.); m.kuzmina@spbu.ru (M.A.K.); ofrank-kam@mail.ru (O.V.F.-K.)
2. Grebenshchikov Institute of Silicate Chemistry of the Russian Academy of Sciences, St. Petersburg 199034, Russia
3. Microbiology Department, S.M. Kirov Military Medical Academy, St. Petersburg 194044, Russia; vladmal_spb@list.ru
* Correspondence: alina.izatulina@mail.ru; Tel.: +7-911-770-3824

Received: 24 April 2019; Accepted: 16 May 2019; Published: 18 May 2019

Abstract: The effect of bacteria that present in the human urine (*Escherichia coli, Pseudomonas aeruginosa, Klebsiella pneumoniae,* and *Staphylococcus aureus*) was studied under the conditions of biomimetic synthesis. It was shown that the addition of bacteria significantly affects both the phase composition of the synthesized material and the position of crystallization boundaries of the resulting phosphate phases, which can shift toward more acidic (struvite, apatite) or toward more alkaline (brushite) conditions. Under conditions of oxalate mineralization, bacteria accelerate the nucleation of calcium oxalates by almost two times and also increase the amount of oxalate precipitates along with phosphates and stabilize the calcium oxalate dihydrate (weddellite). The multidirectional changes in the pH values of the solutions, which are the result of the interaction of all system components and the crystallization process, were analyzed. The obtained results are the scientific basis for understanding the mechanisms of bacterial involvement in stone formation within the human body and the creation of biotechnological methods that inhibit this process.

Keywords: pathogen crystallization; biomimetic synthesis; renal stone; calcium oxalate; apatite; brushite; struvite; octocalcium phosphate; whitlockite; *Escherichia coli; Klebsiella pneumoniae; Pseudomonas aeruginosa; Staphylococcus aureus*

1. Introduction

Urolithiasis is an example of pathogenic mineral formation in the human body. Various exogenous and endogenous factors are considered among the reasons for the development of urolithiasis [1,2]. The more factors act simultaneously, the more difficult the pathogenesis of urolithiasis and the worse its prognosis, which is due to frequent recurrence of the disease and the rapid growth of stones.

Currently, there are many theories explaining the causes and mechanisms of pathogenic stone formation in the human urinary system [3–10]. All theories are based on the complex interaction of biogenic and abiogenic substances, but none of them are exhaustive. The least studied is the bacterial theory [4].

It is well known that the presence of a variety of bacteria in the urine is very likely and bacterial inflammation often accompanies stone formation [11]. Assumptions about the significant effect of microorganisms on the processes of lithiasis in the human urinary system have been made in a number of works [2–13]. The crystallization system (urine) contains about a dozen bacteria species. Microbiological examination of removed urinary stones' microflora shows that more than half of urinary stones are infected, in most cases by several types of bacteria [2,14]. Infectious diseases of the urinary tract are direct or indirect provocateurs of stone formation in the human urinary system. According to the observations of practicing urologists, infectious sequelae after lithotripsy are rather frequent, even against the background of sanitized urine, which indicates that the stones are infected by bacteria during the formation [2]. The results of urine stone sowing showed that *Enterococcus faecalis*, *Enterococcus faecium*, *Staphylococcus epidermidis*, *Staphylococcus haemolyticus*, *Pseudomonas aeruginosa*, *Klebsiella pneumoniae*, *Proteus mirabilis*, and *Escherichia coli*, as well as *Streptococcus spp*, *Staphylococcus aureus*, *Acinetobacter baumanii*, *Candida albicans*, and *Morganella morganii* were among the most frequently excreted microorganisms [14]. The presence of pathogens in the urine affects the parameters and composition of urine, which in turn should affect the crystallization of urinary stones' mineral phases. A number of studies have shown that bacteria can form biofilms on the surface of a stone, which leads to the formation of chronic infection during diseases of the urinary system [2,15,16].

A substantial portion of papers on the effect of bacteria on the stone formation in the human urinary system is devoted to the so-called infectious renal stones, consisting mainly of struvite ($(NH_4)MgPO_4·6H_2O$), and sometimes containing hydroxylapatite ($Ca_5(PO_4)_3(OH)$) and brushite ($Ca(HPO_4)·2H_2O$) [12]. The bacteria that cause the secondary phosphate stone formation belong to the urease-forming microflora [17]. Infectious stones are formed as a result of urea hydrolysis to ammonium ions and bicarbonate, increasing the urine pH to normal or alkaline values and binding to available cations to produce magnesium ammonium phosphate (struvite) and carbonate apatite [12]. Struvite stones are found only in a small number of patients susceptible to urinary tract infections. Thus, in our collection of renal stones of St. Petersburg and the Leningrad region residents, which consists of more than 2000 samples, only 27 belong to this "infectious" type (Figure 1). It is assumed that oxalate stones may also have an infectious origin [2,3,17]. The data on the initiation of the crystallization and aggregation of calcium oxalates in the presence of *E. coli* [18], as well as the work on the crystallization of weddellite ($CaC_2O_4·2H_2O$) in the presence of *E. coli* [19], favor of this assumption. In addition, a number of papers suggest that bacteria can serve as centers of crystallization and the subsequent growth of renal stones, forming a phosphate shell around itself [20].

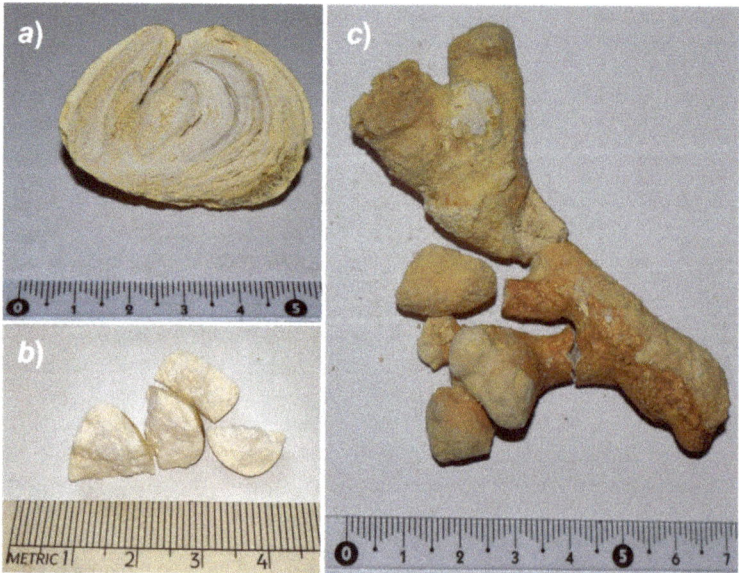

Figure 1. Infectious renal stones: (**a**) Apatite–struvite–brushite, (**b**) struvite, and (**c**) struvite–brushite.

The results of model experiments on the crystallization of pathogenic phase analogs in the presence of bacteria have shown that bacteria change the pH of solutions and can increase the amount and alter the morphology of the resulting oxalate and phosphate crystals [3,6,7,13,21]. Unfortunately, the currently available data are insufficient to characterize the effect of the bacterial presence in the urine on the phase composition of the resulting renal stones.

In order to advance in this direction, we conducted a synthesis experiment using solutions that simulate the composition of human urine, including containing bacteria common for human urine, and revealed their role in the crystallization of urinary phosphate and oxalate stones.

2. Materials and Methods

Biomimetic syntheses in the presence of bacteria were carried out by precipitation at 37 °C from solutions that simulated the composition of human urine and its inorganic components, where the content corresponded to their minimum or maximum values (Table 1). The volume of the solution after mixing of initial components in accordance with Figure 2 was 500 mL. The content of calcium cations in solutions ranged from 5 to 7.7 mmol/L, which is due to the fact that in small volumes of solution (0.2 L) and with a limited time to carry out the synthesis (1–2 days) the formation of a crystalline precipitate at lower calcium concentrations does not occur. To accelerate the crystallization of calcium and magnesium phosphates, oxalate ions (in the form of ammonium oxalate) were also added to the initial solution in a low concentration (0.1 mmol/L). Also, experiments in the so-called "oxalate system" containing only calcium ions and oxalate ions were conducted (calcium oxalate supersaturation is equal to 7, which corresponds to the physiological values of urine), since calcium oxalate does not crystallize in the system simulating the composition of urine. Ovalbumin was added to the experiments at a concentration of 10 mmol/L [22]. Syntheses were carried out by precipitation in an aqueous solution or in solutions of protein-containing nutrient media, the Müller Hinton Broth (MHB) nutrient medium or the Meat-Peptone Broth (MPB), which were prepared according to standard techniques [23,24]. In addition, bacteria associated with inflammatory processes and present in significant quantities both in the environment and in the human body were added to each of the protein media and to the model media in an amount of 10^6 particles per liter: *Escherichia coli* («e»), *Klebsiella pneumoniae* («kl»), *Pseudomonas aeruginosa* («ps»), and *Staphylococcus aureus* («s»). The following bacterial American Type Culture Collection (ATCC) strains were used in the experiments: 25922 («e»), 70060325922 («kl»), 27853 («ps»), and 29213 («s»). The pH of the solutions varied between 5.77–7.26 (minimum concentrations of inorganic components) and 6.10–8.07 (maximum concentrations of inorganic components). The acidity of the initial solutions was adjusted using aqueous solutions of HCl and NaOH. The crystallization start time (clouding of the solution) and phase composition of the obtained precipitates were recorded during experiments. Clouding of the solution was recorded visually. The precipitate obtained a day later was filtered, washed with distilled water, and dried at room temperature; at least three iterations were performed for each experiment.

Table 1. Elemental composition (mmol/L) of model solutions and urine.

Component	Model Solution		Human Urine [22,25]
	Min Concentration	Max Concentration	
Na^+	60	73	67–133
K^+	21.7	102	33–47
Ca^{2+}	5–7.7	5–7.7	1.7–5
Mg^{2+}	5.3	11	5.3–11
NH_4^+	20.8	49.4	20–50
Cl^-	67	80	67–167
CO_3^{2-}	0	33	0–33
PO_4^{3-}	13	33	13–33
SO_4^{2-}	21.7	69	27–80

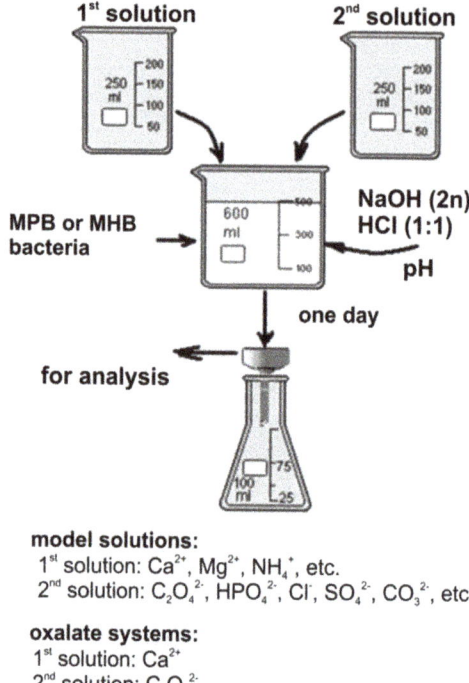

Figure 2. The scheme of the synthesis experiment in system which simulated the composition of urine by inorganic components.

The phase composition of precipitate products was determined by means of powder X-ray diffraction method (PXRD). The measurements were performed using a Rigaku «MiniFlex II» powder diffractometer (CuKα radiation, λ = 1.54178 Å; 30 kV/15 mA; Bragg–Brentano geometry; PSD D-Tex Ultra detector). X-ray diffraction patterns were collected at room temperature in the range of 3–60 °2θ with a step of 0.02 °2θ. Phase identification was carried out using the ICDD PDF-2 Database (release 2016). The unit cell parameters were refined by the Pawley method using TOPAS 4.2 software [26]. The background was modeled using a Chebychev polynomial of 12th order. The peak profile was described using the fundamental parameters approach.

3. Results

3.1. pH Changes of the Medium

The pH of the solutions in the crystallization process of the phosphate phases always decreased in experiments without organic additives and with the addition of nutrient media and bacteria, it either increased or decreased (Table 2). As can be seen from Table 2, the nutrient media and bacteria affected the pH values, which can be explained both by the influence of crystallization processes and bacterial activity. For instance, interaction of the solution with MHB media slightly reduced the pH value of the solution in the case of the minimum concentrations of inorganic components and in the case of maximum concentrations the pH of the solution increased. Addition of *Pseudomonas aeruginosa* bacteria to the MHB medium slightly increased the pH value of the solution (by 0.4), while addition of the same bacteria to the MPB medium increased the pH value of the solution by much more (by 0.6).

Table 2. The change in pH of the solutions during the experiment in nutrient media with the addition of bacteria.

Nutrient Medium	Additives Bacteria	Minimum Concentration Initial pH	Minimum Concentration Final pH	Maximum Concentration Initial pH	Maximum Concentration Final pH
none	none	5.95–7.54	5.94–6.71	5.81–7.73	5.75–7.50
Müller–Hinton Broth	none	5.81–7.15	5.84–6.25	6.10–8.07	6.27–7.39
	Escherichia coli ("e")	5.81–7.15	5.27–6.39	6.10–8.07	5.85–7.35
	Klebsiella pneumoniae ("kl")	5.81–7.15	5.51–6.40	6.10–8.07	5.84–7.50
	Pseudomonas aeruginosa ("ps")	5.81–7.15	6.21–6.51	6.10–8.07	6.16–7.90
	Staphylococcus aureus ("s")	5.81–7.15	5.06–6.45	6.10–8.07	5.66–7.54
Meat–Peptone Broth	none	5.77–7.26	5.78–7.08	6.10–8.03	6.18–7.90
	Escherichia coli ("e")	5.77–7.26	5.94–7.02	6.10–8.03	6.07–7.67
	Klebsiella pneumoniae ("kl")	5.77–7.26	5.90–7.00	6.10–8.03	6.03–7.80
	Pseudomonas aeruginosa ("ps")	5.77–7.26	6.39–7.40	6.10–8.03	6.18–7.90
	Staphylococcus aureus ("s")	5.77–7.26	6.25–7.23	6.10–8.03	6.14–8.05

3.2. Model Solutions with Minimum Concentrations of Additional Ions Characteristic of a Healthy Person's Urine Composition

In syntheses of phosphates with minimum concentrations of inorganic impurities without additives, formation of the following crystalline phases was observed: Brushite ($Ca(HPO_4) \cdot 2H_2O$), octacalcium phosphate ($Ca_8(HPO_4)_2(PO_4)_4 \cdot 5H_2O$), and whitlockite ($Ca_9Mg(HPO_4)(PO_4)_6$) [27,28]. Brushite formed in synthetic experiments when the initial pH of the solution ranged from 6.46 to 6.86. Octacalcium phosphate was usually observed together with brushite (less often with whitlockite) in the pH range of 6.46 to 6.95. Whitlockite was obtained in the pH range of 6.95 to 7.54.

Addition of MHB medium to the model solution changed the phase composition of the sediment (Figure 3). In the pH range 6.75–7.3, the brushite phase was detected. Brushite also formed after addition of various bacteria to the solution. Moreover, the whitlockite phase was detected in the syntheses that were carried out in the presence of "kl" at a pH of 7.15. In addition, in the experiments with "e" and "ps" bacteria at pH 7.05–7.15, formation of struvite was identified (together with brushite).

Addition of the MPB medium to the model solution also led to changes in the phase composition of the sediment (Figure 3). In this case, brushite was detected at pH ~7.06. Brushite did not crystallize at such a high pH in the experiments without additives. Another difference in the phase composition of the precipitate was the formation of struvite at a pH of 7.26, which is absent in the products of syntheses without additives. The brushite phase was detected in the sediments of all syntheses, which were carried out in the presence of bacteria. Whitlockite was formed only in the synthesis in which the E. coli bacteria were present at a pH of 7.07. Struvite was formed in the syntheses with bacteria, except those experiments with the addition of Staphylococcus aureus, at a pH of 7.0 or higher. In all the syntheses with bacteria, the formation of apatite was observed at a pH of 6.72 or higher.

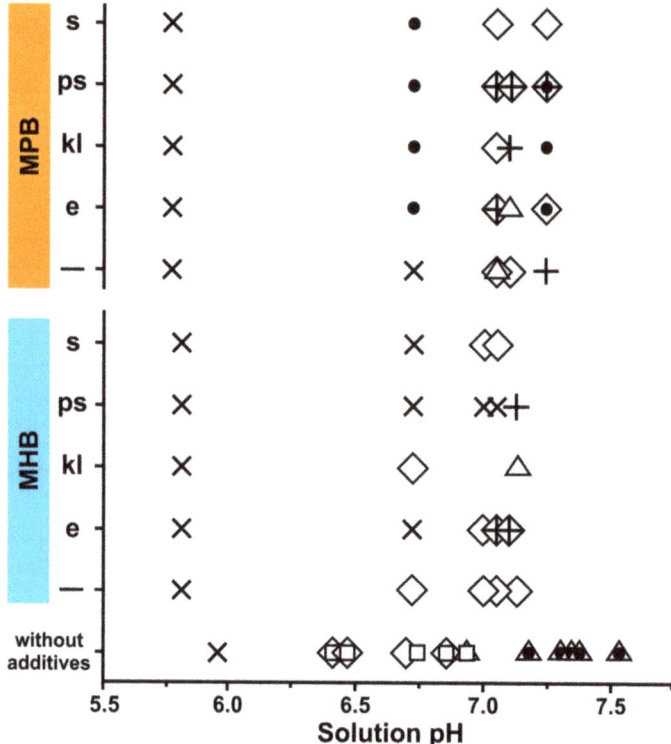

Figure 3. Phase composition of synthesized products from model solutions with minimum concentrations of additional ions characteristic of a healthy person's urine composition. Legend: ◊—brushite, +—struvite, Δ—whitlockite, •—apatite, □—octacalcium phosphate, ×—no precipitation; *Escherichia coli* —«e», *Klebsiella pneumoniae*—«kl», *Pseudomonas aeruginosa*—«ps», and *Staphylococcus aureus*—«s».

3.3. Model Solutions with Maximum Concentrations of Additional Ions Characteristic of a Healthy Person's Urine Composition

In the phosphate syntheses with maximum concentrations of inorganic impurities without additives, the formation of brushite and struvite was observed (Figure 4). Brushite was formed in a wide range of pH values of the initial solution from 5.81 to 7.63. Struvite growth occurred at higher pH values (from 7.23 to 7.73) and usually along with brushite.

When MHB medium was added to the model solution, hydroxylapatite was clearly observed in the precipitate composition, in addition to common brushite and struvite. Brushite and struvite phases also formed when various bacteria were added to the solution, while apatite was detected only in syntheses with *E. coli* and *Pseudomonas aeruginosa*. In all the systems, except for the synthesis in the presence of *Klebsiella pneumoniae*, there was a significant shift in the beginning of the precipitate formation toward higher pH values. Thus, brushite was obtained in syntheses within the pH range 7.0–7.03 (in the system with "kl" at a pH of 6.10), apatite was observed only at a pH of 7.0, and struvite at a pH of 7.0 or higher.

Addition of MPB medium to the model solution did not lead to changes in the phase composition of the sediment. Brushite and struvite were formed when various bacteria were added to the solution (Figure 4). The brushite phase was found in all systems, but at different pH values: Between 6.10 and

7.06 (syntheses with "st", "kl", and "ps") or 6.96–7.06 (syntheses with "e"). Struvite was formed in all the systems at a pH of about 7 and higher.

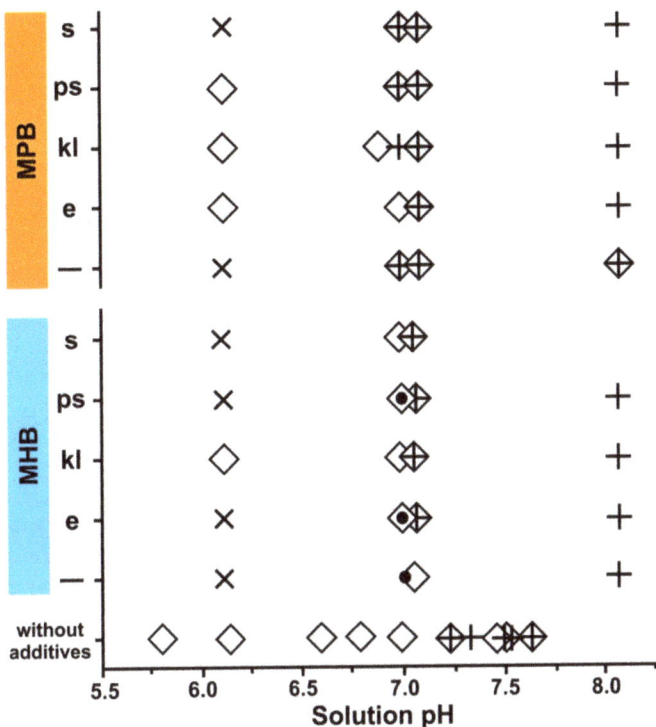

Figure 4. Phase composition of synthesized products from model solutions with maximum concentrations of additional ions characteristic of a healthy person's urine composition. Legend as in Figure 3.

3.4. Crystallization in the Oxalate System

As the result of the biomimetic syntheses, it was found that the presence of bacteria accelerates nucleation within the oxalate system (Table 3). Perhaps, bacteria can act as nucleation centers. The greatest effect (more than twice) in accelerating the crystallization rates of the calcium oxalates was observed in the presence of *Pseudomonas aeruginosa*.

Table 3. Nucleation of calcium oxalates in the presence of bacteria at various supersaturations (γ).

Bacteria	Nucleation Time, s		
	$\gamma = 3$	$\gamma = 7$	$\gamma = 10$
None	More than 2400 (>40 min)	840	140
Staphylococcus aureus	1500	510	30–50
Klebsiella pneumoniae	1290	470	30–50
Escherichia coli	1200	420	30–50
Pseudomonas aeruginosa	1140	370	30–50

PXRD analyses of the precipitates obtained in the presence of bacteria within the oxalate system showed formation of calcium oxalate mono and dihydrate (whewellite and weddellite, respectively), while in the syntheses without bacteria only whewellite was formed. According to the PXRD data, the whewellite/weddellite ratio in precipitants was determined along with unit cell parameters, from which the content of "zeolite" water, x, in the structure of weddellite ($CaC_2O_4 \cdot (2 + x)H_2O$) was calculated (Table 4) [29]. The presence of bacteria did not practically affect the whewellite/weddellite ratio, as well as the content of "zeolite" water (x). Moreover, the results obtained for the syntheses with bacteria were close to the effect of proteins that stabilize calcium oxalate dihydrate crystallization [1,30].

Table 4. Characteristics of phases synthesized within the oxalate system in the presence of bacteria and protein additives.

Additives	Whewellite/Weddellite Ratio	Selected Crystallographic Data for the Weddellite Phase	
		Unit Cell Parameter, Å	Amount of "Zeolite" Water (x), p.f.u. *
None	whewellite	–	–
Ovalbumin	5:2	12.349(1)	0.26
Escherichia coli	5:2	12.344(1)	0.23
Pseudomonas aeruginosa	5:2	12.341(2)	0.21
Staphylococcus aureus	5:2	12.346(2)	0.24

* Per formula unit; calculations were made with regard to the *a* unit cell parameter, using the regression equation reported in [29].

4. Discussion

Perhaps, the most important results of the study are that in systems with minimum and maximum concentrations of inorganic ions, only analogs of the phosphate renal stone mineral phases were observed, while calcium oxalates were obtained under given conditions only with an increase in the concentration of oxalate ions up to the oxalatouria values, both in experiments with bacteria and without them. This result is in general agreement with the literature data on model crystallization experiments in the human urinary system [21,27,31]. Thus, according to the thermodynamic calculations and experiments in systems that simulate composition of the physiological liquid, calcium oxalates are formed in much smaller quantities than what is actually observed during pathogenic processes in the human body. Moreover, the weddellite phase (calcium oxalate dihydrate) does not form at all [31]. Introduction of bacteria and protein (ovalbumin) to the system leads to a similar result in all the experiments, increasing the portion of weddellite and increasing the amount of calcium oxalates in general. It should be also noted that, according to the unit cell parameters of weddellite crystals which are formed in the presence of bacteria, the amount of "zeolite" H_2O molecules (x) falls into a rather narrow range of values, whereas those in the structures of weddellite crystals from human renal stones vary much more (from 0.13 to 0.37 p.f.u.).

According to our data, all bacteria initiate the nucleation of calcium oxalates and promote the crystallization of metastable calcium oxalate dihydrate (weddellite) in the oxalate system (containing only Ca^{2+} and $[C_2O_4]^{2-}$ ions). The initiation of calcium oxalate nucleation in the presence of bacteria is in agreement with the results of some recent studies, which describe an increase in the number of calcium oxalate crystals and their size in the presence of bacteria [32].

As it was shown by the results of phosphate crystallization experiments, the addition of bacteria and nutrient media leads to a change in the phase composition of the precipitate and to the shift of the phosphate phase's formation boundaries (Figures 3 and 4). The addition of the MHB medium to the model solution with the minimum concentration of inorganic impurities led to the disappearance of octacalcium phosphate and whitlockite, followed by the formation of brushite and rare struvite occurrences. The same addition to the model solution with the maximum concentration of inorganic impurities led to the crystallization of apatite, along with brushite and struvite, and to the significant shift of brushite and apatite formation areas toward higher pH values of the solution (~7.0).

The addition of the MPB medium to the model solution with a minimum concentration of inorganic impurities led to the formation of brushite and whitlockite and, in addition, crystallization of struvite was detected at a pH of 7.26, so the shift in the struvite phase formation boundary in this system moved toward being significantly more acidic. The brushite phase was observed in this system in a narrower pH range around 7.06. Since brushite did not form in experiments without organic additives at such a high pH, this suggests that the boundary of its formation expanded to the more alkaline side. The same addition to the model solution with the maximum concentration of inorganic impurities did not lead to any change in the phase composition of the synthesized products. At the same time, it can be stated that the boundary of the brushite formation area has shifted to the more alkaline region of solutions and the boundary of the struvite formation area shifted to the more acidic region (pH of 6.96).

The addition of bacteria to the appropriate media led to additional changes in the composition of the precipitates (Figures 3 and 4). Thus, in the syntheses with minimal concentrations of inorganic impurities, the appearance of *Escherichia coli* and *Pseudomonas aeruginosa* in an MHB medium led to the formation of struvite and shifted its starting crystallization boundary to the more acidic region (pH of 7.05). Although struvite was initially present in the synthetic products, the appearance of bacteria in the MHB medium contributed to the displacement of its crystallization area to the more acidic region. The effect of *Escherichia coli* and *Staphylococcus aureus* bacteria on the crystallization of brushite was also well demonstrated in systems containing MHB; the shift of the brushite initial crystallization area occurred toward the more alkaline region. The effect of the bacteria addition on the crystallization of apatite was clearly visible in the MPB medium; the appearance of bacteria promoted crystallization of apatite and shifted its formation boundary to the more acidic region (pH of 6.72).

The change in the pH values of the solution during the biomimetic syntheses process occurred in different directions, due to both the crystallization process of various phases and the effect of a certain protein medium type and all types of bacteria addition. The decrease in pH in systems that modeled urine using inorganic components can be explained by the result of phosphate phase crystallization, while an increase in systems with bacteria can be explained by the influence of metabolic products. The presence of urease-producing bacteria such as *Pseudomonas aeruginosa*, *Klebsiella pneumoniae* and *Staphylococcus aureus* in urea led to an increase in pH [11,12].

The displacement of struvite crystallization boundaries obtained in the experiments, which led to its intensive formation, once again underlines the involvement of bacteria in the formation of "infectious" renal stones, described in a number of works [3,12]. At the same time, the expansion of the brushite crystallization area boundaries and the crystallization of apatite, as well as the formation of weddellite in the oxalate system, shows that the influence of the presence and function of bacteria in the crystallization medium was not only limited to the alkalization of the urine and the formation of ammonium ions, but significantly affected the types of growing mineral phases and the size of their crystallization areas with natural variations in urine pH.

5. Conclusions

Under the conditions of model experiments, the effect of bacteria that are present in human urine (*Escherichia coli*, *Pseudomonas aeruginosa*, *Klebsiella pneumoniae* and *Staphylococcus aureus*) on the formation of the renal stone mineral phases, such as brushite, struvite, vitlocite, octacalcium phosphate, apatite, whewellite, and weddellite, was studied in systems simulating the composition of human urine and using two types of nutrient media (Muller–Hinton Broth and Meat–Peptone Broth). Multidirectional changes in the pH values of the solutions were analyzed, which are the result of all system components' interactions with the crystallization process.

It was shown that the presence of bacteria has a different effect on the phosphate and oxalate phases' formation. The presence of pathogens and nutrient media significantly affect the precipitant phase composition and the position of the resulting phosphate phase's crystallization boundaries, which can shift both to more acidic (struvite, apatite) and more alkaline (brushite) areas. Under conditions of oxalate mineralization, bacteria accelerate the nucleation of calcium oxalates by almost two times and also increase the amount of oxalate precipitates along with phosphates and stabilize the calcium oxalate dihydrate to weddellite.

As it can be seen from the reported results and the available literature data, the bacterial effect on oxalate and phosphate phase formation is different. Thus, in the case of oxalate mineralization, primarily (most likely), the inflammatory process will contribute to the decrease of oxalate supersaturation in urine due to calcium oxalate crystallization. In the case of phosphate mineralization, the change in urine pH and the products of bacterial metabolism will be of major importance. Studies aimed at identifying the specific action of certain microorganisms on the crystallization of certain mineral phases should serve to develop individual methods of treatment and prevention of urolithiasis.

The obtained results could be regarded as the scientific basis for understanding the mechanisms of bacterial participation stone formation in the human urinary system and the creation of biotechnological methods for the prevention of this disease.

Author Contributions: Conceptualization, A.R.I., O.V.F.-K., and V.V.M.; methodology, A.R.I., M.A.K., and O.V.F.-K.; investigation, A.R.I., A.M.N., M.A.K., and V.V.M.; writing—original draft preparation, A.R.I., M.A.K., and A.M.N.; writing—review and editing, A.R.I., O.V.F.-K., and V.V.M.; visualization, A.R.I. and A.M.N.

Funding: This research was funded by the Russian Science Foundation (grant 18-77-00026 to A.R.I.).

Acknowledgments: The XRD have been performed at the X-ray Diffraction Centre of the St. Petersburg State University. We are grateful to reviewers for useful comments.

Conflicts of Interest: The authors declare no conflict of interest.

References

1. Khan, S.R.; Kok, D.J. Modulators of urinary stone formation. *Frontiers Biosci.* **2004**, *9*, 1482. [CrossRef]
2. Seregin, A.V.; Mulabaev, N.S.; Tolordava, E.R. Urolithiasis: Current aspects of etiology and pathogenesis. *Med. Buisiness* **2012**, *4*, 4–10.
3. Balaji, K.C.; Menon, M. Mechanism of stone formation. *Urol. Clin. N. Am.* **1997**, *24*, 1–11. [CrossRef]
4. Bauza, J.L.; Pieras, E.C.; Grases, F.; Tubau, V.; Guimerà, J.; Sabaté, X.A.; Pizà, P. Urinary tract infection's etiopathogenic role in nephrolithiasis formation. *Med. Hypotheses* **2018**, *118*, 34–35. [CrossRef] [PubMed]
5. Evan, A.P. Physiopathology and etiology of stone formation in the kidney and the urinary tract. *Pediatr Nephrol* **2010**, *25*, 831–841. [CrossRef] [PubMed]
6. Fleming, D.E.; Bronswijk, W.V.; Ryall, R.L. A comparative study of the adsorption of amino acids on to calcium minerals found in renal calculi. *Clin Sci* **2001**, *101*, 159–168. [CrossRef] [PubMed]
7. Grases, U.F.; Costa-Bauzá, A.; García-Ferragut, L. Biopathological crystallization: A general view about the mechanisms of renal stone formation. *Adv. Colloid. Interface. Sci.* **1998**, *74*, 169–194. [CrossRef]
8. Korago, A.A. *Introduction to Biomineralogy*; Nedra: St. Petersburg, Russia, 1992; p. 280.
9. Tiktinskiy, O.L.; Aleksandrov, V.P. *Urolithiasis*; Liter: St. Peterburg, Russia, 2000; p. 384.
10. Waltona, R.C.; Kavanagha, J.P.; Heywoodb, B.R.; Rao, P.N. Calcium oxalates grown in human urine under different batch conditions. *J. Cryst. Growth* **2005**, *284*, 517–529. [CrossRef]
11. Jan, H.; Akbar, I.; Kamran, H.; Khan, J. Frequency of renal stone disease in patients with urinary tract infection. *J. Ayub Med. Coll.* **2008**, *20*, 60–62.
12. Bichler, K.-H.; Eipper, E.; Naber, K.; Braun, V.; Zimmermann, R.; Lahme, S. Urinary infection stones. *Int. J. Antimicrob. Ag.* **2002**, *19*, 488–498. [CrossRef]
13. Borghi, L.; Nouvenne, A.; Meschi, T. Nephrolithiasis and urinary tract infections: 'The chicken or the egg' dilemma? *Nephrol. Dial. Transplant.* **2012**, *27*, 3982–3985. [CrossRef]
14. Romanova, Y.M.; Mulabaev, N.S.; Tolordava, E.R.; Seregin, A.V.; Seregin, I.V.; Alexeeva, N.V.; Stepanova, T.V.; Levina, G.A.; Barhatova, O.I.; Gamova, N.A.; et al. Microbial communities on kidney stones. *Mol. Genet. Microbiol. Virol.* **2015**, *2*, 20–25. [CrossRef]
15. Costerton, J.W.; Stewart, P.S.; Greenberg, E.P. Bacterial biofilm: A common cause of persistent infections. *Science* **1999**, *284*, 1318–1322. [CrossRef]
16. Musk, D.J.; Hergenrother, P.J. Chemical countermeasures for the control of bacterial biofilms: Effective compounds and promising targets. *Cur. Med. Chem.* **2006**, *18*, 2163–2177. [CrossRef]
17. Cógain, M.R.; Lieske, J.C.; Vrtiska, T.J.; Tosh, P.K.; Krambeck, A.E. Secondarily infected non-struvite urolithiasis: A prospective evaluation. *Urology* **2014**, *84*, 1295–1300. [CrossRef]

18. Chen, L.; Shen, Y.; Jia, R.; Xie, A.; Huang, B.; Cheng, X.; Zhang, Q.; Guo, R. The role of escherichia coliform in the biomineralization of calcium oxalate crystals. *Eur. J. Inorg. Chem.* **2007**, *20*, 3201–3207. [CrossRef]
19. Zhao, Z.; Xia, Y.; Xue, J.; Wu, Q. Role of E. coli-secretion and melamine in selective formation of $CaC_2O_4 \cdot H_2O$ and $CaC_2O_4 \cdot 2H_2O$. *Cryst. Growth Des.* **2014**, *14*, 450–458. [CrossRef]
20. Jamshaid, A.; Ather, M.N.; Hussain, G.; Khawaja, K.B. Single center, single operator comparative study of the effectiveness of electrohydraulic and electromagnetic lithotripters in the management of 10- to 20-mm single upper urinary tract calculi. *Urol.* **2008**, *72*, 991–995. [CrossRef]
21. Izatulina, A.R.; Yelnikov, V. Structure, chemistry and crystallization conditions of calcium oxalates— The main components of kidney stones. In *Minerals as Advanced Materials I*; Krivovichev, S., Ed.; Springer: Berlin, Germany, 2008; pp. 231–241.
22. Borodin, E.A. *Biochemical Diagnosis*; V. 1,2; Amuruprpoligraphizdat: Blagoveschensk, Russia, 1989; p. 77.
23. GOST 20730-75. *Nutrient Media. Meat-Peptone Broth (for Veterinary Purposes)*; № 899; Standards publishing house: Moscow, Russia, 1975. (in Russian)
24. Mueller, J.H.; Hinton, J. A protein free medium for primary isolation of gonococcus and meningococcus. *Proc. Soc. Exp. Biol. Med.* **1941**, *48*, 330–333. [CrossRef]
25. Moskalev, Y.I. *Mineral Exchange*; Medicine: Moscow, USSR, 1985; p. 288.
26. *Topas V4.2: General Profile and Structure Analysis Software for Powder Diffraction Data*; Bruker AXS: Karlsruhe, Germany, 2009.
27. Kuz'mina, M.A.; Nikolaev, A.M.; Frank-Kamenetskaya, O.V. The formation of calcium and magnesium phosphates of the renal stones depending on the composition of the crystallization medium. In *Processes and Phenomena on the Boundary between Biogenic and Abiogenic Nature*; Springer: Basel, Switzerland, 2019. (In press)
28. Nikolaev, A.M.; Kuz'mina, M.A.; Izatulina, A.R.; Frank-Kamenetskaya, O.V.; Malyshev, V.V. Influence of the albumin substance and bacteria on formation of urinal phosphate stones (According to results of the modelling experiment). *Zapiski RMO (Proceedings of the Russian Mineralogical Society, in Russian)* **2014**, *143*, 120–133.
29. Izatulina, A.R.; Gurzhiy, V.V.; Frank-Kamenetskaya, O.V. Weddellite from renal stones: Structure refinement and dependence of crystal chemical features on H_2O content. *Amer. Min.* **2014**, *99*, 2–7. [CrossRef]
30. Wong, T.Y.; Wu, C.Y.; Martel, J.; Lin, C.W.; Hsu, F.Y.; Ojcius, D.M.; Lin, P.Y.; Young, J.D. Detection and characterization of mineralo-organic nanoparticles in human kidneys. *Sci. Rep.* **2015**, *5*, 15272. [CrossRef] [PubMed]
31. El'nikov, V.Yu.; Rosseeva, E.V.; Golovanova, O.A.; Frank-Kamenetskaya, O.V. Thermodynamic and experimental modeling of the formation of major mineral phases of uroliths. *Russ. J. Inorg. Chem.* **2007**, *52*, 50–157. [CrossRef]
32. Chutipongtanate, S.; Sutthimethakorn, S.; Chiangjong, W.; Thongboonkerd, V. Bacteria can promote calcium oxalate crystal growthand aggregation. *J. Biol. Inorg. Chem.* **2013**, *18*, 299–308. [CrossRef] [PubMed]

© 2019 by the authors. Licensee MDPI, Basel, Switzerland. This article is an open access article distributed under the terms and conditions of the Creative Commons Attribution (CC BY) license (http://creativecommons.org/licenses/by/4.0/).

Article

Synthesis and Characterization of (Ca,Sr)[C$_2$O$_4$]·nH$_2$O Solid Solutions: Variations of Phase Composition, Crystal Morphologies and in Ionic Substitutions

Aleksei V. Rusakov *, Mariya A. Kuzmina, Alina R. Izatulina and Olga V. Frank-Kamenetskaya *

Crystallography Dept., Institute of Earth Sciences, St. Petersburg State University, 7/9, University emb., 199034 St. Petersburg, Russia; m.kuzmina@spbu.ru (M.A.K.); alina.izatulina@spbu.ru (A.R.I.)
* Correspondence: alex.v.rusakov@gmail.com (A.V.R.); o.frank-kamenetskaia@spbu.ru (O.V.F.-K.);
Tel.: +7-921-924-2229 (A.V.R.)

Received: 14 November 2019; Accepted: 6 December 2019; Published: 8 December 2019

Abstract: To study strontium (Sr) incorporation into calcium oxalates (weddellite and whewellite), calcium-strontium oxalate solid solutions (Ca,Sr)[C$_2$O$_4$]·nH$_2$O (n = 1, 2) are synthesized and studied by a complex of methods: powder X-ray diffraction (PXRD), scanning electron microscopy (SEM) and energy dispersive X-ray (EDX) spectroscopy. Two series of solid solutions, isomorphous (Ca,Sr)[C$_2$O$_4$]·(2.5 − x)H$_2$O) (space group I4/m) and isodimorphous Ca[C$_2$O$_4$]·H$_2$O(sp.gr. P2$_1$/c)–Sr[C$_2$O$_4$]·H$_2$O(sp.gr. P$\bar{1}$), are experimentally detected. The morphogenetic regularities of their crystallization are revealed. The factors controlling this process are discussed.

Keywords: calcium oxalate; strontium oxalate; solid solutions; ionic substitutions; weddellite; whewellite; X-ray powder diffraction; scanning electron microscopy; EDX spectroscopy

1. Introduction

Natural calcium oxalates dihydrous weddellite Ca[C$_2$O$_4$]·(2.5 − x)H$_2$O and monohydrous whewellite Ca[C$_2$O$_4$]·H$_2$O are widespread biominerals. They often form in the human body, for instance, a major part of the stones in the human urinary system consist of both [1,2]. These minerals also are found in salivary stones and other human pathogenic formations [3–5]. Besides, weddellite and whewellite often occur in crustose and foliose lichen thalli on the surface of Ca-bearing rocks and minerals [6–9]. Oxalate crystallization is a result of the interaction between metabolites of lichens and associated microscopic fungi with an underlying stone substrate in this case [10–14].

The crystal structure of monoclinic whewellite (space group P2$_1$/c) and tetragonal weddellite (sp.gr I4/m) of renal stones are well known [15–21]. Variations of weddellite unit cell parameters are well explained by the variable water content [18–20]. There is no evidence of ionic substitutions occurring at calcium sites of calcium oxalate crystal structures of renal stones. The complex multicomponent composition of the crystallization medium (of urea and of other human physiological liquids), however, allows for the probability of such substitutions [22]. The probability of Cd–Ca substitutions in synthetic whewellite is proposed by McBride et al., 2017 [23].

We found strontium impurities in weddellite and whewellite crystals in lichen thalli on Sr-bearing apatite rock via energy-dispersive X-ray (EDX) spectroscopy [9]. It allows us to suggest that Sr^{2+} ions leach from fluorapatite and substitute Ca^{2+} ions in oxalates. Synthetic strontium oxalates, monohydrate and dihydrate, have been reported by Baran, Sterling and Christensen, Hazen [7,15,24]. The tetragonal Sr [C$_2$O$_4$]·(2.5 − x)H$_2$O is isotypic to weddellite and, similarly, contains a variable number of water molecules. The monohydrate, Sr[C$_2$O$_4$]·H$_2$O, belongs to the triclinic crystal system (sp gr. P$\bar{1}$). This suggests the presence of two series of solid solutions, isomorphous (Ca,Sr)[C$_2$O$_4$]·(2.5 − x)H$_2$O (sp.gr. I4/m) and isodimorphous Ca[C$_2$O$_4$]·H$_2$O(sp.gr. P2$_1$/c)-Sr[C$_2$O$_4$]·H$_2$O (sp.gr. P$\bar{1}$).

To clarify the patterns of Sr^{2+} ion incorporation into calcium oxalates (whewellite and weddellite), we synthesize $(Ca,Sr)[C_2O_4]\cdot nH_2O$ (n = 1, 2) solid solutions and study the variations of their phase composition and crystal morphologies as well as in ion substitutions.

2. Materials and Methods

2.1. Synthesis

Ca–Sr oxalates were crystallized by precipitation from aqueous solutions (0.5 L volume) containing calcium chloride ($CaCl_2$, 99% purity, Vekton), strontium nitrate ($Sr(NO_3)_2$, 99% purity, Vekton), sodium oxalate (1.5 mmol, $Na_2C_2O_4$, 98% purity, Vekton) and citric acid (6.5 mmol, $C_6H_8O_7\cdot H_2O$, 99% purity, Vekton). The atomic ratios of Sr/(Sr + Ca) cations ranged from 0 to 100%. The total amount of Ca^{2+} and Sr^{2+} ions was 5 times higher than the content of oxalate ions in each synthesis.

A non-stoichiometric composition of the solution and the presence of citric acid were used to stabilize weddellite formation [25]. The pH range of the solutions was 4.6–6.1 due to the small additions of solutions of NaOH (NaOH, 99%, Vekton) or HCl (HCl, 35–38 wt.% aqueous solution, 99.9% purity, Vekton). The syntheses were carried out at room temperature (22–25 °C), with the exposure of the solution for five days until complete precipitation. The resulting precipitate was filtered off, washed with distilled water and dried in air at room temperature. The precipitate consisted of fine-grained white crystalline powder, each 0.5 L solution volume provided around 200 mg of it.

2.2. Methods

2.2.1. X-Ray Powder Diffraction (PXRD)

The method was used to determine the phase composition of the precipitates. The measurements were performed using a Bruker « D2 Phaser » powder diffractometer (CuKα radiation of wavelength λ = 1.54178 Å). X-ray diffraction patterns were collected at room temperature in the range of 2θ = 3–60° with a step of 0.02°. Phase identification was carried out using the ICDD PDF-2 Database (release 2016). The unit cell parameters and coherently scattering domain (CSD) size were refined by using TOPAS 4.2 software [26].

2.2.2. Scanning Electron Microscopy (SEM) and Energy-Dispersive X-Ray (EDX) Spectroscopy

Scanning electron microscopy and energy-dispersive X-ray spectroscopy were used for the identification of calcium oxalates and for estimating calcium and strontium content in their crystals. Tetragonal weddellite and monoclinic whewellite crystals and their intergrowths were identified on SEM images by their previously described morphological features [12,27].

The study was carried out by means of a Zeiss Supra 40MP electron microscope, equipped with a variable-pressure secondary electron (VPSE) detector and Hitachi S-3400N with energy dispersive attachment AzTec Energy 350, at an accelerating voltage of 2 or 5 kV (depending on the image resolution). Magnification range varied from 100x to 1000x. Two SE detectors (secondary electron Everhart–Thornley), as well as a BSE detector (scintillation detector based on the highly sensitive YAG crystal with the resolution of 0.1Z of the atomic number) were used. The specimens were applied on two sided conducting tape and were coated with carbon (~15 nm). EDX analysis was performed by a standardless method that is generally reliable for elements with Z > 10. The mineral standards used were diopside (Ca) and celestine (Sr).

3. Results

3.1. X-Ray Powder Diffraction

The results of PXRD (Table 1, Figure 1) showed that, in the absence of strontium in the solution, the obtained precipitate was represented almost exclusively by weddellite (Figure 1a).

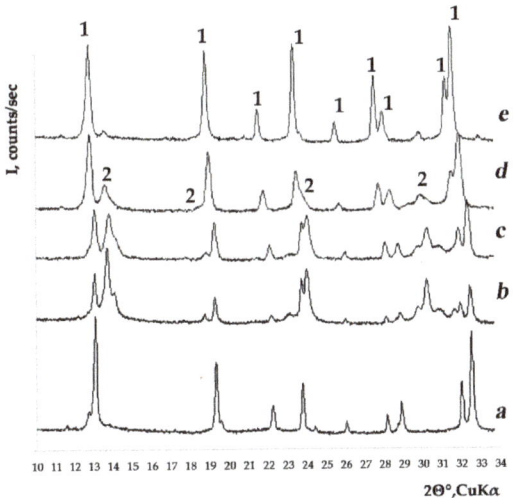

Figure 1. Typical XRD patterns of precipitates (1—weddellite, 2—whewellite), obtained from solutions with different atomic ratio Sr/(Sr + Ca), %: **a**—0; **b**—5; **c**—35; **d**—70; **e**—100.

Accompanying the addition of a small amount of strontium to the solution (Sr/(Sr + Ca) ≤ 30%), the precipitate became dominated by calcium oxalate monohydrate whewellite (Figure 1b). Weddellite content ranged from 23 to 34%. Concurrently, at Sr/(Sr + Ca) = 25–30%, traces of calcium oxalate trihydrate caoxite ($Ca[C_2O_4] \cdot 3H_2O$) were recorded in the precipitate (Table 1). Along with an increase of strontium content in the solution, the amount of whewellite in the precipitate decreased, while weddellite content increased, and at a ratio of Sr/(Ca + Sr) ~40%, these oxalates were present in the precipitate in almost equal amounts (Figure 1c). A further increase of strontium content in the solution (40% < Sr/(Sr + Ca) ≤ 80%) saw weddellite gradually start to prevail (Figure 1d). Accompanying a higher strontium content in the solution, the precipitate became close to monophasic—it is almost solely represented by weddellite (Figure 1e).

Along with the increase of strontium in the solution (and, consequently, in the formed crystals), the parameters of the weddellite tetragonal unit cell increased: *a* from 12.341 to 12.770 Å, *c* from 7.356 to 7.529 Å (Table 2). Linear parameters of monoclinic whewellite also increased: *a* from 6.289 to 6.396 Å, *b* from 14.576 to 14.860 Å, *c* from 10.120 to 10.367 Å (Table 2). The irregular fluctuations of the angle *β* value increased after Sr/(Sr + Ca) ≥ 40% (Table 2).

The average CSD size in Sr-containing weddellites varied from 162 to 71 nm and was smaller than 269 Å in Ca-weddellite. The minimum values of average CSD size (smaller than 100 Å) were observed at intermediate Sr/(Sr + Ca) ratios in solutions of 50–70%. Regarding Sr-containing whewellite, the average CSD sizes were from 57 to 28 Å, with an increase of strontium in the solution (Sr/(Sr + Ca) from 15 to 75), i.e., no less than two-times less than in weddellite (Table 2). Generally, the CSD size of the whewellite crystals gradually decreased with an increase in Sr content in the solution (up to the ratio Sr/(Sr + Ca) = 75%).

Table 1. Weddellite content in oxalate precipitate and strontium concentration in solution and in synthesized oxalates (via EDX).

Sample	Sr/(Ca + Sr) in Solution, %	Wd Content in Oxalate Precipitate, wt%	Sr Content in Crystal Phase, wt%		Sr/(Sr + Ca) in Crystal Phase, %	
			Wd	Wh	Wd	Wh
1	0.0	99.9	0.00	0.00	0.0	0.0
2	5.0	28.0	8.71	4.44	4.2	2.1
3	10.0	23.0	14.45	7.84	7.2	3.8
4	15.0	24.0	18.47	11.17	9.4	5.4
5	20.0	30.0	26.75	15.58	12.6	7.8
6	25.0	34.0 *	26.39	17.31	13.8	8.7
7	30.0	32.0 *	29.40	22.63	16.0	11.8
8	35.0	57.0	35.50	21.25	20.1	11.0
9	40.0	51.0	35.32	24.07	20.0	12.7
10	45.0	66.0	42.70	25.41	29.7	17.4
11	50.0	81.0	48.03	31.55	34.2	24.0
12	60.0	90.0	56.77	42.14	37.5	25.0
13	65.0	93.0	57.03	45.78	37.8	27.9
14	70.0	78.0	64.29	49.58	45.2	31.0
15	75.0	86.0	72.76	54.31	55.0	35.5
16	80.0	96.0	77.70	No data	61.4	No data
17	85.0	99.8	83.72	No data	70.2	No data
18	90.0	99.8	87.09	No data	75.5	No data
19	95.0	99.8	92.81	No data	85.5	No data
20	100.0	99.5	100.00	No data	100.0	No data

* Caoxite (2 wt%) is present in the precipitate.

Table 2. Oxalate unit cell parameters and average coherent scattered domain (CSD) size.

Sample	Sr/(Sr + Ca) in Crystal Phase, %		X-Ray Powder Diffraction Data							
			Weddellite (sp gr I4/m)			Whewellite (sp gr P2$_1$/c)				
	Wd	Wh	a, Å	c, Å	CSD, nm	a, Å	b, Å	c, Å	β, deg	CSD, nm
1	0.00	0.00	12.341(1)	7.356(1)	269(5)	Whewellite not detected				
2	4.2	2.1	12.3568(10)	7.3624(6)	121(3)	6.2889(6)	14.5762(14)	10.1200(12)	109.568(8)	43(1)
3	7.2	3.8	12.3772(14)	7.3708(8)	93(3)	6.2942(8)	14.5936(18)	10.1312(12)	109.533(6)	47(1)
4	9.4	5.4	12.3968(8)	7.3810(4)	162(5)	6.2999(5)	14.6134(12)	10.1415(10)	109.563(6)	47(1)
5	12.6	7.8	12.4027(9)	7.3873(5)	110(2)	6.3026(5)	14.6166(12)	10.1482(10)	109.544(6)	46(1)
6	13.8	8.7	12.4211(5)	7.3927(3)	155(4)	6.3072(4)	14.6313(1)	10.1615(10)	109.546(7)	43(1)
7	16.0	11.8	12.4360(7)	7.4009(3)	140(4)	6.3122(5)	14.6491(11)	10.1716(11)	109.527(1)	43(1)
8	20.1	11.0	12.4629(8)	7.4033(4)	103(3)	6.3122(7)	14.6538(15)	10.1761(15)	109.545(1)	37(1)
9	20.0	12.7	12.4784(7)	7.4090(3)	110(3)	6.3169(7)	14.6666(15)	10.1836(16)	109.535(1)	40(1)
10	29.7	17.4	12.5112(5)	7.4255(2)	107(1)	6.3362(8)	14.7075(17)	10.2130(2)	109.579(2)	35(1)
11	34.2	24.0	12.5301(5)	7.4303(3)	95(1)	6.344(2)	14.725(3)	10.243(4)	109.65(3)	32(1)
12	37.5	25.0	12.5571(9)	7.4382(4)	60(1)	6.333(4)	14.769(7)	10.258(9)	109.39(8)	37(3)
13	37.8	27.9	12.5902(7)	7.4529(3)	91(2)	6.339(3)	14.817(6)	10.229(9)	109.11(6)	57(7)
14	45.2	31.0	12.5907(8)	7.4583(4)	71(1)	6.351(2)	14.790(5)	10.301(6)	109.56(4)	28(1)
15	55.0	35.5	12.6315(6)	7.4762(3)	102(2)	6.396(3)	14.860(8)	10.367(6)	110.07(5)	30(2)
16	61.4	No data	12.6540(9)	7.4839(3)	106(2)	No data				
17	70.2	No data	12.7070(15)	7.4979(9)	184(7)					
18	75.5	No data	12.7384(5)	7.5133(3)	136(3)					
19	85.5	No data	12.7698(6)	7.5286(3)	127(3)					
20	100.0	No data	12.8247(4)	7.5377(2)	194(5)					

3.2. Scanning Electron Microscopy (SEM) and Energy-Dispersive X Ray (EDX) Spectroscopy

SEM data on the phase composition of the synthesized calcium oxalates support the XRD data (Table 1, Figure 2).

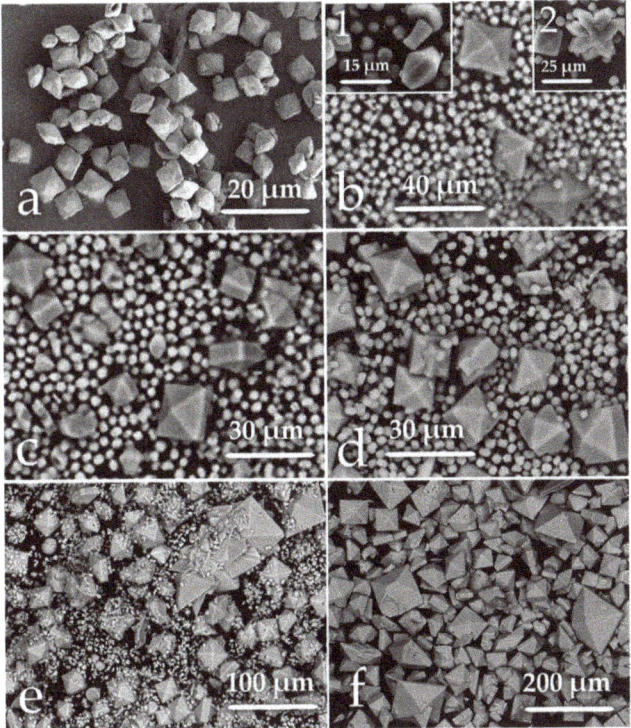

Figure 2. SEM images of formed (Ca,Sr)[C$_2$O$_4$]nH$_2$O (n = 1, 2) crystals, synthesized from solutions with different Sr/(Sr + Ca), % ratio: **a**: 0, **b**: 5, **c**: 20, **d**: 35, **e**: 70, **f**: 80.

SEM images clearly show that weddellite dipyramidal crystals represent calcium oxalates synthesized from strontium-free solutions, the dipyramidal base edge (**Dp**) of which does not exceed 5–6 microns (Figure 2a).

Found in the precipitate obtained from a solution with a ratio of Sr/(Sr + Ca) = 5%, small spherical spherulites of calcium oxalate monohydrate whewellite (diameter ~4–5 µm) prevail, among which there are individual weddellite crystals and their intergrowths (Figure 2b). According to EDX spectroscopy, strontium content in weddellite (Sr/(Sr + Ca) = 4.2%) is close to that in the solution and is two times greater than in whewellite (Sr/(Sr + Ca) = 2.1%). Weddellite crystals are defined by dipyramidal habit (**Dp** ~10–25 µm). Not very well-developed prism faces appear on smaller crystals (rib length between prism faces **Pr** = 3–4 µm, Figure 2b inset 1). The value of the average **Pr/Dp** ratio (which describes the degree of prism face development compared to dipyramid) is only 0.1. There also are twin intergrowths which form "quadruplets" consisting of two intergrown tetragonal dipyramids with a common fourth-order symmetry axis, with each dipyramid being rotated relative to each other around this axis of symmetry by 45° (Figure 2b inset 2).

Accompanying an increase in—ratio in the solution up to 30%, whewellite continues to prevail over weddellite (Figure 2c). According to EDX, Sr content increases in both phases (up to ratio Sr/(Sr + Ca) = 16.0% in weddellite and up to 11.8% in whewellite), but it decreases relative to the solution. The size of whewellite spherulites increases (average diameter ~6 µm, maximum 15 µm). Weddellite average crystal size also increases (**Dp** ~20–25 µm). While the number of weddellite crystals of dipyramidal-prismatic pattern increases, the prism face continues to develop: **Pr** ~10 µm, the **Pr/Dp** ~0.35.

When the Sr/(Sr + Ca) ratio in the solution reached 35–40%, the number of crystals of whewellite and weddellite became comparable (Figure 2d). According to EDX, Sr content continued to increase in both phases (up to Sr/(Sr + Ca) = 20% in weddellite and up to 11–13% in whewellite) and decreased relative to the solution. The size of whewellite spherulites was relatively small (5–6 microns in diameter, but reached 20 microns). Weddellite crystals also increased in size (**Dp** ~15–30 μm). All weddellite crystals had prism faces and the prism faces continued to develop (**Pr** ~10–20, **Pr/Dp** ratio ~0.40–0. 45).

When the ratio in the solution Sr/(Sr + Ca) ≤ 40%, weddellite began to prevail over whewellite. Two generations of weddellite crystals were observed in the precipitate: large dipyramidal (**Dp** up to 80 μm) and smaller dipyramidal-prismatic (**Dp** ~25 μm, **Pr** ~15–20 μm) (Figure 2e). The **Pr/Dp** ratio in weddellite crystals continued to increase and reached 0.56 at Sr/(Sr + Ca) = 50% in solution, and then began to decrease to 0.33 at Sr/(Sr + Ca) = 70%. The diameter of spherulites of whewellite decreased to 3–5 microns. According to EDX, the Sr/(Sr + Ca) ratio reached 55.0% in weddellite and 35.5% in whewellite.

Reaching a Sr/(Sr + Ca) ratio of 80% and more in the solution, weddellite crystallized almost solely, and was represented by large dipyramidal (200–250 μm) and smaller dipyramidal-prismatic (30–50 μm) crystals (Figure 2f). Attaining Sr/(Sr + Ca) = 85–95%, the development of the prism face slowed (**Pr** remained ~20 μm, the **Pr/Dp** ratio decreased to ~0.1). The Sr/(Sr + Ca) ratio in weddellite via EDX increased from 78 to 100%, approaching the value of this ratio in solution.

According to the EDX data, the Sr/(Sr + Ca) ratios of 29.7–45.2 in weddellite and of 17.4–31.0 in whewellite correspond to the minimum CSD size for both phases.

4. Discussion

The results of the study show that in all syntheses, solid solutions $(Ca,Sr)[C_2O_4] \cdot nH_2O$ (n = 1, 2) with a variable ratio of dihydrate and monohydrate phases are obtained. Phase and elemental composition of synthesized solid solutions, as well as the morphology of their crystals, is strongly relevant to strontium concentration in the solution.

4.1. The Effect of Strontium Concentration in Solution on Phase Composition of the Precipitate

The strontium content in the solution significantly affects the phase composition of the precipitate. Considering the absence of Sr^{2+} ions in the solution of non-stoichiometric composition (Ca/C_2O_4 = 5) containing citric acid, tetragonal calcium oxalate dihydrate weddellite (Figures 1a and 2a) is obtained, which is unstable in the crystallization field of monoclinic calcium oxalate monohydrate whewellite [17,21]. The addition of a small amount of Sr^{2+} ions (Sr/(Sr + Ca) = 5%) to this solution violates the conditions favorable for weddellite crystallization and leads to intensive crystallization of whewellite, the content of which in the precipitate exceeds the amount of weddellite by 2.6 times (Figures 1b and 2b, Table 1). Accompanying an increase in the strontium content in the solution, the amount of weddellite in the precipitate gradually increases and, at a ratio of Sr/(Sr + Ca) = 35–45%, matches the amount of whewellite (Figures 1c and 2d). A further increase in strontium content again creates conditions favorable for the crystallization of weddellite, the content of which in the precipitate begins to prevail (Figures 1d and 2e). Reaching an Sr/(Sr + Ca) ≥ 80% ratio in solution, the precipitate consists almost exclusively of weddellite. Thus, the presence of strontium in the crystallization medium can be favorable for the crystallization of both weddellite and whewellite, depending on the concentration.

4.2. Sr-Ca Ionic Substitutions in Solid Solutions

According to EDX, strontium is present in crystals of both calcium oxalate monohydrate and dihydrate (Table 1). The ratio Sr/(Sr + Ca) varies from 0 to 100% in weddellite and from 2.1 to 35.5% in whewellite. A narrower range of observed strontium concentrations in whewellite is explained by the synthesis conditions under which whewellite, unlike weddellite, precipitates only in solutions with an Sr/(Sr + Ca) ratio varying from 5 to 75% and always alongside weddellite. Altogether, strontium content in weddellite is always higher than in whewellite by 1.4–2 times (Table 1), which indicates that the entry of Sr into whewellite is difficult compared to weddellite.

It is known that, despite the significant difference in ionic radii ($r_{Ca}^{VIII} = 1.12$ Å, $r_{Sr}^{VIII} = 1.26$ Å, [28]), Sr^{2+} ions easily replace Ca^{2+} ions which occupy large cavities coordinated by 7–9 cations in various crystal structures, like in apatite [29,30]. This suggests that all (or almost all) strontium recorded via EDX isomorphically substitutes calcium in its positions in weddellite and whewellite. A simultaneous increase of the unit cell parameters of weddellite and whewellite with the increasing Sr content (Figures 3 and 4) confirms this assumption.

An increase in the unit cell parameters of weddellite, primarily parameter a, also occurs with the increase of zeolite water (W_z) [18,19]. An increase of zeolite water amount in weddellite (from 0.13 to 0.37 apfu), however, leads to an increase in the a parameter by only 0.049 Å [18]. Conversely, the contribution of strontium to the increase of the weddellite parameter a is an order of magnitude greater; the change in parameter a reaches 0.469 Å (Figure 3a). The dependence of unit cell parameters of weddellite on strontium content is not linear and follows a second degree polynomial function (Figure 3). Violation of the linear correlation between the strontium content and the parameter values can be due, first, to the fact that when the Sr/(Sr + Ca) ratio reaches ~40% in weddellite (~60% in solution) Sr^{2+} ion entry into weddellite slows. Second, it is possible that at this strontium content, the amount of zeolite water in weddellite ceases to increase and a further slower increase in parameters occurs only due to an increase in strontium.

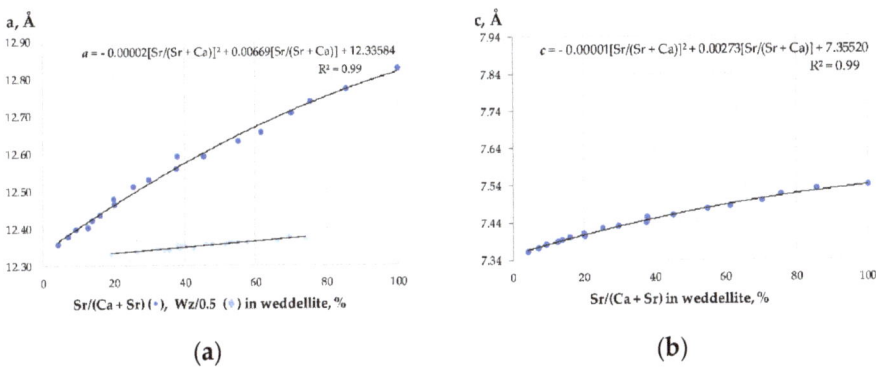

Figure 3. The increase of weddellite unit cell parameters: a unit cell parameter versus strontium (present study) and zeolite water [18] content, % (**a**); c unit cell parameter versus strontium content, % (**b**).

Seen at low strontium concentrations in whewellite (with Sr/(Sr + Ca) ratios in crystals varying from 2.1 to 12.7%; and in solution from 5 to 40%), there is a well-defined linear relationship between the values of linear parameters (a, b, c) and the content of strontium, while the angle β does not change significantly (Figure 4). The fluctuations of both linear parameters and angle β at higher strontium concentrations can be explained by desymmetrization (lowering the symmetry from monoclinic to triclinic), since the crystal structure of strontium oxalate monohydrate ($Sr[C_2O_4] \cdot H_2O$) is triclinic [7,24]. Desymmetrization may be associated with reciprocal turns of calcium polyhedra and oxalate groups, associated with the partial ordering of calcium and strontium atoms at the crystallographic sites. To the question of why desymmetrization is not at maximum in the middle of the series, further studies should be made.

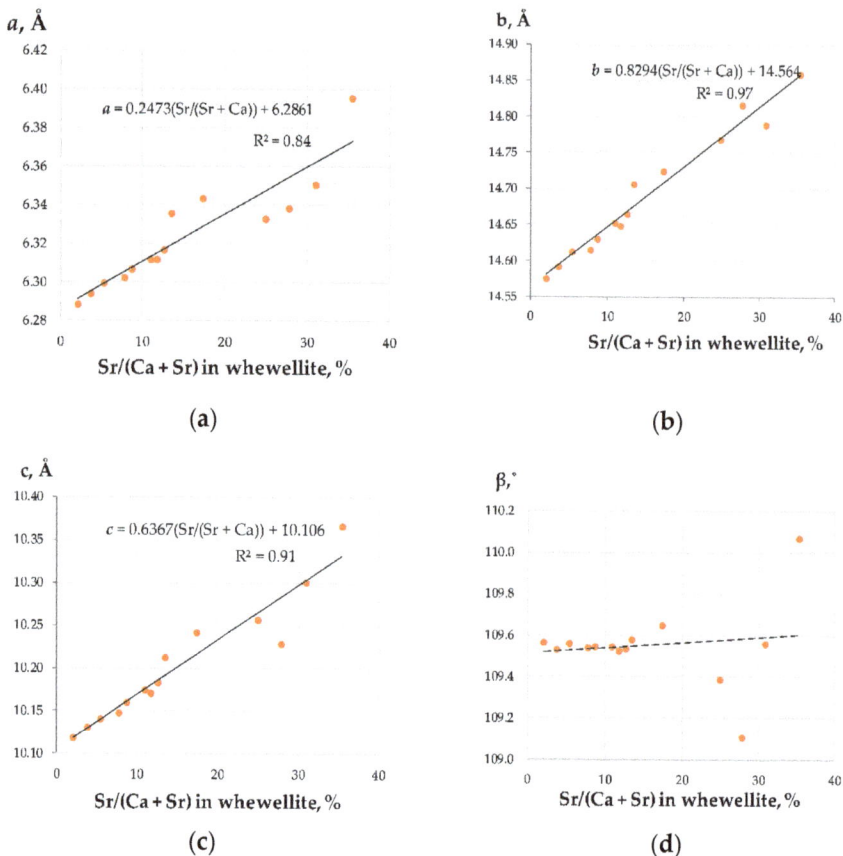

Figure 4. Variations of whewellite unit cell parameters versus strontium content, %: a (**a**); b (**b**); c (**c**); β (**d**).

Thus, assumption of the presence of two series of solid solutions, isomorphous $(Ca,Sr)[C_2O_4]\cdot(2.5-x)H_2O$ (sp.gr. I4/m) and isodimorphous $Ca[C_2O_4]\cdot H_2O$(sp.gr. $P2_1/c$)–$Sr[C_2O_4]\cdot H_2O$(sp.gr. $P\bar{1}$), received experimental confirmation. A detailed study of the desymmetrization pattern of the monoclinic whewellite structure with strontium incorporation via single-crystal X-ray diffraction analysis is now in progress.

4.3. Morphogenetic Patterns of the Formation of Solid Solutions $(Ca,Sr)[C_2O_4]\cdot nH_2O$ (n = 1, 2)

4.3.1. The Effect of Sr Concentration in Solutions on Crystal Morphology

Regarding all variations of strontium content in solution ($0 \leq Sr/(Sr + Ca) \leq 100\%$), weddellite is represented by dipyramidal and/or dipyramidal-prismatic crystals, and whewellite by round spherulites (Figure 2).

As strontium content in the solution increases, the size of the weddellite dipyramidal crystals grows as well: **Dp** varies from ~5–6 μm in pure calcium crystals to 30 μm at $Sr/(Sr + Ca) = 35\%$ (Figure 2a–d). Accompanying a further increase of strontium concentration in the crystallization medium, the formation of two weddellite crystal generations is clearly observed. Larger ones are mainly of dipyramidal pattern and smaller ones of dipyramidal-prismatic configuration (Figure 2e). The size of larger weddellite crystals increases with the increase of strontium concentration in the solution, from **Dp** = 30 μm at $Sr/(Sr + Ca) = 35\%$ to **Dp** = 200 μm at $Sr/(Sr + Ca) = 80$–95% (Figure 2f).

The size of smaller weddellite crystals also increases with the increase of strontium content, from $Dp = 15$ μm at Sr/(Sr + Ca) = 35% in solution to $Dp = 50$–60 μm at Sr/(Sr + Ca) = 80–95% in solution. Tetragonal prism faces of smaller weddellite crystals are well developed and the size of these faces along the prism edge also gradually increase with the increase of strontium content in the solution, from $Pr = 3$–4 μm at Sr/(Sr + Ca) = 5% (Figure 2b) to $Pr = 20$ μm at Sr/(Sr + Ca) = 70–80% (Figure 2e,f). The aspect ratio between the prism and dipyramid faces with the increase of strontium content in the solution increases from 0.1 (at Sr/(Sr + Ca) = 5%) to 0.56 (at Sr/(Sr + Ca) = 50%) and then it starts to decrease to 0.1 (at Sr/(Sr + Ca) = 95%).

Shown above, the first and last syntheses of the studied series (purely calcium and purely strontium solutions) are characterized by the formation of weddellite crystals in the form of a tetragonal dipyramid, which indicates optimal conditions for the precipitation of calcium and strontium oxalate dihydrate. The middle members in the weddellite series exhibit tetragonal prism faces in their pattern. The ratio between linear sizes of the prism and dipyramidal faces increases with the increase in strontium content in solution from 0.1 (at Sr/(Sr + Ca) = 5%) to 0.56 (at Sr/(Sr + Ca) = 50%), and then decreases to 0.1 (at Sr/(Sr + Ca) = 95%). Accordingly, with an increase in strontium in the crystallization medium from 5 to 50%, the initial prism growth rate increases significantly and then (at a ratio Sr/(Sr + Ca) ≥ 65%) begins to decrease. The slowdown in the growth rate of weddellite prism faces in the middle of the calcium-strontium series is most likely due to the deviation of the crystallization conditions from optimal.

Kuzmina et al [25] demonstrates that the crystallization of whewellite in the form of spherulites is common for solutions containing citrate ions and indicates a more rapid growth of whewellite (compared with weddellite) under the conditions of a higher supersaturation of the solution. The average diameter of whewellite spherulites is dependent on the content of Sr^{2+} cations in the solution. It first increases from 4 μm (at Sr/(Sr + Ca) = 5%, Figure 2b) to 6 μm (at Sr/(Sr + Ca) = 35–50%, Figure 2d), then it decreases to 1–2 μm (at Sr/(Sr + Ca) = 70–80%, Figure 2e,f) amid the decrease in the total amount of whewellite.

4.3.2. Sr-distribution in Oxalate System «Solution–Crystal»

The strontium concentration in oxalate monohydrate and oxalate dihydrate (0 ≤ Sr/(Ca + Sr) ≤ 100%) depends on strontium concentration in the solution following a third degree polynomial function (Figure 5):

$$Sr/(Sr + Ca)_{Solution} = 0.0006[Sr/(Sr + Ca)_{wh}]^3 - 0.0815[Sr/(Sr + Ca)_{wh}]^2 + 4.3286[Sr/(Sr + Ca)_{wh}] - 5.4929,$$

$$Sr/(Sr + Ca)_{Solution} = 0.0001[Sr/(Sr + Ca)_{Wd}]^3 - 0.0266[Sr/(Sr + Ca)_{Wd}]^2 + 2{,}67[Sr/(Sr + Ca)_{Wd}] - 7.1757.$$

Fitting with a third degree polynomial function is necessary due to the fact that the difference between strontium content in the crystallization medium and in synthesized solid solutions increases in the middle of the corresponding series.

The described distribution of strontium in the oxalate "crystal-solution" system indicates that the incorporation of strontium ions into calcium oxalates from solutions with close to equal amounts of Ca and Sr is difficult, which can explain the decrease of weddellite prism face growth rate and the lowering of CSD sizes for both weddellite and whewellite.

The difference in strontium content between the solution and whewellite is greater than between the solution and weddellite (Figure 5), which supports a proposal for a more difficult incorporation of Sr^{2+} cations into whewellite than into weddellite. This effect can most likely be explained by the specific features of weddellite and whewellite crystal structures. This result also agrees well with the above conclusion that whewellite crystals grow more rapidly than weddellite crystals.

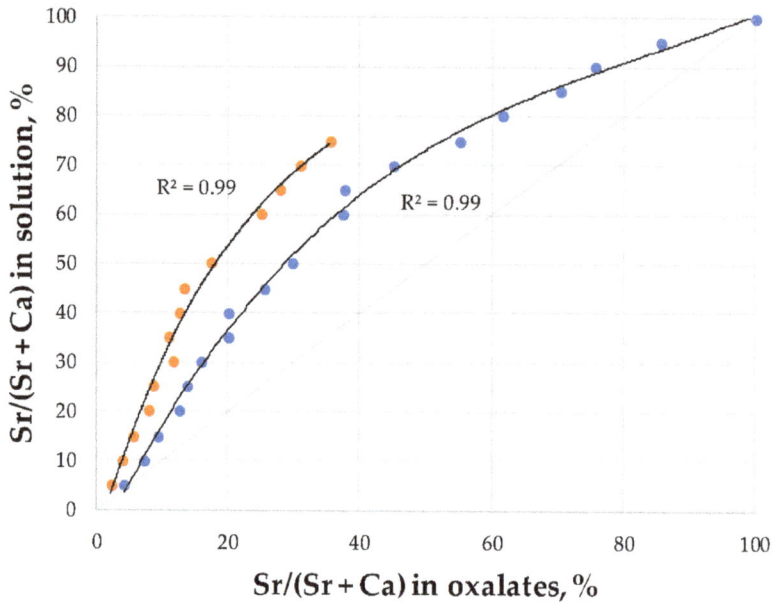

Figure 5. Sr/(Ca + Sr) ratio in the solution versus Sr/(Ca + Sr) ratio in the weddellite (•) and whewellite (•).

5. Conclusions

To clarify the patterns of Sr incorporation into calcium oxalates (whewellite and weddellite) the $(Ca,Sr)[C_2O_4] \cdot nH_2O$ (n = 1, 2) solid solutions were synthesized and studied by complex methods (PXRD, SEM and EDX). It was shown that phase and elemental composition of synthesized solid solutions, as well as the morphology of their crystals, is strongly relevant to Sr concentration in the solution. The presence of two series of solid solutions, isomorphous $(Ca,Sr)[C_2O_4] \cdot (2.5 - x)H_2O$ (sp.gr. I4/m) and isodimorphous $Ca[C_2O_4] \cdot H_2O$(sp.gr. $P2_1/c$)–$Sr[C_2O_4] \cdot H_2O$(sp.gr. $P\bar{1}$) was experimentally proven for the first time. The causes of a difficult incorporation of strontium ions into calcium oxalates (especially into whewellite) were discussed. Morphogenetic regularities of their formation were revealed.

The results of this work show the possibility of diverse ionic substitutions occurring at calcium sites of calcium oxalate crystal structures, which opens a new page in the crystal chemistry of oxalic acid salts. Among the oxalates found in nature, biofilm biominerals formed as a result of the interaction of metabolism products of lithobiotic community with bedrock are highly favorable to exhibit ionic substitutions at a global scale. The regularities of these ionic substitutions are determined by both the mineral and elemental composition of the underlying rock and the species composition of the microorganisms inhabiting them. The applied value of the obtained patterns lies in the field of the development of biotechnologies which use the oxalate microbial crystallization as a source of bioremediation for environments contaminated with toxic elements.

Author Contributions: Conceptualization (A.V.R. and O.V.F.-K.); Investigation (A.V.R., M.A.K. and A.R.I.); Methodology (A.V.R., O.V.F.-K. and M.A.K.); Visualization (A.V.R. and A.R.I.); Writing—original draft (A.V.R., M.A.K. and A.R.I.); Writing—review & editing (A.V.R. and O.V.F.-K.).

Funding: This research was funded by the Russian Science Foundation (grant 19-17-00141 to A.V.R., O.V.F.-K., M.A.K., A.R.I.).

Acknowledgments: The laboratory researches were carried out in Research Resource Centers of Saint Petersburg State University: XRD measurements—in the X-ray Diffraction Centre; SEM investigations—in the Interdisciplinary Resource Center for Nanotechnology and Centre for Geo-Environmental Research and Modelling (Geomodel).

Conflicts of Interest: The authors declare no conflict of interest.

References

1. Balaji, K.C.; Menon, M. Mechanism of stone formation. *Urolithiasis* **1997**, *24*, 1–11. [CrossRef]
2. Grases, F.; Costa-Bauza, A.; Garcia-Ferragut, L. Biopathological crystallization: A general view about the mechanisms of renal stone formation. *Adv. Colloid Interface Sci.* **1998**, *74*, 169–194. [CrossRef]
3. Kraaij, S.; Brand, H.S.; van der Meij, E.H.; de Visscher, J.G. Biochemical composition of salivary stones in relation to stone and patient-related factors. *Med. Oral Patol. Oral Cir. Bucal* **2018**, *23*, e540–e544. [CrossRef] [PubMed]
4. Nishizawa, Y.; Higuchi, C.; Nakaoka, T.; Omori, H.; Ogawa, T.; Sakura, H.; Nitta, K. Compositional Analysis of Coronary Artery Calcification in Dialysis Patients in vivo by Dual-Energy Computed Tomography Angiography. *Ther. Apher. Dial.* **2018**, *22*, 365–370. [CrossRef] [PubMed]
5. Kuranov, G.; Nikolaev, A.; Frank-Kamenetskaya, O.; Gulyaev, N.; Volina, O. Physicochemical characterization of human cardiovascular deposits. *J. Biol. Inorg. Chem.* **2019**, *24*, 1047–1055. [CrossRef]
6. Rusakov, A.V.; Frank-Kamenetskaya, O.V.; Zelenskaya, M.S.; Vlasov, D.Y.; Gimelbrant, D.E.; Knauf, I.V.; Plotkina, Y.V. Calcium oxalates in bio-films on surface of the chersonesus archaeological limestone monuments (Crimea). *Zap. Rmo (Proc. Russ. Mineral. Soc. Russ.)* **2010**, *5*, 96–104.
7. Baran, E.J. Review: Natural oxalates and their analogous synthetic complexes. *J. Coord. Chem.* **2014**, *67*, 3734–3768. [CrossRef]
8. Marques, J.; Gonçalves, J.; Oliveira, C.; Favero-Longo, S.E.; Paz-Bermúdez, G.; Almeida, R.; Prieto, B. On the dual nature of lichen-induced rock surface weathering in contrasting micro-environments. *Ecology* **2016**, *97*, 2844–2857. [CrossRef]
9. Frank-Kamenetskaya, O.V.; Ivanyuk, G.Y.; Zelenskaya, M.S.; Izatulina, A.R.; Kalashnikov, A.O.; Vlasov, D.J.; Polyanskaya, E.I. Calcium Oxalates in Lichens on Surface of Apatite-Nepheline Ore (Kola Peninsula, Russia). *Minerals* **2019**, *9*, 656. [CrossRef]
10. Ríos de los, A.; Cámara, B.; Cura del, M.Á.G.; Rico, V.J.; Galván, V.; Ascaso, C. Deteriorating effects of lichen and microbial colonization of carbonate building rocks in the Romanesque churches of Segovia (Spain). *Sci. Total Environ.* **2009**, *407*, 1123–1134. [CrossRef]
11. Sayer, J.A.; Gadd, J.M. Solubilization and transformation of insoluble inorganic metal compounds to insoluble metal oxalates by *Aspergillus niger*. *Mycol. Res.* **1997**, *6*, 653–661. [CrossRef]
12. Sturm, E.V.; Frank-Kamenetskaya, O.V.; Vlasov, D.Y.; Zelenskaya, M.S.; Sazanova, K.V.; Rusakov, A.V.; Kniep, R. Crystallization of calcium oxalate hydrates by interaction of calcite marble with fungus *Aspergillus niger*. *Am. Mineral.* **2015**, *100*, 2559–2565. [CrossRef]
13. Rusakov, A.V.; Vlasov, A.D.; Zelenskaya, M.S.; Frank-Kamenetskaya, O.V.; Vlasov, D.Y. The Crystallization of Calcium Oxalate Hydrates Formed by Interaction Between Microorganisms and Minerals. In *Biogenic—Abiogenic Interactions in Natural and Anthropogenic Systems*; Frank-Kamenetskaya, O., Panova, E., Vlasov, D., Eds.; Lecture Notes in Earth System Sciences; Springer: Cham, Switzerland, 2016; pp. 357–377. [CrossRef]
14. Zelenskaya, M.S.; Rusakov, A.V.; Frank-Kamenetskaya, O.V.; Vlasov, D.Y.; Izatulina, A.R.; Kuz'mina, M.A. Crystallization of Calcium Oxalate Hydrates by Interaction of Apatites and Fossilized Tooth Tissue with Fungus *Aspergillus niger*. In *Processes and Phenomena on the Boundary Between Biogenic and Abiogenic Nature*; Frank-Kamenetskaya, O., Vlasov, D., Panova, E., Lessovaia, S., Eds.; Lecture Notes in Earth System Sciences; Springer: Cham, Switzerland, 2020; pp. 581–603. [CrossRef]
15. Sterling, C. Crystal-structure of Tetragonal Strontium Oxalate. *Nature* **1965**, *205*, 588–589. [CrossRef]
16. Tazzoli, V.; Domeneghetti, C. The crystal structures of whewellite and weddellite: Reexamination and comparison. *Am. Mineral.* **1980**, *65*, 327–334.
17. Izatulina, A.R.; Yelnikov, V.Y. Structure, chemistry and crystallization conditions of calcium oxalates—the main components of kidney stones. In *Minerals as Advanced Materials I*; Krivovichev, S.V., Ed.; Springer: Berlin/Heidelberg, Germany, 2008; pp. 231–241.
18. Izatulina, A.R.; Gurzhiy, V.V.; Frank-Kamenetskaya, O.V. Weddellite from renal stones: Structure refinement and dependence of crystal chemical features on H_2O content. *Am. Mineral.* **2014**, *99*, 2–7. [CrossRef]

19. Rusakov, A.V.; Frank-Kamenetskaya, O.V.; Gurzhiy, V.V.; Zelenskaya, M.S.; Izatulina, A.R.; Sazanova, K.V. Refinement of the crystal structures of biomimetic weddellites produced by microscopic fungus *Aspergillus niger*. *Crystallogr. Rep.* **2014**, *59*, 362. [CrossRef]
20. Mills, S.J.; Christy, A.G. The Great Barrier Reef Expedition 1928–29: The crystal structure and occurrence of weddellite, ideally $CaC_2O_4 \cdot 2.5H_2O$, from the Low Isles, Queensland. *Mineral. Mag.* **2016**, *80*, 399–406. [CrossRef]
21. Izatulina, A.R.; Gurzhiy, V.V.; Krzhizhanovskaya, M.G.; Kuz'mina, M.A.; Leoni, M.; Frank-Kamenetskaya, O.V. Hydrated Calcium Oxalates: Crystal Structures, Thermal Stability and Phase Evolution. *Cryst. Growth Des.* **2018**, *18*, 5465–5478. [CrossRef]
22. Frank-Kamenetskaya, O.V.; Izatulina, A.R.; Kuz'mina, M.A. Ionic substitutions, non-stoichiometry, and formation conditions of oxalate and phosphate minerals of the human body. In *Biogenic—Abiogenic Interactions in Natural and Anthropogenic Systems*; Frank-Kamenetskaya, O., Panova, E., Vlasov, D., Eds.; Lecture Notes in Earth System Sciences; Springer: Cham, Switzerland, 2016; pp. 425–442. [CrossRef]
23. McBride, M.B.; Frenchmeyer, M.; Kelch, S.E.; Aristilde, L. Solubility, structure, and morphology in the co-precipitation of cadmium and zinc with calcium-oxalate. *J. Colloid Interface Sci.* **2017**, *486*, 309–315. [CrossRef]
24. Christensen, N.; Hazell, R.G. Thermal analisys and crystal structure of tetragonal strontium oxalate dihydrate and of triclinic strontium oxalate hydrate. *Acta Chem. Scand.* **1998**, *52*, 508. [CrossRef]
25. Kuz'mina, M.A.; Rusakov, A.V.; Frank-Kamenetskaya, O.V.; Vlasov, D.Y. The influence of inorganic and organic components of biofilms with microscopic fungi on the phase composition and morphology of crystallizing calcium oxalates. *Crystallogr. Rep.* **2019**, *64*, 161. [CrossRef]
26. Bruker, A.X.S. Topas V4.2: General Profile and Structure Analysis Software for Powder Diffraction Data. Available online: http://www.topas-academic.net/ (accessed on 6 December 2019).
27. Zuzuk, F.V. *Urinary Calculus Mineralogy, 2*; Volynsk State University: Luzk, Ukraine, 2003; p. 507. (In Ukranian)
28. Shannon, R.D. Revised effective ionic radii and systematic studies of interatomic distances in halides and chalcogenides. *Acta Crystallogr.* **1976**, *32*, 751–767. [CrossRef]
29. Heijligers, H.J.M.; Driessens, F.C.M.; Verbeeck, R.M.H. Lattice parameters and cation distribution of solid solutions of calcium and strontium hydroxyapatite. *Calcif. Tissue Int.* **1979**, *29*, 127–131. [CrossRef] [PubMed]
30. Nikolaev, A.; Kuz'mina, M.; Frank-Kamenetskaya, O.; Zorina, M. Influence of carbonate ion in the crystallization medium on the formation and chemical composition of CaHA–SrHA solid solutions. *J. Mol. Struct.* **2015**, *1089*, 73–80. [CrossRef]

© 2019 by the authors. Licensee MDPI, Basel, Switzerland. This article is an open access article distributed under the terms and conditions of the Creative Commons Attribution (CC BY) license (http://creativecommons.org/licenses/by/4.0/).

Article

Characterization of Biominerals in Cacteae Species by FTIR

Alejandro De la Rosa-Tilapa [1], Agustín Maceda [2] and Teresa Terrazas [1,*]

[1] Instituto de Biología, Universidad Nacional Autónoma de México, México City 04510, Mexico; aledrosa17@gmail.com
[2] Programa de Botánica, Colegio de Postgraduados en Ciencias Agrícolas, Texcoco, Estado de México 56230, Mexico; biologoagustin@hotmail.com
* Correspondence: tterrazas@ib.unam.mx

Received: 7 May 2020; Accepted: 27 May 2020; Published: 29 May 2020

Abstract: A biomineral is a crystalline or amorphous mineral product of the biochemical activity of an organism and the local accumulation of elements available in the environment. The cactus family has been characterized by accumulating calcium oxalates, although other biominerals have been detected. Five species of Cacteae were studied to find biominerals. For this, anatomical sections and Fourier transform infrared, field emission scanning electron microscopy and energy dispersive X-ray spectrometry analyses were used. In the studied regions of the five species, they presented prismatic or spherulite dihydrate calcium oxalate crystals, as the predominant biomineral. Anatomical sections of *Astrophytum asterias* showed prismatic crystals and *Echinocactus texensis* amorphous silica bodies in the hypodermis. New findings were for *Ariocarpus retusus* subsp. *trigonus* peaks assigned to calcium carbonate and for *Mammillaria sphaerica* peaks belonging to silicates.

Keywords: anatomy; Cactaceae; calcium carbonate; oxalate; silica; stem; weddellite

1. Introduction

A biomineral is a crystalline or amorphous mineral product of the biochemical activity of an organism and the local accumulation of elements available in the environment [1]. In plants, the most common biominerals are amorphous silica, calcium oxalate and calcium carbonate salts [2–7]. Some species of the Cactaceae family accumulate up to 85% of their dry weight in calcium oxalate crystals [8,9]. These calcium oxalate crystals may have one of two states of hydration: monohydrate ($CaC_2O_4 \cdot H_2O$; whewellite) or dihydrate ($CaC_2O_4 \cdot 2H_2O$; weddellite) [10–17]. In addition, other biominerals, such as magnesium oxalate ($MgC_2O_4 \cdot 2H_2O$; Glushinkite) [17], amorphous silica bodies ($SiO_2 \cdot nH_2O$; opal) [18] and silica in crystalline form (SiO_2; α-quartz) [19], have been identified in cacti.

Like other plants, biominerals in Cactaceae develop mainly in the cellular vacuole of different epidermal, fundamental or vascular stem tissues. The accumulation of biominerals in a given tissue is usually highly specific in some species or genera [10,18,20–23]. Thirty-four species from the Cacteae tribe have been studied with techniques such as X-ray diffraction to detect calcium oxalates [11], by Raman spectroscopy in *Ferocactus latispinus* and *Coryphantha clavata* [24] and by Fourier transform infrared (FTIR) only in one species, *Mammillaria uncinata* [25]. In these techniques, the tissues were blended or macerated together, so they have not been analyzed individually in the Cacteae species.

For this reason, the FTIR technique can be a good option to study biominerals due to the minimum amount of sample and the speed of data acquisition. Therefore, the aims of this study were to characterize the biominerals and determine their hydration state in the different tissues of five Cacteae species. With this, it will be possible to identify other biominerals that are not calcium oxalate, magnesium oxalate or amorphous silica, in addition to the state of hydration of biominerals in the different tissues.

2. Material and Methods

Two adult and healthy plants of five species classified within the Cacteae tribe were collected in their natural habitats (Table 1). For each species, a portion of the plant was prepared as a voucher, which was deposited in the National Herbarium (MEXU). In one of the two plants per species, spines were removed. Then, using a dissecting microscope at different magnifications, the stem of each sample was dissected into the epidermis plus hypodermis (EH), cortex (C), vascular cylinder (VC) and pith (P) (Figure 1A,B). Each tissue was blended with distilled water and filtered with a mesh of approximately 300 µm pore diameter to separate biominerals from cell debris. The biominerals were precipitated, washed with distilled water several times until no residue was observed under a stereomicroscope, and finally, dried at room temperature. Small samples from pith to epidermis were prepared for FE-SEM-EDS (see 2.3, JEOL Ltd., Akishima, Tokyo, Japan).

Table 1. Species analyzed and the state of Mexico where they were collected. Vouchers deposited at MEXU.

Species	Collection Number	Location
Astrophytum asterias (Zucc.) Lem.	TT1020	La Esperanza, Tamaulipas
	TT846	San Carlos Tamaulipas
Ariocarpus retusus subsp. *trigonus* (F.A.C.Weber) E.F.Anderson & W.A.Fitz Maur.	TT1005	La Soledad, Tamaulipas
	TT879	Moctezuma, San Luis Potosí
Echinocactus texensis Hopffer	TT1021	La Esperanza, Tamaulipas
	TT851	Tula, Tamaulipas
Mammillaria melanocentra subsp. *rubrograndis* (Repp. & A.B. Lau) D.R. Hunt	TT1050	Ejido Huizache, Tamaulipas
Mammillaria sphaerica A. Dietr.	TT1051	Ejido Huizache, Tamaulipas

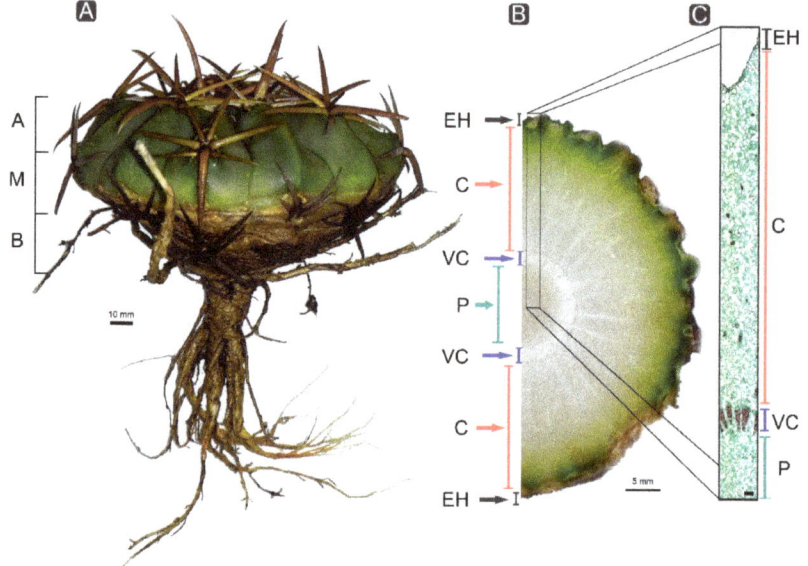

Figure 1. Stem and macro and microscopic tissues studied. (**A**) *Echinocactus texensis* (TT1021), showing the globose depressed stem. (**B**) *Echinocactus texensis*, transverse section illustrating the stem tissues. (**C**) *Mammillaria sphaerica* (TT1051), microscopic transverse section showing the tissues. A = apical, M = medium, B= basal, EH = epidermis-hypodermis, C = cortex, VC = vascular cylinder, P = pith. Bar is 300 µm in C.

The second individual of each species was divided into three parts (apical, median and basal) no larger than 1.5 cm length that included epidermal, cortical, vascular and pith tissues (Figure 1A,C).

Each part was immediately fixed in a formalin-acetic acid-ethanol solution (10:5:85) [26] and processed for paraffin embedding according to Loza-Cornejo and Terrazas [27]. Transverse and tangential sections 14 µm thick were made with a rotary microtome, stained with Safranin-fast green, and mounted with synthetic resin.

2.1. Fourier Transform Infrared (FTIR) Spectroscopy Analysis

Approximately 0.1 g of biomineral dry sample of each species was used to obtain the infrared spectra. The spectra were obtained with an Attenuated Total Reflectance Fourier Transform Infrared Spectrometer (ATR-FTIR) (Agilent Cary 630 FTIR, Agilent Technologies, Santa Clara, CA, USA) equipped with an ATR diamond unit [28]. The samples were processed with two wavenumber ranges of 4000–400 cm^{-1} and 4000–650 cm^{-1} (30 scans with a resolution of 4 cm^{-1}, 15 seconds per sample and three replicates per sample) with the program MicroLab PC (Agilent Technologies, Santa Clara, CA, USA). The baseline correction, ATR correction (applied only in 4000 to 400 cm^{-1}) and the spectra average were performed with the Resolution Pro FTIR Software program (Agilent Technologies, Santa Clara, CA, USA). With the spectra obtained, the peaks corresponding to the mineral components were identified and compared with those reported in the literature [29,30].

2.2. Polarized and Brightfield Microscope Analysis

For each species, permanent slides were observed with both types of lighting in an Olympus BX 51 microscope (Olympus, Tokyo, Japan) and photographs obtained with Image Pro Plus 7.1 software (Media Cybernetics, MD, USA) to characterize the morphology and distribution of biominerals in the different tissues of the three regions studied (Figure 1). Additionally, photographs of isolated biominerals from each tissue were also obtained with the two types of lighting.

2.3. Field Emission Scanning Electron Microscopy (FE-SEM) and Energy Dispersive X-ray Spectrometry (EDS) Analysis

For FE-SEM observations, samples from pith to epidermis of stems (<1 cm) per species were placed between to coverslips and dried in an oven (56 °C) overnight. The dried samples were fixed to aluminum specimen holders with double-sided tape and coated with gold in a Hitachi-S-2460N sputter coater. The coated samples were then observed using a Field Emission Scanning Electron Microscope (FE-SEM; JEOL-JSM7800, JEOL Ltd., Akishima, Tokyo, Japan) (20 Kv) coupled to Energy Dispersive X-ray Spectrometry (EDS; Oxford X-Maxn^2 50 mm^2, Oxford Instruments, Tubney Woods, Abingdon, UK) at the Physics Institute, UNAM. It was calibrated with copper and standards were: C (Cvit), O (SiO_2), Mg (MgO), Al (Al_2O_3), Si (SiO_2), K (KBr) and Ca (Wollastonite). The relative concentration of the element is given in percentages of weight.

3. Results

3.1. Analysis by FTIR

Figure 2 shows the spectra of the four tissues analyzed from the five species. In all spectra, the characteristic calcium oxalate crystals peaks were present (Table S1) in all species and in the four tissues. Furthermore, in all the tissues of the five species, a weak peak at 915 cm^{-1} and a wide peak ranging from 3000 to 3600 cm^{-1} due to the OH stretching were present and reflected the occurrence of dihydrate calcium oxalate ($CaC_2O_4·2H_2O$). This dihydrate calcium oxalate was detected in *Astrophytum asterias*, *Echinocactus texensis* and *Mammillaria sphaerica* with the peak 517 cm^{-1} (Figure 2A,E,I), although there is noise on the spectra. The peaks between 1014 and 1048 cm^{-1} in some tissues of the five species studied were assigned to opal and silicates (Figure 2B,D,F,H,J), whereas some peaks in the pith of *Ariocarpus retusus* subsp. *trigonus* were assigned to calcium carbonate (Figure 2C,D). The peaks at 555 and 560 cm^{-1} could represent some contaminants or unknown biomineral residue (Figure 2A–J). The occurrence of primary cell wall debris was assigned to hemicellulose residues in the pith (4) of *Ariocarpus retusus* subsp. *trigonus* (Figure 2D) and to cellulose residues in the pith (4) and epidermis-hypodermis (1) of *Astrophytum asterias* (Figure 2A,B), *Ariocarpus retusus* subsp. *trigonus* (Figure 2C,D) and *Mammillaria sphaerica* (Figure 2I,J).

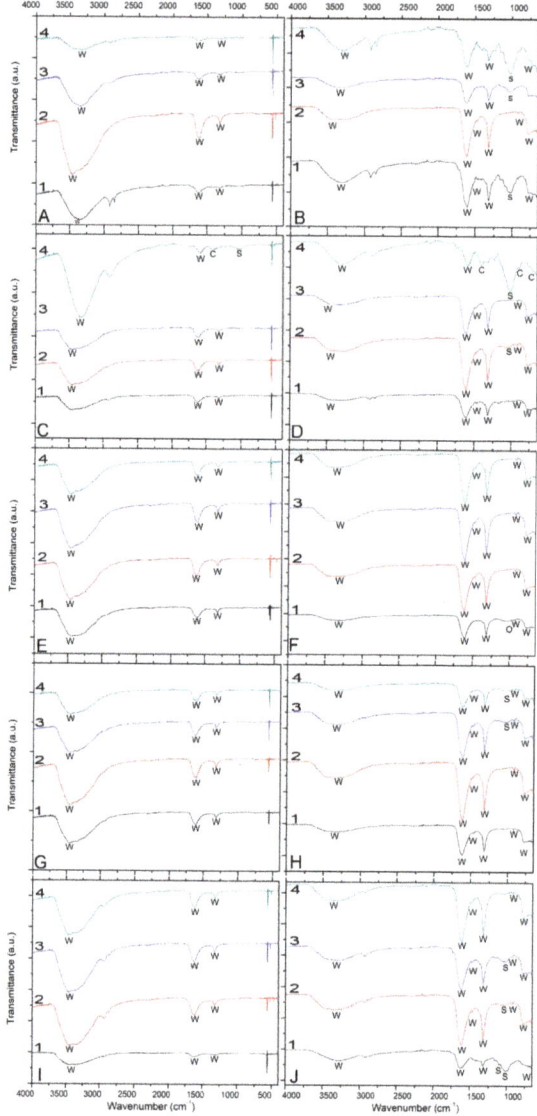

Figure 2. Spectra of biominerals in each tissue studied. The numbers below each spectra (line) indicate the wavenumbers of the peaks. A, C, E, G, I 4000–400 cm^{-1} spectra. B, D, F, H, J 4000–650 cm^{-1} spectra. (**A,B**) *Astrophytum asterias*. (**C,D**) *Ariocarpus retusus* subsp. *trigonus*. (**E,F**) *Echinocactus texensis*. (**G,H**) *Mammillaria melanocentra* subsp. *rubrograndis*. (**I,J**) *Mammillaria sphaerica*. Tissues: 1(black line) = Epidermis-hypodermis, 2(red) = Cortex, 3(purple) = Vascular cylinder, 4(green) = Pith, W = weddellite (calcium oxalate dihydrate), O = opal (amorphous silica), C = Calcium carbonate, S = silicates.

3.2. Anatomy with Light and Polarized Microscopy

The five species studied have the typical stem anatomy for the Cacteae. There is an epidermis with straight or convex outer cell wall and a hypodermis of one to five strata of parenchyma or collenchyma (Figure 3A,E and Figure 4A). Both cortex and pith have exclusively parenchyma cells (Figure 4A) and in the vascular cylinder, nonfibrous wood with wide medullary unlignified rays (Figure 4F–H). The studied species do not form visible biominerals in the epidermis (Figure 3). Two of the five species studied here had biominerals in the hypodermal cells. *Astrophytum asterias* presented prismatic biominerals (Figure 3A,C) that were birefringent under polarized light (Figure 3B,D) and *Echinocactus texensis* had amorphous silica bodies that lacked birefringence under polarized light (Figure 3E,F). Biominerals in the hypodermal cells of *Ariocarpus retusus* subsp. *trigonus* and both *Mammillaria* species were not detected in the anatomical sections.

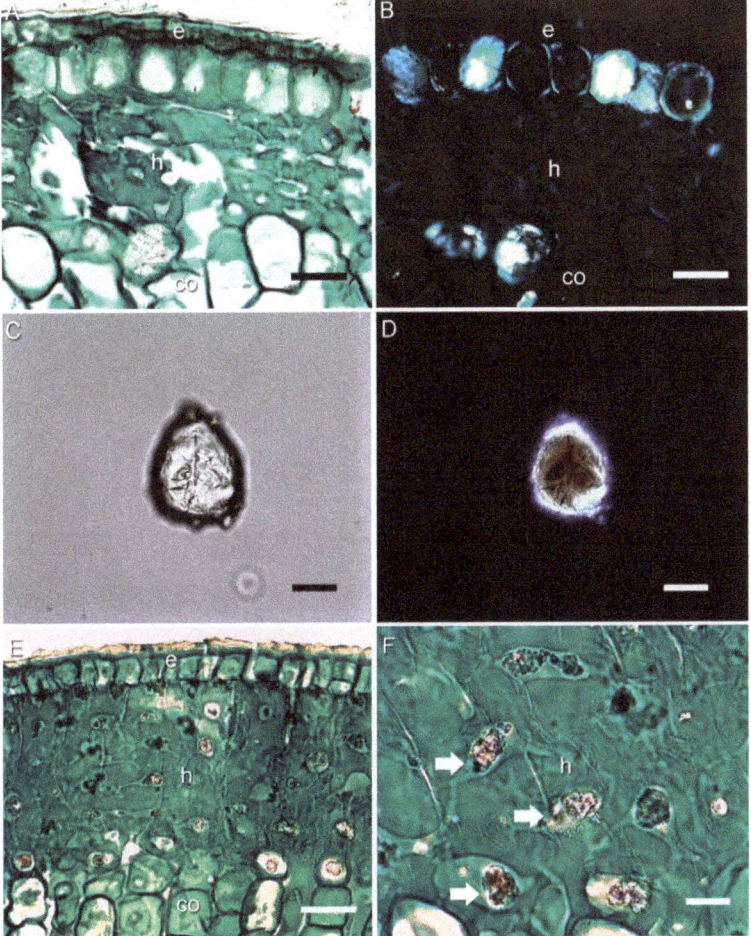

Figure 3. Biominerals in the hypodermis of stem in Cacteae, transverse sections (TS) and isolated (IS). (A, C, D, E, F) Bright field. (B, D) Polarized light. (**A,B**) *Astrophytum asterias* (TT846), TS. (**C,D**) *Astrophytum asterias* (TT1020), IS. (**E,F**) *Echinocactus texensis* (TT859), TS. Bar is 50 μm in A, B, E; 20 μm in C, D, F. e = epidermis, co = cortex, h = hypodermis, white arrows = amorphous silica bodies.

In the cortex (Figure 4A,B,D), vascular cylinder (Figure 4F–H) and pith (Figure 4I,J), all species had spherulites or prisms, which were birefringent with polarized light (Figure 4C,E,G,L,N,P). In the vascular tissue, the biominerals were deposited in the ray cells (Figure 4F,H).

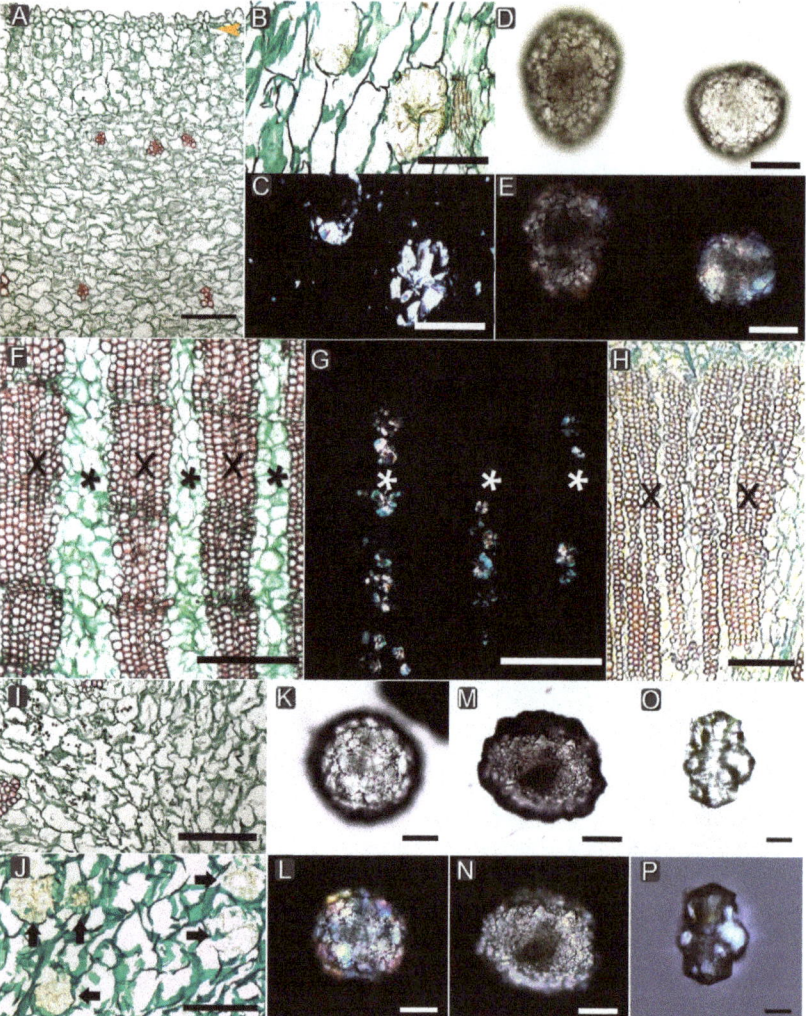

Figure 4. Biominerals in the stem tissues of transverse sections (TS) and isolated (IS) using bright field (b) or polarized light (p). (**A**) *Ariocarpus retusus* subsp. *trigonus* (TT1005), TSbAR, epidermis, hypodermis (orange arrowhead) and cortical tissue. (**B,C**) *Astrophytum asterias* (TT846), TSbpMR, spherulites in cortical cells. (**D,E**) *Astrophytum asterias* (TT1020), IS, spherulites of cortical tissue. (**F,G**) *Echinocactus texensis* (TT859), TSbpBR, spherulites in vascular cylinder. (**H**) *Mammillaria sphaerica* (TT1051), TSbBR, vascular cylinder. (**I**) *Ariocarpus retusus* subsp. *trigonus* (TT879), TSbAR, pith. (**J**) *Echinocactus texensis* (TT859), TSbBR, spherulites (black arrows) in pith. (**K,L**) *Ariocarpus retusus* subsp. *trigonus* (TT1005), ISbp, spherulites of cortical tissue. (**M,N**) *Mammillaria melanocentra* subsp. *rubrograndis* (TT1050), ISbp, spherulites of pith. (**O,P**) *Mammillaria sphaerica* (TT1051), ISbp, spherulites of cortical tissue. Bar is 300 µm in A, B, C, F–J; 100 µm in M, N; 50 µm in D, E, K, L; 20 µm in O, P. AR = apical region; MR = medium region; BR = basal region; X = tracheary cells; * = rays.

3.3. FE-SEM-EDS Analysis

Figure 5 shows the morphology and elements detected by EDS analysis. Differences between hypodermal and cortical element concentrations were observed. *Astrophytum asterias* has the highest concentration Ca (49.53%) and Al (0.87%) in the prismatic crystals of the hypodermis, whereas in *Echinocactus texensis*, hypodermis showed the lowest concentrations of Ca (5.08%) and Al (0.15%), but the highest of Si (25.34%). No traces of Mg, K and Si were detected in *Astrophytum asterias* spherulites (Figure 5F), but traces of Si (0.91%), K (0.79%), Mg (0.17%) were detected in *Ariocarpus retusus* subsp. *trigonus* (Figure 5H).

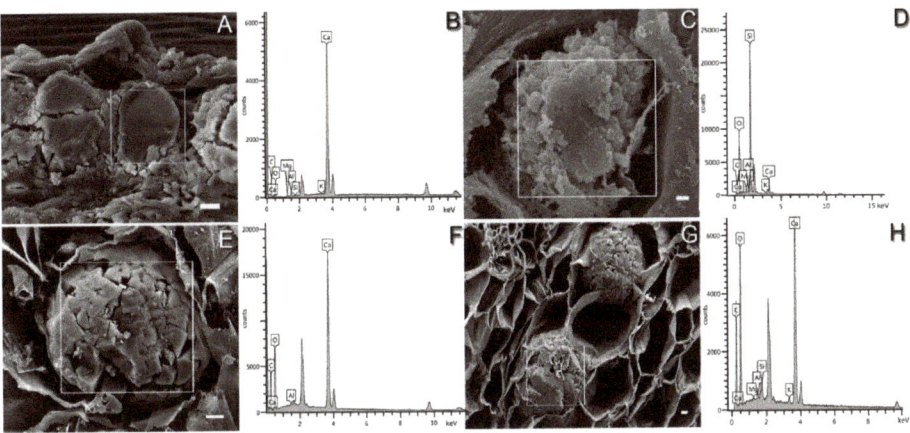

Figure 5. SEM images and EDS spectra for stem tissues of Cacteae species. (**A,B**) *Astrophytum asterias* (TT1020), prisms in hypodermis. (**C,D**) *Echinocactus texensis* (TT1021), amorphous silica bodies in hypodermis. (**E,F**) *Astrophytum asterias*, spherulite in cortex. (**G,H**) *Ariocarpus retusus* subsp. *trigonus* (TT1005), spherulites in cortex. The white square on images was the area used to analyze the elements. Bar is 10 µm in A, E, G; 1 µm in C.

4. Discussion

Calcium oxalates and amorphous silica bodies turned out to be the main biominerals in the Cacteae species studied. Both are not exclusive to the cactus family, as calcium oxalates have been described in at least 215 plant families [6] and amorphous silica bodies in 56 [31]. However, the presence of these biominerals in Cactaceae has had great relevance in its systematics, because they are not deposited randomly in the stem tissues [20,23]. Moreover, it has been suggested the possibility of identifying biominerals other than calcium oxalate [10,21,23] and different crystalline structures have been described even within the stem of the same species [23] as we found in the studied species. Calcium carbonate is poorly described for Cactaceae, possibly due to the fixation technique of the samples, so that unfixed samples combined with the FTIR analysis favor the recognition of calcium carbonate as well as the presence of other compounds.

4.1. Biominerals Identification

4.1.1. Calcium Oxalate

The differences between the monohydrate (whewellite, Wh) and dihydrate (weddellite, W) spectra of calcium oxalate have been documented by Conti et al. [32] and Petit et al. [33]. These authors point out that in the infrared spectrum, the calcium oxalate monohydrate (Wh) is characterized in the elongation zone of water molecules (3000–3600 cm^{-1}) with several peaks at 3058, 3258, 3336, 3429 and 3483 cm^{-1}. Furthermore, the pattern of calcium oxalate dihydrate has a more intense peak close to 3469 cm^{-1}. In the studied species, there was an intense peak that varies from 3466 to 3327 cm^{-1} (Figure 1A,C,E,G,I). Other characteristic vibrations of calcium oxalate dihydrate are a weak signal at 1475 and 915 cm^{-1} that belongs to the symmetric vibrations of CO and CO + H$_2$O, respectively [33], and were present from 1476 to 1466 and 913 to 910 cm^{-1} (Figure 1B,D,F,H,J) in most tissues of the species studied. In the species studied, calcium dihydrate vibrations ranged from 774 to 762 cm^{-1}, very close to the reported vibrations (770 cm^{-1}), while the calcium oxalate monohydrate is assigned to 782 cm^{-1} [33]. Based on these results, we consider that calcium oxalate dihydrate is the dominant biomineral in the studied species and is mainly distributed in the three tissues. Hartl et al. [12] documented the presence of this state of hydration for other species of *Ariocarpus* (W), *Astrophytum* (W), *Echinocactus* (W) and *Mammillaria* (Wh or W). *Mammillaria* is the largest genus of the Cacteae tribe with more than 100 species; to date, only six have been studied, Wh (3) and W (3) (12, 25), plus the two here reported. The study of other *Mammillaria* species is needed to understand the variability of calcium oxalate in terms of the state of hydration as well as the identification of other elements.

4.1.2. Silica–Amorphous Hydrated Silica

Zancajo et al. [34] and Corrales-Ureña et al. [35] identified the vibration of the Si-O-Si bonds at 1090 and 1020 cm^{-1}, in amorphous silica bodies (phytoliths) from *Sorghum* and *Ananas comosus*. In *Echinocactus texensis*, amorphous silica bodies were found in the EH spectrum region, with the vibration of the Si-O-Si bond at 1014 cm^{-1} (Figure 2D). These silica bodies were distinctive in the anatomical sections of the hypodermis (Figure 3E,F). Anatomically, silica bodies were described and characterized from the visualization of stem sections [20,21,36]. In their chemical composition, just one previous study analyzed with Raman spectroscopy the presence of amorphous silica bodies in three *Opuntia* species and *Stenocereus thurberi* (Engelm.) Buxb. [36] and identified differences in their structural composition, mainly in the SiH bonds. Therefore, with Raman spectroscopy, the structure of the phytoliths could be analyzed in future studies. We consider that these amorphous silica bodies are the product of controlled biomineralization [36,37] and should be considered as a taxonomic character.

The EDS analysis supports the presence of Si in the hypodermis of *Echinocactus texensis* with the highest concentration, while the concentrations of Ca and Mg were lower. Although the precipitation phases of hypodermic biominerals were not studied in *Echinocactus texensis*, Si and Mg may be the first phase of biomineral precipitation along with some form of calcium carbonate or amorphous calcium oxalate (both have carbon, oxygen, and calcium), as in *Ficus microcarpa* L.f and *Morus alba* L. cystoliths. In both Moraceae, Si and Mg precipitate in the first phase of biomineral development, and amorphous calcium carbonate is deposited later [38]. Biominerals of *Echinocactus texensis* precipitate in cells with a small lumen (20 μm on average) compared to *Ficus* idioblasts that exceed 100 μm [38]. Density was not quantified in *Echinocactus texensis* but they are quite abundant in hypodermis strata. For these reasons, we consider that the silica bodies differ from the cystoliths and the calcium carbonate does not precipitate.

4.1.3. Calcium Carbonate

The pith of *Ariocarpus retusus* subsp. *trigonus* presented weak peaks belonging to calcium carbonate as reported by Palacio et al. [39] for some gypsophilic plants. Monje and Baran [40] identified on stems of another cactus, *Cylindropuntia kleiniae* (DC.) F.M. Knuth, vibration in the infrared between 1415 and 1422 cm^{-1}; they were assigned to the carbonate antisymmetric stretching mode of calcite. Anatomical techniques allow the identification of biominerals in certain plant tissues, however, during fixation with substances such as acetic acid, some calcium salts and phosphates could be lost [41–43]. This may be the case detected here in the pith of *Ariocarpus retusus* subsp. *trigonus*, since no previous report of calcium carbonate for any member of Cacteae is known. Probably calcium carbonate is rare in the group because they were not detected in the other species studied, even when the same method to separate biominerals were applied.

4.1.4. Unknown Biominerals

In *Astrophytum asterias*, the spectra in the EH region showed a very broad peak in the 1027 cm^{-1} region, which unlike the biominerals of *Echinocactus texensis*, they showed birefringence and were named here as prismatic crystals (see 4.1.1). However, in *Astrophytum asterias*, there was a high intensity peak in the vascular cylinder at 1028 cm^{-1} and an intense peak at 1026 cm^{-1} in the pith. These three peaks in *Astrophytum asterias* suggest a different biomineral composition, which according to Palacio et al. [39], the spectrum of the regions that presents the strongest link patterns could be assigned to some types of silicates (1100–950 cm^{-1}) or phosphates (1100–1000 cm^{-1}). We consider that these vibrations belong to silicates, probably aluminosilicates due to the presence of silicon and aluminum in biominerals detected by EDS in the hypodermis or aluminum oxides in the cortex (Figure 5). To support this finding, new FTIR spectra at 400–100 cm^{-1} are needed [44]. It is important to mention that these aluminosilicate-weddellite prismatic crystals appear to be conservative for the genus since they have been observed in the other four *Astrophytum* species studied by EDS [45]. However, they were not detected by EDS in the other stem tissues as FTIR did here for *Astrophytum asterias*.

Here, we report for the first time the presence of silicates in *Mammillaria* biominerals. In *Mammillaria melanocentra* subsp. *rubrograndis*, the peaks assigned to silicates occurred in the vascular cylinder and pith. It should be noted that these biominerals were also birefringent. Peaks assigned to silicates were also present in three of the four *Mammillaria sphaerica* tissues (Figure 2). The surprising results were the peaks in the epidermis-hypodermis of *M. sphaerica* because no vacuolar biominerals were detected in this region by microscopy. Furthermore, no biominerals other than calcium oxalate have been described for the genus [12,25]. Our results suggest that other *Mammillaria* species should be studied as mentioned above to confirm biomineral diversity.

Magnesium was detected by EDS spectra (Figure 5) but not with FTIR (Figure 2). The characteristic peaks assigned to magnesium oxalate reported by Monje and Baran [17] for other cacti such as *Opuntia* were not detected in our study by FTIR but Mg is present in traces.

4.2. Possible Functions of Biominerals in Different Tissues

Different ecological functions are attributed to biominerals in plant families [37]. Pierantoni et al. [46] showed in *Abelmoschus esculentus* that the calcium oxalates can scatter light through photosynthetic tissue and the amorphous silica bodies have a protective effect against UV radiation. The occurrence of biominerals in these species that grow in the Chihuahuan desert could be a mechanism for protecting the photosynthetic tissue and against excessive solar radiation in *Astrophytum asterias* and *Echinocactus texensis*. The presence of calcium oxalates in the cortex, vascular cylinder and pith of the stem could be explained as calcium deposits in the cellular vacuoles within the tissues, as suggested by Volk et al. [47] for *Pistia stratiotes*. The crystalline structure of calcium oxalate dihydrate has zeolitic channels that allow the adsorption of large quantities of water molecules that can diffuse "freely" in the structure [33], suggesting that these biominerals function as small water reserves. This water would be available together with the water stored in the vacuoles of cortical and pith cells, allowing plants to withstand the periods of greatest drought.

On the other hand, silicates in plants may have a role of improving the aluminum tolerance capacity, so the presence of silicon transporters shows that the deposition of phytoliths is an active and regulated process by the plant [36]. Further to this, the presence of phytoliths in plant tissues works as a defensive method against herbivores by abrasion of the teeth and reduction in the absorption of nitrogen during digestion [48], so it is possible that in cacti, the presence of silicates works in this way. These silicates were detected in experiments with *Sorghum bicolor* [49] and four cacti species [36]. In *Astrophytum asterias* and *Mammillaria melanocentra* subsp. *rubograndis*, these silicates could be present in the soil; however, the metabolic uptake capacity of silicon in the soil depends on the species. Silicon accumulation in different species could be related to the presence of silicon transporters (Lsi proteins), belonging to the Nod26-like major protein (NIP) in the plasma membranes of root cells [4,48]. In Cacteae, the occurrence of these transporters has never been evaluated and future studies are needed.

5. Conclusions

There were vibrations associated with calcium carbonate in the P of *Ariocarpus retusus* subsp. *trigonus*. The calcium carbonate reported here increases the diversity of biominerals in cactus. The hydration state of calcium oxalate is conserved in the different tissues that were studied in the stems of the Cacteae species. Both calcium oxalate and silica bodies were present in the same species but in different tissues for *Echinocactus texensis*. The presence of silicate peaks belonging to species such as *Astrophytum asterias* and *Mammillaria sphaerica* opens the opportunity to study the role of silicates in the physiology of Cacteae species.

Supplementary Materials: The following are available online at http://www.mdpi.com/2073-4352/10/6/432/s1, Table S1. Assignment of FTIR absorption bands of biominerals extracted from tissues of Cacteae species.

Author Contributions: T.T. and A.D.l.R.-T. designed the work, T.T. collected the plants, A.D.l.R.-T. and A.M. performed the lab work and prepared the figures, A.D.l.R.-T., T.T. and A.M. analyzed the data, A.D.l.R.-T. wrote the original draft. All authors reviewed and edited the manuscript. All authors have read and agreed to the published version of the manuscript.

Funding: Funding for this research was provided by Programa de Apoyo a Proyectos de Investigación e Innovación Tecnológica, Universidad Nacional Autónoma de México (PAPIIT-UNAM); grant no. IN205419 to TT and Consejo Nacional de Ciencia y Tecnología (CONACyT), grant no. 703332 to ADR-T.

Acknowledgments: ADR-T thanks Posgrado en Ciencias Biológicas (UNAM) and AM & TT thank to Rubén San Miguel-Chávez for helping us to use the FTIR.

Conflicts of Interest: The authors declare no conflict of interest.

References

1. Skinner, H.C.W. Biominerals. *Mineral. Mag.* **2005**, *69*, 621–641. [CrossRef]
2. Arnott, H.J. Studies of Calcification in Plants. In *Calcified Tissues*; Fleish, H., Blackwood, H.J.J., Owen, M., Eds.; Springer: Berlin/Heidelberg, Germany, 1966; pp. 152–157, ISBN 978-3-642-85843-7.
3. Arnott, H.J. Three Systems of Biomineralization in Plants with Comments on the Associated Organic Matrix. In *Biological Mineralization and Demineralization*; Nancollas, G.H., Ed.; Springer: Berlin, Germany, 1982; pp. 199–218, ISBN 978-3-642-68574-3.
4. Bauer, P.; Elbaum, R.; Weiss, I.M. Calcium and Silicon Mineralization in Land Plants: Transport, Structure and Function. *Plant Sci.* **2011**, *180*, 746–756. [CrossRef] [PubMed]
5. Franceschi, V.R.; Nakata, P.A. Calcium Oxalate in Plants: Formation and Function. *Annu. Rev. Plant Biol.* **2005**, *56*, 41–71. [CrossRef]
6. McNair, J.B. The Interrelation between Substances in Plants: Essential Oils and Resins, Cyanogen and Oxalate. *Am. J. Bot.* **1932**, *19*, 255–272. [CrossRef]
7. Skinner, H.C.W.; Jahren, A.H. Biomineralization. In *Treatise on Geochemistry*; Holland, D.H., Turekian, K.K., Eds.; Elsevier Sciences: Amsterdam, The Netherlands, 2003; Volume 8, pp. 1–69, ISBN 9780080437514.
8. Cheavin, W.H.S. The Crystals and Cystolites Found in Plant Cells. Part 1: Crystals. *Microscope* **1938**, *2*, 155–158.
9. Schleiden, M.J. Beiträge zur Anatomie der Cacteen. *Mem l'Acadèmie Imp des Sci St. Pétersbg* **1845**, *4*, 335–380.
10. Bárcenas-Argüello, M.L.; Gutiérrez-Castorena, M.C.-D.-C.; Terrazas, T. The Polymorphic Weddellite Crystals in Three Species of *Cephalocereus* (Cactaceae). *Micron* **2015**, *77*, 1–8. [CrossRef] [PubMed]
11. Hartl, W.P.; Barbier, B.; Klapper, H.M.; Barthlott, W. Dimorphism of Calcium Oxalate Crystals in Stem Tissues of RHIPSALIDEAE (Cactaceae)—A Contribution to the Systematics and Taxonomy of the Tribe. *Bot. Jahrb. Syst. Pflanzengesch. Pflanzengeogr.* **2003**, *124*, 287–302. [CrossRef]
12. Hartl, W.P.; Klapper, H.; Barbier, B.; Ensikat, H.J.; Dronskowski, R.; Müller, P.; Ostendorp, G.; Tye, A.; Bauer, R.; Barthlott, W. Diversity of Calcium Oxalate Crystals in Cactaceae. *Can. J. Bot.* **2007**, *85*, 501–517. [CrossRef]

13. Monje, P.V.; Baran, E.J. On the Formation of Weddellite in *Chamaecereus Silvestrii*, a Cactaceae Species from Northern Argentina. *Z. Naturforsch. C J. Biosci.* **1996**, *51*, 426–428. [CrossRef]
14. Monje, P.V.; Baran, E.J. On the Formation of Whewellite in the Cactaceae Species *Opuntia Microdasys*. *Z. Naturforsch. C J. Biosci.* **1997**, *52*, 267–269. [CrossRef]
15. Monje, P.V.; Baran, E.J. Characterization of Calcium Oxalates Generated as Biominerals in Cacti. *Plant Physiol.* **2002**, *128*, 707–713. [CrossRef] [PubMed]
16. Rivera, E.R.; Smith, B.N. Crystal Morphology and ^{13}Carbon/^{12}Carbon Composition of Solid Oxalate in Cacti. *Plant Physiol.* **1979**, *64*, 966–970. [CrossRef] [PubMed]
17. Monje, P.V.; Baran, E.J. Evidence of Formation of Glushinskite as a Biomineral in a Cactaceae Species. *Phytochemistry* **2005**, *66*, 611–614. [CrossRef] [PubMed]
18. Jones, J.G.; Bryant, V.M. Phytolith Taxonomy in Selected Species of Texas Cacti. In *Phytolith Systematics*; Springer: Boston, MA, USA, 1992; Volume 1, pp. 215–238, ISBN 9781489911575.
19. Monje, P.V.; Baran, E.J. First Evidences of the Bioaccumulation of α-Quartz in Cactaceae. *J. Plant Physiol.* **2000**, *157*, 457–460. [CrossRef]
20. Loza-Cornejo, S.; Terrazas, T. Epidermal and Hypodermal Characteristics in North American Cactoideae (Cactaceae). *J. Plant Res.* **2003**, *116*, 27–35. [CrossRef]
21. Terrazas, T.; Loza-Cornejo, S.; Arreola-Nava, H.J. Anatomía Caulinar de las Especies del Género *Stenocereus* (Cactaceae). *Acta Botánica Venez.* **2005**, *28*, 321–336.
22. Gibson, A.C.; Horak, K.E. Systematic Anatomy and Phylogeny of Mexican Columnar Cacti. *Ann. Mo. Bot. Gard.* **1978**, *65*, 999. [CrossRef]
23. De La Rosa-Tilapa, A.; Vázquez-Sánchez, M.; Terrazas, T. Stem Anatomy of *Turbinicarpus* s.l. (Cacteae, Cactaceae) and Its Contribution to Systematics. *Plant Biosyst.* **2019**, *153*, 600–609. [CrossRef]
24. Frausto-Reyes, C.; Loza-Cornejo, S.; Terrazas, T.; De La Luz Miranda-Beltrán, M.; Aparicio-Fernández, X.; López-Mací, B.M.; Morales-Martínez, S.E.; Ortiz-Morales, M. Raman Spectroscopy Study of Calcium Oxalate Extracted from Cacti Stems. *Appl. Spectrosc.* **2014**, *68*, 1260–1265. [CrossRef]
25. López-Macías, B.M.; Morales-Martínez, S.E.; Loza-Cornejo, S.; Reyes, C.F.; Terrazas, T.; Patakfalvi, R.J.; Ortiz-Morales, M.; Miranda-Beltrán, M.D.l.L. Variability and Composition of Calcium Oxalate Crystals in Embryos-Seedlings-Adult Plants of the Globose Cacti *Mammillaria Uncinata*. *Micron* **2019**, *125*, 102731. [CrossRef]
26. Ruzin, S.E. *Plant Microtechnique and Microscopy*; Oxford University Press: Oxford, UK, 1999; ISBN 0195089561.
27. Loza-Cornejo, S.; Terrazas, T. Anatomía del tallo y de la raíz de dos Especies de *Wilcoxia* Britton & Rose (Cactaceae) del Noreste de México. *Bot. Sci.* **1996**, *59*, 13–23. [CrossRef]
28. Durak, T.; Depciuch, J. Effect of Plant Sample Preparation and Measuring Methods on ATR-FTIR Spectra Results. *Environ. Exp. Bot.* **2020**, *169*, 103915. [CrossRef]
29. Stuart, B.H. *Infrared spectroscopy: Fundamentals and Applications*; ANTS (Analytical Techniques in the Sciences); John Wiley & Sons, Ltd.: Chichester, UK, 2004; ISBN 978-0-470-85428-0.
30. Mascarenhas, M.; Dighton, J.; Arbuckle, G.A. Characterization of Plant Carbohydrates and Changes in Leaf Carbohydrate Chemistry due to Chemical and Enzymatic Degradation Measured by Microscopic ATR FT-IR Spectroscopy. *Appl. Spectrosc.* **2000**, *54*, 681–686. [CrossRef]
31. Sharma, R.; Kumar, V.; Kumar, R. Distribution of Phytoliths in Plants: A Review. *Geol. Ecol. Landsc.* **2019**, *3*, 123–148. [CrossRef]
32. Conti, C.; Casati, M.; Colombo, C.; Possenti, E.; Realini, M.; Gatta, G.D.; Merlini, M.; Brambilla, L.; Zerbi, G. Synthesis of Calcium Oxalate Trihydrate: New Data by Vibrational Spectroscopy and Synchrotron X-Ray Diffraction. *Spectrochim. Acta Part A Mol. Biomol. Spectrosc.* **2015**, *150*, 721–730. [CrossRef]
33. Petit, I.; Belletti, G.D.; Debroise, T.; Llansola-Portoles, M.J.; Lucas, I.T.; Leroy, C.; Bonhomme, C.; Bonhomme-Coury, L.; Bazin, D.; Daudon, M.; et al. Vibrational Signatures of Calcium Oxalate Polyhydrates. *Chem. Sel.* **2018**, *3*, 8801–8812. [CrossRef]
34. Zancajo, V.M.R.; Diehn, S.; Filiba, N.; Goobes, G.; Kneipp, J.; Elbaum, R. Spectroscopic Discrimination of *Sorghum* Silica Phytoliths. *Front. Plant Sci.* **2019**, *10*, 1571. [CrossRef]
35. Corrales-Ureña, Y.R.; Villalobos-Bermúdez, C.; Pereira, R.; Camacho, M.; Estrada, E.; Argüello-Miranda, O.; Vega-Baudrit, J.R. Biogenic Silica-Based Microparticles Obtained as a Sub-Product of the Nanocellulose Extraction Process from Pineapple Peels. *Sci. Rep.* **2018**, *8*, 1–9. [CrossRef]

36. Wright, C.R.; Waddell, E.A.; Setzer, W.N. Accumulation of Silicon in Cacti Native to the United States: Characterization of Silica Bodies and Cyclic Oligosiloxanes in *Stenocereus Thurberi*, *Opuntia Littoralis*, *Opuntia Ficus-Indica*, and *Opuntia Stricta*. *Nat. Prod. Commun.* **2014**, *9*, 873–878. [CrossRef]
37. He, H.; Veneklaas, E.J.; Kuo, J.; Lambers, H. Physiological and Ecological Significance of Biomineralization in Plants. *Trends Plant Sci.* **2014**, *19*, 166–174. [CrossRef] [PubMed]
38. Gal, A.; Hirsch, A.; Siegel, S.; Li, C.; Aichmayer, B.; Politi, Y.; Fratzl, P.; Weiner, S.; Addadi, L. Plant Cystoliths: A Complex Functional Biocomposite of Four Distinct Silica and Amorphous Calcium Carbonate Phases. *Chem. Eur. J.* **2012**, *18*, 10262–10270. [CrossRef] [PubMed]
39. Palacio, S.; Aitkenhead, M.; Escudero, A.; Montserrat-Martí, G.; Maestro, M.; Robertson, A.H.J. Gypsophile Chemistry Unveiled: Fourier Transform Infrared (FTIR) Spectroscopy Provides New Insight into Plant Adaptations to Gypsum Soils. *PLoS ONE* **2014**, *9*, e107285. [CrossRef] [PubMed]
40. Monje, P.V.; Baran, E.J. Complex Biomineralization Pattern in Cactaceae. *J. Plant Physiol.* **2004**, *161*, 121–123. [CrossRef]
41. Berg, R.H. A Calcium Oxalate-Secreting Tissue in Branchlets of the Casuarinaceae. *Protoplasma* **1994**, *183*, 29–36. [CrossRef]
42. Lersten, N.R.; Horner, H.T. Unique Calcium Oxalate "Duplex" and "Concretion" Idioblasts in Leaves of Tribe Naucleeae (Rubiaceae). *Am. J. Bot.* **2011**, *98*, 1–11. [CrossRef]
43. He, H.; Bleby, T.M.; Veneklaas, E.J.; Lambers, H.; Kuo, J. Morphologies and Elemental Compositions of Calcium Crystals in Phyllodes and Branchlets of *Acacia Robeorum* (Leguminosae: Mimosoideae). *Ann. Bot.* **2012**, *109*, 887–896. [CrossRef]
44. Vahur, S.; Teearu, A.; Peets, P.; Joosu, L.; Leito, I. ATR-FT-IR Spectral Collection of Conservation Materials in the Extended Region of 4000–80 cm^{-1}. *Anal. Bioanal. Chem.* **2016**, *408*, 3373–3379. [CrossRef]
45. De la Rosa-Tilapa, A. Structure and Composition of Biominerals in the Stem of the Cacteae Tribe (Cactaceae). Master's Thesis, National Autonomous University of Mexico, Mexico City, Mexico, 2020.
46. Pierantoni, M.; Tenne, R.; Brumfeld, V.; Kiss, V.; Oron, D.; Addadi, L.; Weiner, S. Plants and Light Manipulation: The Integrated Mineral System in Okra Leaves. *Adv. Sci.* **2017**, *4*, 1600416. [CrossRef]
47. Volk, G.M.; Lynch-Holm, V.J.; Kostman, T.A.; Goss, L.J.; Franceschi, V.R. The Role of Druse and Raphide Calcium Oxalate Crystals in Tissue Calcium Regulation in *Pistia Stratiotes* Leaves. *Plant Biol.* **2002**, *4*, 34–45. [CrossRef]
48. Nawaz, M.A.; Zakharenko, A.M.; Zemchenko, I.V.; Haider, M.S.; Ali, M.A.; Imtiaz, M.; Chung, G.; Tsatsakis, A.; Sun, S.; Golokhvast, K.S. Phytolith Formation in Plants: From Soil to Cell. *Plants* **2019**, *8*, 249. [CrossRef] [PubMed]
49. Hodson, M.J.; Sangster, A.G. The interaction between silicon and aluminium in *Sorghum bicolor* (L.) Moench: Growth analysis and x-ray microanalysis. *Ann. Bot.* **1993**, *72*, 389–400. [CrossRef]

© 2020 by the authors. Licensee MDPI, Basel, Switzerland. This article is an open access article distributed under the terms and conditions of the Creative Commons Attribution (CC BY) license (http://creativecommons.org/licenses/by/4.0/).

Review

Synthesis Methods and Favorable Conditions for Spherical Vaterite Precipitation: A Review

Donata Konopacka-Łyskawa

Department of Process Engineering and Chemical Technology, Faculty of Chemistry,
Gdańsk University of Technology, Narutowicza 11/12, 80-233 Gdańsk, Poland;
donata.konopacka-lyskawa@pg.edu.pl; Tel.: +48-58-347-2910

Received: 15 March 2019; Accepted: 18 April 2019; Published: 25 April 2019

Abstract: Vaterite is the least thermodynamically stable anhydrous calcium carbonate polymorph. Its existence is very rare in nature, e.g., in some rock formations or as a component of biominerals produced by some fishes, crustaceans, or birds. Synthetic vaterite particles are proposed as carriers of active substances in medicines, additives in cosmetic preparations as well as adsorbents. Also, their utilization as a pump for microfluidic flow is also tested. In particular, vaterite particles produced as polycrystalline spheres have large potential for application. Various methods are proposed to precipitate vaterite particles, including the conventional solution-solution synthesis, gas-liquid method as well as special routes. Precipitation conditions should be carefully selected to obtain a high concentration of vaterite in all these methods. In this review, classical and new methods used for vaterite precipitation are presented. Furthermore, the key parameters affecting the formation of spherical vaterite are discussed.

Keywords: vaterite; calcium carbonate; polymorph; precipitation; synthesis; carbonation

1. Introduction

Vaterite is the least thermodynamically stable anhydrous calcium carbonate polymorph and it easily transforms into more stable calcite or aragonite in the presence of water. This form of calcium carbonate mineral was named to honor the German chemist and mineralogist, Heinrich Vater, in 1903.

Because of its instability, the existence of vaterite is very rare in nature. It have been found in some sediments and rocks [1], e.g., as a major constituent of a carbonated calcium silicate hydrogel complex formed from larnite in Ballycraigy, Ireland [2]. Vaterite can be precipitated in some mineral springs when specific glacial conditions take place [3]. Also, vaterite crystals have been identified in materials produced by living organisms, e.g., otolith organs of fishes [4–6], spicules of the ascidian Herdmania momus [6,7], freshwater pearls, crustacean tissues, or bird eggs [6,8] as well as the chalky crust on the surface of leaves of the alpine plant, *Saxifraga scardica* [9].

Synthetic vaterite particles have been used as a carrier of active compounds for medical treatments [10–14]. They have been tested as a template for biodegradable polymer capsules, which can be used for applications in nanomedicine [10,11,15]. Also, vaterite particles are added to personal care products as abrasives, adsorbents, anticaking agents, buffers, or dyes [15]. Due to their unique optical properties, vaterite microspheres has been useful in microrheology and microfluidics [16]. Spherical vaterite particles have been used to generate flow within microfluidic channels that has allowed the creation of an optical driven pump [17]. This polymorphic $CaCO_3$ form has been proposed as a coating pigment for ink jet paper [18]. The main advantages of vaterite particles are their easy and low-cost preparation, ability to design particles with defined characteristics, porous structure, mild conditions for decomposition, non-toxicity, and biocompatibility [11,15].

Recently, this unstable mineral was widely investigated to identify the favorable conditions for its production as well as to verify its usefulness for various applications. Therefore, the current opinions on vaterite synthesis as well as a discussion on the variables affecting its formation are overviewed. The issues raised in this review are presented in the diagram in Figure 1.

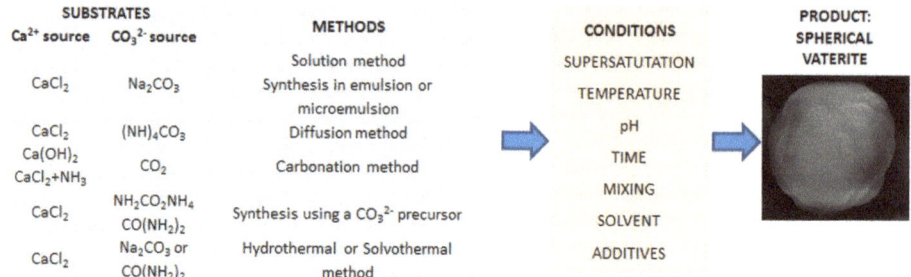

Figure 1. Diagrammatic representation of raised issues.

2. Vaterite Properties

Synthetic vaterite particles are usually produced as polycrystalline spheres. The main advantages of such particles' morphology are the porous structure, large surface area, and greater hydrophilicity in comparison to more stable calcite or aragonite [10,19,20]. Other morphological forms of vaterite can be also obtained, i.e., plates [21], hexagonal crystals [22], lenses [23], lamellar aggregates [24], florets, or rosettes [21,25] as well as microtablets [26]. Examples of vaterite particles are shown in Figure 2.

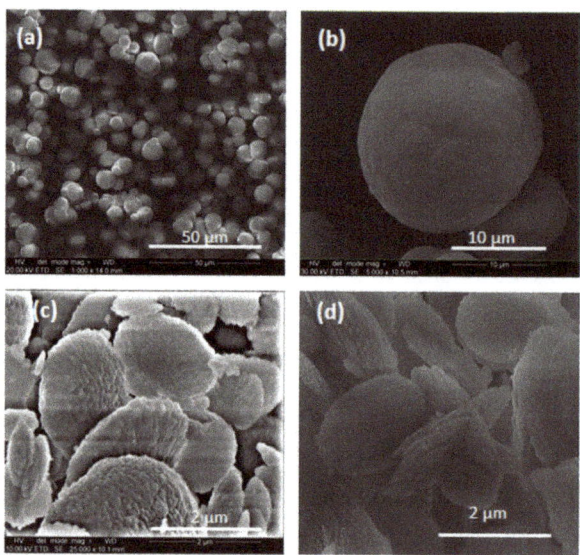

Figure 2. Vaterite particles: (a) and (b) typical spherical particles; (c) spherical and lens-like particles; (d) deformed lens-like and crossed lens-like particles.

Vaterite has a hexagonal crystal system, but the exact crystal structure of vaterite is still under discussion. The analysis of experimental data is consistent in that all vaterite structures belong to the order-disorder (OD) family [27]. It means that the occurrence of multiple polytypes on the micro- to macroscopic scale, as well as considerable stacking disorder, are both to be expected. Recently Burgess and Bryce [28] used the combined ^{43}Ca solid-state nuclear magnetic resonance spectroscopic and computational method to indicate two crystal structures, i.e., the hexagonal lattice, P3$_2$21, and monoclinic lattice, C2, which have the best agreement between the simulated spectra and diffractograms with the experimental data.

Selected properties of the vaterite are summarized in Table 1.

Table 1. Selected properties of vaterite.

Properties	Values	Ref.
Density	2.54 g/cm^3	[2]
	2.65 g/cm^3	[27]
K_{sp}^{1} at 25°C	1.22·10^{-8}	[29]
K_{sp} for t = 0–90 °C	$K_{sp} = -172.1495 - 0.077993T + 3074.688/T + 71.595 \log T$	[29]
Optical properties	Semitransparent, colorless	[17]
Effective birefringence	$\Delta n = 0.06$–0.1	[17]
Refractive index	$n_\omega = 1.55$, $n_\varepsilon = 1.65$	[27]
α_V^{2} at 25 °C	35.5·10^{-6} K^{-1}	[30]
Surface energy	Calculated: 90 mJ/m^2; experimental 34–73 mJ/m^2	[31]

^{1} Solubility product; ^{2} Volumetric thermal expansion coefficient.

The surface of the vaterite particles is usually hydrophilic. The hydrophobic vaterite can be obtained by the adsorption of amphiphilic molecules, e.g., oleic acid, at the interface of the produced vaterite [32]. The charge of vaterite particles depends on the composition of the solution and its pH. The values of the ζ-potential were negative when the vaterite particles were dispersed in saturated CaCO$_3$ solution at pH 9.0 (−4 mV) and at pH 10.6 (−26 mV) [33]. Another experiment showed that when the solution was composed from 0.01 mol/dm^3 CaCl$_2$, 0.002 mol/dm^3 Na$_2$CO$_3$, and 0.5 mol/dm^3 NaCl, the charge of the vaterite particles was positive in the range of pH 7.5 to 9.9 [34]. The addition of organic compounds, like polypeptides or fulvic acid, can change the charge of the vaterite particles due to its adsorption at the precipitated crystal interface [33,34].

The mechanical properties of synthetic vaterite particles were determined using nanoidentation analysis [30]. The elastic modulus was found to be in the range of 16 to 61 GPa and the calculated mean value of this parameter was 31 GPa. The determined hardness of vaterite was in the range of 4.2 to 0.3 GPa with a mean value of 0.9 GPa.

During the heating of vaterite particles, thermal transformation and decomposition occurs. The exothermic transformation of vaterite into calcite takes place at temperatures between 395 and 540 °C [6,35,36]. The exact transformation temperature of vaterite into calcite depends on the particle characteristics, the presence of additives, and the heating rate. The shift of the vaterite transformation to a lower temperature may be observed when calcite is present in the sample [35,36] or in the case of the incorporation of organic molecules or foreign ions into the vaterite particles [6,37]. Also, the coexistence of pure vaterite and vaterite in contact with the calcite phase (e.g., vaterite particles covered by a calcite layer) [36] can cause the appearance of a broad range of transformation temperatures.

Recently, a report on the pressure-induced phase transition of vaterite was presented [38]. With increasing pressure, vaterite transformed to high-pressure vaterite forms (vaterite II, vaterite III, and vaterite IV) or partially to calcite. All phase transitions related to vaterite were reversible, except for vaterite II to calcite III.

The presence of vaterite in calcium carbonate samples can be determined using Fourier transformed infrared spectroscopy (FTIR) [39], powder X-ray diffraction (PXRD), Raman spectroscopy [40], or ^{43}Ca solid state nuclear magnetic resonance (^{43}Ca ssNMR) [41]. Characteristic peaks of calcite, aragonite, and vaterite, obtained in spectra or diffractograms that allow these polymorphic forms to be distinguished by FTIR, XRD, Raman spectroscopy, and ^{43}Ca ssNMR, are listed in Table 2.

Table 2. Characteristic peaks of CaCO$_3$ polymorphs in FTIR, XRD, Raman spectroscopy, and ^{43}Ca ssNMR analysis.

Analytical technique	Vaterite	Aragonite	Calcite	Ref.
FTIR, Wave number, cm^{-1}	745	710, 713	713	[39]
XRD, 2Θ°	29.5	45.9	25.0	[40]
Raman, Wave number, cm^{-1}	750	705	711	[40]
^{43}Ca ssNMR, δ_{iso} ^{1}, ppm	−3	−34	6	[41]

^{1} The isotropic chemical shift.

3. Mechanisms of Spherical Vaterite Formation

The crystallization of calcium carbonate crystalline polymorphs often occurs via amorphous calcium carbonate (ACC), which is initially formed in the solution [42]. The structure of the ACC precursor consists of a porous calcium-rich nanoscale framework containing water and carbonate ions. ACC is transformed into vaterite due to the rapid dehydration and internal structure reorganization. It is suggested that the possibility of crystallization of the vaterite particles occurs, as the initially generated ACC exhibits proto-vaterite features. [43]. In the final stage, the slow transformation of vaterite into calcite takes place via a dissolution and recrystallization process [42].

There are two main concepts proposed for the explanation of spherical vaterite formation. The first one is based on the aggregation of nanoparticles and the second mechanism is a development of the classical theory of crystal growth [44]. According to the first concept, the production of polycrystalline vaterite particles is a result of the assembly of nano-sized crystals by oriented or not-oriented attachment [44,45]. The aggregation requires the production of many small particles at the beginning of the reaction, which is supported by a high supersaturation. The aggregation mechanisms were applied to interpret both the formation of core-shell microspheres [45] and hollow vaterite particles [46]. It was suggested that the core-shell vaterite structure was a result of a successive aggregation and coverage of formed spheres with hexagonal plates [45] in the presence of poly(styrene sulfonic acid) sodium salt, while the precipitation with the addition of ethylene glycol led to the formation of an initial shell structure from the primary nanoparticles and then hollow vaterite particles [46].

However, the spherulitic growth mechanism is based on the classical theory of crystal growth [44]. Spherulites are produced when a new nucleus arises on the surface of the growing crystal. Based on the experiments, two concepts of spherical growth are proposed: (i) Spherulites arise from a central precursor via multidirectional growth of crystalline fibers, and (ii) spherules grow from a precursor via low angle branching starting on the edges [44,47]. The spherulitic growth mechanism was used to explain the formation of vaterite particles with spherical and dumbbell morphology [47,48].

4. Methods of the Synthesis of Vaterite Particles

4.1. Solution Route (L-L)

In this method, a solution containing calcium salt is mixed with a solution of carbonate salt [49]. When calcium chloride and potassium carbonate solutions are used, the reaction is as follows:

$$CaCl_2 + K_2CO_3 \rightarrow CaCO_3 + 2KCl \qquad (1)$$

The total volume of the first solution used can be added immediately to the second solution [50,51] or it can be injected to another one with a controlled rate [52]. Different types of stirring are used to produce a homogeneous reaction mixture. Stirrers applied in laboratories are usually mechanical and magnetic, although other types are also proposed, e.g., ultrasound or microwave [13,53]. Recently, a "dropwise precipitation" has been adopted to calcium carbonate precipitation [54]. In this method, a calcium ion solution is added in very small portions to a carbonate solution.

4.2. Carbonation Route (G-L)

Vaterite particles can be synthesized using carbon dioxide as a reagent. The reaction may be carried out using calcium hydroxide or calcium salt as a source of Ca^{2+} ions. The reaction of calcium hydroxide and carbon dioxide can be written as:

$$Ca(OH)_2 + CO_2 \rightarrow CaCO_3 + H_2O \tag{2}$$

When gaseous CO_2 is introduced into the aqueous solution of calcium salt (e.g., $CaCl_2$), the overall following reactions can be presented as:

$$CaCl_2 + CO_2 + 2OH^- \leftrightarrow CaCO_3 + H_2O + 2Cl^- \tag{3}$$

The detailed mechanisms of calcium carbonate precipitation are complex and include the transfer of carbon dioxide from the gas phase into the liquid phase, the formation of a carbonic acid, and its hydrolysis to produce carbonate ions that are the reagent for $CaCO_3$ precipitation. The basic pH of the reactive mixture favors both CO_2 absorption and carbonate ion formation. In case of calcium hydroxide slurry, the initial pH is about 12.4, and therefore the CO_2 absorption is relatively easy. However, when calcium salts (e.g., calcium chloride, $CaCl_2$, or calcium nitrate, $Ca(NO_3)_2$) are used, the pH of solution is more acidic and less advantageous to the formation of carbonate ions. Therefore, CO_2 absorption promoters, like ammonia [19,20,55,56] or amines [56,57], are added to the initial solution. Moreover, carbamate ions that form during carbon dioxide absorption in ammonia and primary or secondary amine aqueous solutions may stabilize the vaterite phase [56,58]. A modification of this approach is the use of extracts obtained by the leaching of calcium to reach minerals or solid waste with ammonium salt solutions [59]. The formed leachates contain calcium ions and ammonia and can be utilized in the carbonation process.

Recently, an interesting synthesis method using solid CO_2 (dry ice) was proposed [60]. The reaction was carried out at a minimum temperature of $-50°C$ in a water-methanol solution containing calcium oxide. The addition of dry ice pellets into a water-methanol solution kept the temperature low and the use of methanol allowed the liquid state of the reaction system to be maintained. Such conditions promoted the CO_2 solubility and enabled the synthesis of vaterite nanoparticles.

4.3. Diffusion Method

Calcium carbonate in vaterite form can be produced by a diffusion method, in which ammonium carbonate [52] or ammonium bicarbonate [61] is used. These compounds slowly decompose the formed ammonia gas and carbon dioxide. The precipitation is carried out in a closed vessel containing a container with a calcium salt solution and a container with solid carbonates. The formed gases diffuse into the calcium salt solution. After the absorption of ammonia and carbon dioxide, the calcium carbonate is formed in the liquid phase according to the reaction described by Equation (3).

4.4. Synthesis in Emulsions and Microemulsions

The reaction between calcium ions and carbonate ions may take place in the emulsion system. A water-in-oil emulsion containing a single reagent solution as the dispersed phase can be used for the reaction. A second reactant solution is added to the emulsion and precipitation occurs in the dispersed aqueous phase [62]. A stable double water-in-oil-in-water emulsion can be used for vaterite synthesis as well. In this case, calcium ions dissolved in the external phase are extracted into the organic phase containing oil-soluble extractant and stripped into the internal aqueous phase, where the reaction with carbonate ions occurs [63]. Also, the use of microemulsions is proposed for the production of vaterite. For this purpose, two microemulsions containing Ca^{2+} and CO_3^{2-} ions are prepared. Then, these microemulsions are mixed together to produce calcium carbonate [64].

Another approach to calcium carbonate precipitation in a microemulsion system is the decomposition of calcium bicarbonate:

$$Ca(HCO_3)_2 \leftrightarrow CaCO_3 + CO_2 + H_2O \qquad (4)$$

When a saturated solution of calcium bicarbonate is a water phase in a water-in-oil emulsion, vaterite is formed during the slow desorption of carbon dioxide from an aqueous phase. Sponge-like vaterite spheroids were produced using this method [65].

4.5. Synthesis Using a Precursor of Carbonate Ions

The synthesis of calcium carbonate can be carried out using a substance that forms carbonate ions in the reaction environment. Such carbonate ion promoters may be ammonium carbamate [58] or urea [66,67]. Ammonium carbamate hydrolyzes in aqueous solutions to form ammonium carbonate:

$$NH_2CO_2NH_4 + H_2O \leftrightarrow (NH_4)_2CO_3 \qquad (5)$$

Ammonium carbonate is also the product in the reaction of urea with water:

$$CO(NH_2)_2 + 2H_2O \leftrightarrow (NH_4)_2CO_3 \qquad (6)$$

A high concentration of both soluble calcium salts and carbonate ions precursors is necessary to precipitate vaterite polymorph [58,66]. Successful vaterite precipitation has been carried out at temperatures of 15, 25, and 50 °C, when ammonium carbamate has been used [58]. While, the reaction in the solution containing soluble calcium salt and urea required a temperature of 90 °C [66].

4.6. Hydrothermal and Solvothermal Methods

A hydrothermal or solvothermal process is a method used to create ceramic materials at elevated temperatures and pressure. Hydrothermal conditions were applied to precipitate vaterite-reach particles using diethylenetriaminepentaacetic acid as an additive [68]. The first step of this process was $CaCO_3$ precipitation using $CaCl_2$ and Na_2CO_3 solution at a temperature of 120 °C. Then, the hydrothermal process was carried out using the filtrate from the first step as a reactive mixture. The filtrate was placed in an autoclave at temperatures between 130 and 230 °C for 48 hours and the highest concentration of vaterite (90%) was obtained at the highest tested temperature, i.e., 230 °C.

The solvothermal method for vaterite synthesis was proposed by Li et al. [23]. Calcium chloride and urea (a precursor of carbonate ions) were reagents dissolved in ethylene glycol, 1,2-propanediol, or glycerol. Carbonate ions were formed by decomposition of urea in solvothermal conditions at temperatures from 100 to 190 °C for 12 hours. No additives or pH control were needed to precipitate the pure vaterite phase. The pure vaterite particles were produced in ethylene glycol at 100 °C, in 1,2-propanediol at 130 °C, and in glycerol at 150 °C.

5. Factors Influencing Vaterite Formation

Calcium carbonate can form several polymorphs, therefore, its crystallization requires careful control of the process parameters to obtain the desired product. Factors affecting the crystallization of the preferred polymorphic form have been grouped by Kitamura in a set of primary and secondary variables [69]. Primary factors include supersaturation, temperature, stirring rate, and seed crystals. However, the solvent composition, additives, and pH are secondary factors. The main investigated factors influencing the precipitation of vaterite are presented below. The effect of solvents and other organic substances is discussed together because organic solvents are treated as additives in many studies on calcium carbonate precipitation.

5.1. Supersaturation

Supersaturation is defined as:

$$S = \left(\frac{a_{Ca^{2+}} a_{CO_3^{2-}}}{K_{sp}}\right)^{1/2} \qquad (7)$$

where $a_{Ca^{2+}}$ and $a_{CO_3^{2-}}$ are activities of a calcium and carbonate ions, respectively, and Ksp is a solubility product. In the liquid-liquid systems, vaterite particles precipitate at room temperature from moderately supersaturated aqueous solutions, i.e., when S < 6.5 [21,33,69,70]. The supersaturation can influence the size of the crystallite forming vaterite particles, and a smaller crystal subunit was observed when supersaturation increased [69,70]. When precipitation is carried out by a carbonation route, the composition of the gas stream can impact on the supersaturation, i.e., an increase in the CO_2 concentration in the gas phase resulted in the higher supersaturation. The high supersaturation in the buffered pH range (from about 9.5 to 7.7) helps to trap metastable vaterite, preventing its transformation in calcite [19]. Also, more regular spherical particles are produced at higher concentrations of CO_2 in feed gas mixtures [71].

5.2. Temperature

Usually, vaterite particles can be formed in a broad range of temperatures using a solution method of calcium carbonate precipitation [24,42,58,72]. The comparison of vaterite concentrations in calcium carbonate samples precipitated at various conditions is shown in Figure 3. The favored range of temperatures for vaterite precipitation in various experiments is up to 40 °C. This range is also valid when the carbonation route is applied to produce $CaCO_3$ in the vaterite form [59,71].

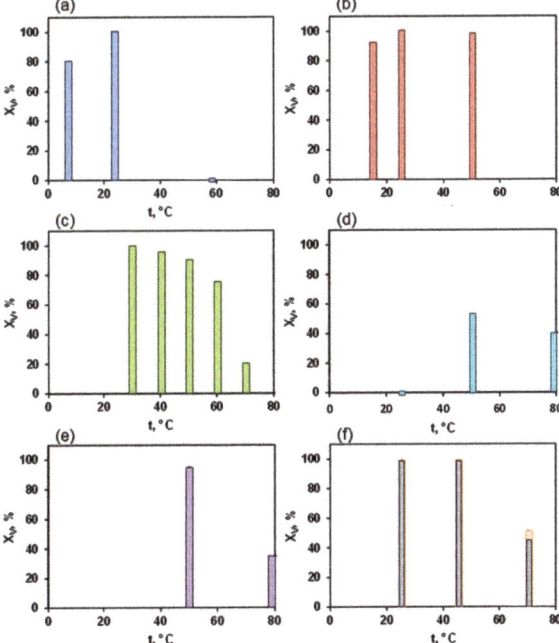

Figure 3. Influence of temperature on vaterite concentration in $CaCO_3$ samples; (a) L-L, S = 6.5, pH = 9, based on data from [21]; (b) with carbamate, [Ca^{2+}] = 1.5 M, based on data from [58]; (c) L-L, [Ca^{2+}] = 0.25 M, based on data from [24]; (d) L-L, [Ca^{2+}] = 0.015 M [73]; (e) L-L, [Ca^{2+}] = 0.015 M with ethylene glycol based on data from [73]; (f) G-L, pH = 9–10, x_{CO_2} = 1, V_G = 50 dm^3/h (gray), V_G = 100 dm^3/h (orange) based on data from [59].

The application of the hydrothermal or solvothermal method allowed the precipitation of vaterite at temperatures above 100 °C [23,68]. Some additives, e.g., ethylene glycol, can promote vaterite formation at temperatures above 40 °C [42] A higher concentration of vaterite was observed in the CaCO$_3$ product precipitated in an ethylene glycol-water solution at a temperature of 50 °C compared to the reaction carried out in an aqueous solution (see Figure 3d,e).

5.3. pH

The pH value of the reaction mixture is an important parameter during calcium carbonate precipitation. It was reported that vaterite is a dominant polymorph in solutions with an initial basic pH [21,42,74]. However, Han et al. [55] studied a carbonation system with ammonia as an absorption promoter and found that pure vaterite is formed when the pH is 8, but an increase in the pH value of the reactive mixture resulted in a decrease in the vaterite concentration. Almost pure vaterite produced in a gas-liquid system with the addition of ammonia and ammonium chloride was obtained when the pH decreased from 9.7 to 7.7 during the carbonation process [19]. The influence of pH on the vaterite content in selected processes is shown in Figure 4. Vaterite precipitation was promoted in the pH range from 8 to 10 for all compared processes. An extension of this range was possible by conducting precipitation in the presence of ethylene glycol [74] or ionic liquid surfactant [75].

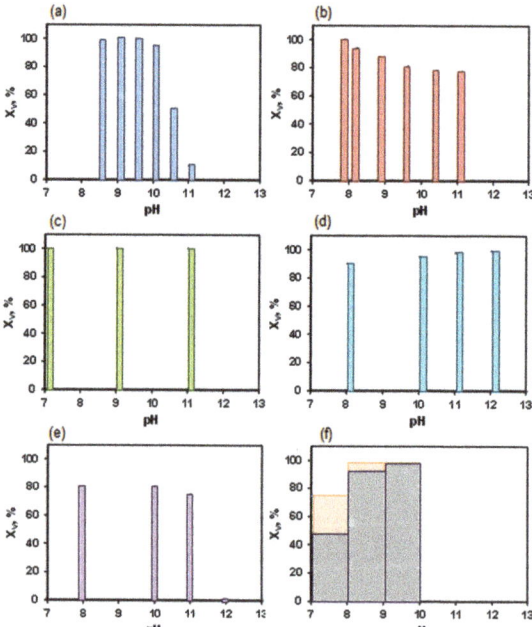

Figure 4. Influence of pH on vaterite concentration in CaCO$_3$ samples; (**a**)) L-L, S = 6.5, t = 24 °C, based on data from [21]; (**b**) G-L, [Ca^{2+}] = 0.1 M, t = 20 °C, V$_G$ = 18 dm^3/h, x$_{CO_2}$ = 033, based on data from [55]; (**c**) L-L, [Ca^{2+}] = 0.5 M, t$_a$, with ILS, based on data from [75]; (**d**) L-L, [Ca^{2+}]:[CO$_3^{2-}$] = 1:3, t = 23 °C, with ethylene glycol, based on data from [74]; (**e**) L-L, [Ca^{2+}]:[CO$_3^{2-}$] = 1:1, t = 23 °C, with ethylene glycol, based on data from [74]; (**f**) G-L, t = 25 °C, x$_{CO_2}$ = 1, V$_G$ = 50 dm^3/h (gray), V$_G$ = 100 dm^3/h (orange), based on data from [59].

5.4. Time

Because vaterite is a metastable calcium carbonate polymorph, prolonging the reaction time results in a reduction of the vaterite content when an aqueous solution without the presence of organic additives is the medium of CaCO$_3$ synthesis [13,49]. The remaining precipitated vaterite particles in the aqueous solutions lead its recrystallization to the more stable calcium carbonate polymorph, i.e., aragonite and calcite [58,69]. The time needed for the transformation of vaterite to calcite is usually in the range of a few minutes to several hours [13,58,69,76]. However, a longer time (up to 20 h) of vaterite stability in aqueous solutions was also reported when calcium carbonate was synthesized at a temperature of 7 °C [48]. Slight acceleration of the conversion of vaterite into calcite with an increase of the ionic strength was observed [77]. Some organic compounds used as additives can prevent the transformation of vaterite into calcite (see Section 5.6. Additives).

5.5. Mixing

The mixing rate of the reactant solutions is an important factor of the precipitation process [69]. The stirring intensity can affect the activation energy of nucleation of calcite and vaterite in the aqueous system [52]. Local non-homogeneities of the supersaturation can affect the creation of conditions conducive to the formation of one of the polymorphic forms. Homogeneous, high-shear, and constant agitation of reactant solutions enables precipitation of pure vaterite particles [49]. Also, ultrasonic agitation can be used for this purpose [13]. Ultrasounds provide a large amount of energy into the reaction systems, which produces mechanical and thermal effects and facilitates mass transfer. When CaCO$_3$ precipitation was carried out using ultrasounds, the vaterite concentration was higher compared to the process using the same reagent mixed by a magnetic stirrer. In the carbonation method using gaseous CO$_2$, mixing is generated by a gas flow. Therefore, an increase in the gas flow rate resulted in an increase in the vaterite precipitation [19,59,78,79].

5.6. Additives

The presence of additives in the reaction mixture may affect the precipitation by changing the solubility of a forming substance, influencing the reactive crystallization rate, nucleation, and crystal growth; the selective stabilization of a less stable polymorph; and the morphology of the forming crystals. The selected compounds that were additives in the reaction mixture used for the precipitation of calcium carbonate and their effect on the formation of vaterite are presented in Table 3.

The selection of the solvent for the precipitation of calcium carbonate has been intensively investigated. The addition of an organic solvent may change supersaturation [80], because usually both the solubility of each polymorphs and the activities of ions decreases. Therefore the promotion of vaterite precipitation in aqueous solutions of organic solvents is frequently reported [10,31,73,74]. Moreover, organic solvent and water molecules may be inhomogeneously dispersed. Such phenomena are observed, e.g., in ethanol-water mixtures, and are enhanced by an increase of the ethanol concentration and the addition of an inorganic salt [81]. Another feature of an aqueous ethanol solution is the ability to perform selective solvation as confirmed by a molecular dynamic simulation [81]. In this case, carbonate ions were mostly solvated by water and hardly solvated by ethanol, while calcium ions were solvated by both water ethanol molecules. These can lead to changes in the morphology of the precipitated particles of vaterite. As mentioned before, the addition of an organic solvent to the reaction mixture increases the supersaturation of the solution, hence the formation of vaterite particles considerably smaller in size is observed [10,74]. This is a result of the higher nucleation rate that provides a large number of nucleation sites in these systems. Moreover, organic solvent molecules that have negatively charged hydroxyl groups can adsorb at the forming vaterite surface, change the surface energy of the vaterite, and, as a result, stabilize these phases, thus preventing its transformation into more stable forms as has been reported for vaterite precipitation using the solvothermal method [23].

The formation of the adsorption layer at the produced crystals at the early stage of precipitation by different organic molecules added into the solution is often raised in the discussion of the role of organic additives. The adsorbed layer can inhibit the dissolution step of vaterite that is attributed to the remaining more unstable phases. If the adsorption energy of organic molecules at the solid surface interface is not enough to overcome the hydration energy of the hydrophilic part of these molecules, they are removed from the crystal surface. Then, the precipitated vaterite is re-dissolved into water and a more stable calcite is formed by the recrystallization process [82].

Table 3. The influence of selective additives on vaterite precipitation.

Additive	Synthesis Method	Influence				Ref.
		Rate	Stability	Morphology	Size	
Ethanol	L-L, t = 25 °C	+	+	+		[31]
	L-L, t = 25–30 °C			+		[81]
Iso-propanol	L-L, t = 25 °C	+	+	+		[31]
	L-L, t = 25–30 °C			+		[81]
Diethylene glycol	L-L, t = 25 °C	+	+	+		[31]
Ethylene glycol	L-L, t = 2–40 °C		+			[10]
	L-L, t = 25–80 °C				+	[73]
	L-L, t = 25–50 °C			+	+	[47]
	L-L, t = 40, 70 °C			+		[80]
	Solv., t = 100–150 °C	+	+			[23]
Glycerol	L-L, t = 2–40 °C		+		+	[10]
	Solv., t = 100–150 °C		+			[23]
Erythritol	L-L, t = 2–40 °C		+		+	[10]
1,2-propanediol	Solv., t = 100–150 °C		+			[23]
1,8-diaminooctane	G-S; t = 30 °C	+				[82]
Glycine	L-L, Diff., t_a			+		[52]
	G-S; t = 30 °C	+				[82]
4-aminobutyric acid	G-S; t = 30 °C	+				[82]
6-aminohexanoic acid		+				
Poly-glutamic acid	L-L, t = 25 °C	+		+		[33]
Poly-aspartic acid		+		+		
Oleic acid	L-L, t = 30 °C		+			[83]
EDTMPA	G-L, t = 30, 60 °C			+	+	[84]
Sucrose	L-L, t = 30 °C			+		[50]
	G-L, t = 22 °C	+		+		[85]
SDSN	CaCl$_2$+urea; t = 90 °C			+		[66]
SDBS				+		
Tween 20				+		
Tween 40	L-L, t_a			+		[49]
Tween 60				+		
Tween 80				+		
Ionic liquid surfactant	L-L, t = 25 °C	+		+		[47]
Guar gum	L-L, t = 0, 20, 40 °C			+		[12]

Abbreviations: Solv.—a solvothermal method; G-S—a gas-slurry system, t_a—an ambient temperature; EDTMPA—ethylenediamine-tetrakis-N,N,N,N,-(methylenephosphonic acid); SDSN—sodium dodecylsulfonate; SDBS—sodium dodecylbenzenesulfonate.

The stabilization of a vaterite polymorph was reported for alcohols, polyalcohols [10,23,31,81], and aminoacids [82,86]. Vaterite particles were produced in the presence of aminoacids. Both polar interactions from the hydrophilic groups of additives and the hydrophobic interactions due to the van der Waals forces from the hydrophobic alkyl groups play an important role in stabilizing vaterite particles [82]. The stabilizing effect of oleic acid molecules was also demonstrated and the adsorption of oleic acid at the vaterite surface was confirmed by FTIR analysis [83]. Also, tetrazole [61] and fulvic acid [34] were identified as compounds that were able to absorb at the vaterite surface and retard vaterite dissolution. The stabilization of vaterite particles was also observed when polypeptides [33], bovine serum albumin, or soluble starch [87] was added into the reaction mixture. Hydrophilic and hydrophobic parts are present in surfactant molecules, therefore, these molecules can easily adsorb on the hydrophilic surface of vaterite particles. In the same conditions, the vaterite polymorphic form was precipitated in the solution containing surfactant while calcite formation was favored in a medium without the addition of surfactant molecules [75,88]. Moreover, changes in the morphology of vaterite particles precipitated in the presence of surfactant were observed [49,66].

Polymeric substances were tested as additives in vaterite precipitation, as well. Polypeptides adsorbed at the interfaces of forming particles. They decreased in aggregation and changed the electrokinetic and morphological properties of the precipitate [33]. However the addition of the guar gum to the initial calcium chloride solution resulted in the production of hollow spherical vaterite particles [12]. However, core-shell vaterite microspheres composed of nanoparticles in the core and hexagonal nanoplates at the outer layer were precipitated using solutions of calcium chloride and ammonium carbonate with the addition of poly(styrene sulfonic acid) sodium salt [45]. When the synthesis was performed using a sodium carbonate solution as a carbonate source, vaterite microspheres covered by nanorods were precipitated [45].

The influence of additives on the rate of calcium carbonate precipitation was reported for some systems. In the solution method, it was found that the addition of ethylene glycol [80] reduced the precipitation rate. The retardation of crystal growth was observed when poly-aspartic acid and poly-glutamic acid were present in the solution [33]. In these systems, high supersaturation was preserved for a longer time and it resulted in a higher concentration of the vaterite phase in precipitated calcium carbonate. However, an increase in the vaterite growth was observed in the presence of ethanol, isopropanol, and diethylene glycol [31]. However, the stabilization of vaterite particles by organic solvent molecules prevented its transformation to more stable calcite. The opposite effect was found when citric acid was added [34]. Then, the precipitation was also slower, but the creation of calcite was privileged. It seems that the addition of citric acid increased the solubility of calcium carbonate and reduced the supersaturation in the system, which could result in the crystallization of calcite. However, when precipitation was carried out by carbonation of the calcium hydroxide suspension, the addition of amines, diamines, and amino acids resulted in a longer reaction time [82]. Vaterite was created when diaminooctane and amino acids were used. Although amino acids did not promote the CO_2 absorption and the formation of high supersaturation, their stabilizing effect on the vaterite was prevalent. On the other hand, the addition of sucrose into the initial solution of calcium chloride and ammonia reduced the reaction time [85]. Sucrose facilitated the absorption of CO_2 and caused the high supersaturation in the system, which promoted the creation of vaterite.

In summary, according to the Ostwald rule, the least stable vaterite precipitates first and subsequently transforms to the more stable one. In the absence of an additive, the kinetics is a dominant factor influencing the vaterite concentration in the produced calcium carbonate. As additives can affect each stage of crystallization, i.e., nucleation, growth, and transformation, they can therefore change the course of the precipitation process. Comprehensive information on the influence of tested additives on vaterite precipitation is not available. As shown in Table 3, only one study presented the effects of the additive in three of the highlighted areas [31]. Little information is available for carbonation-based precipitation, where additional substances can also affect the rate of CO_2 absorption and the generation of supersaturation in the system.

6. Summary

Vaterite is a polymorphic form of calcium carbonate, which is the subject of many studies due to its unique properties and related potential applications. Especially, spherical polycrystalline particles of vaterite are indicated as the most promising ones for applications. Various utilizations require particles with defined characteristics to be obtained. In this review, methods used for spherical vaterite precipitation were presented. Classical routes and recently new proposed approaches for calcium carbonate precipitation were summarized. Favorable conditions for vaterite particles precipitation were also described.

A variety of proposed methods allows the selection of an approach due to the availability of substrates and equipment, which may have an impact on the cost of the produced calcium carbonate. The problem of separating the obtained particles, e.g., in the emulsion method, is not discussed, but it may decide on the choice of a specific method. Also, the recovery or recirculation of the liquid residue has not been investigated so far.

Spherical vaterite particles can be formed using the presented methods when appropriate process parameters are maintained. The most frequently indicated conditions conducive to the formation of vaterite are the relatively high supersaturation, temperatures up to 40°C, and pH between 8 and 10. However, these parameters can be shifted because there are many relationships between them that are not fully understood. In particular, the presence of additives may affect the range of conditions favorable to the precipitation of vaterite. Therefore, the stabilizing role of additives seems to be a promising research area. Especially, the selection of non-toxic compounds is very important when vaterite particles are used in pharmaceutical and cosmetic preparations or for biomedical applications.

Funding: The research described in this paper was financially supported by Faculty of Chemistry, Gdansk University of Technology, grant number DS 033155.

Conflicts of Interest: The author declares no conflict of interest.

References

1. Friedman, G.M.; Schultz, D.J. Precipitation of vaterite (CaCO$_3$) during oil field drilling. *Mineral. Mag.* **1994**, *58*, 401–408. [CrossRef]
2. Anthony, J.W.; Bideaux, R.A.; Bladh, K.W.; Nichols, M.C. *Handbook of Mineralogy*; MDP Inc.: Chantilly, VA, USA, 2003; Volume 5.
3. Jones, B. Review of calcium carbonate polymorph precipitation in spring systems. *Sediment. Geol.* **2017**, *353*, 64–75. [CrossRef]
4. Berman, A. Biomineralization of Calcium Carbonate. The Interplay with Biosubstrates. In *Metal Ions Life Sciences, Vol. 4*; Sigel, A., Freisinger, E., Sigel, R.K.O., Eds.; J. Wiley & Sons Ltd., 2008; pp. 167–205.
5. Chakoumakos, B.C.; Pracheil, B.M.; Koenigs, R.P.; Bruch, R.M.; Feygenson, M. Empirically testing vaterite structural models using neutron diffraction and thermal analysis. *Sci. Rep.* **2016**, *6*, 36799. [CrossRef]
6. Falini, G.; Fermani, S.; Reggi, M.; Njegić Džakula, B.; Kralj, D. Evidence of structural variability among synthetic and biogenic vaterite. *Chem. Commun.* **2014**, *50*, 15370–15373. [CrossRef] [PubMed]
7. Schenk, A.S.; Albarracin, E.J.; Kim, Y.Y.; Ihli, J.; Meldrum, F.C. Confinement stabilises single crystal vaterite rods. *Chem. Commun.* **2014**, *50*, 4729–4732. [CrossRef]
8. Portugal, S.J.; Bowen, J.; Riehl, C. A rare mineral, vaterite, acts as a shock absorber in the eggshell of a communally nesting bird. *Ibis* **2018**, *160*, 173–178. [CrossRef]
9. Wightman, R.; Wallis, S.; Aston, P. Leaf margin organisation and the existence of vaterite-producing hydathodes in the alpine plant Saxifraga scardica. *Flora Morphol. Distrib. Funct. Ecol. Plants* **2018**, *241*, 27–34. [CrossRef]
10. Trushina, D.B.; Bukreeva, T.V.; Antipina, M.N. Size-Controlled Synthesis of Vaterite Calcium Carbonate by the Mixing Method: Aiming for Nanosized Particles. *Cryst. Growth Des.* **2016**, *16*, 1311–1319. [CrossRef]
11. Volodkin, D. CaCO$_3$ templated micro-beads and -capsules for bioapplications. *Adv. Colloid Interface Sci.* **2014**, *207*, 306–324. [CrossRef] [PubMed]
12. Yang, H.; Wang, Y.; Liang, T.; Deng, Y.; Qi, X.; Jiang, H.; Wu, Y.; Gao, H. Hierarchical porous calcium carbonate microspheres as drug delivery vector. *Prog. Nat. Sci. Mater. Int.* **2017**, *27*, 674–677. [CrossRef]
13. Svenskaya, Y.I.; Fattah, H.; Zakharevich, A.M.; Gorin, D.A.; Sukhorukov, G.B.; Parakhonskiy, B.V. Ultrasonically assisted fabrication of vaterite submicron-sized carriers. *Adv. Powder Technol.* **2016**, *27*, 618–624. [CrossRef]
14. Trofimov, A.D.; Ivanova, A.A.; Zyuzin, M.V.; Timin, A.S. Porous inorganic carriers based on silica, calcium carbonate and calcium phosphate for controlled/modulated drug delivery: Fresh outlook and future perspectives. *Pharmaceutics* **2018**, *10*, 167. [CrossRef] [PubMed]

15. Trushina, D.B.; Bukreeva, T.V.; Kovalchuk, M.V.; Antipina, M.N. CaCO$_3$ vaterite microparticles for biomedical and personal care applications. *Mater. Sci. Eng. C* **2015**, *45*, 644–658. [CrossRef]
16. Vogel, R.; Persson, M.; Feng, C.; Parkin, S.J.; Nieminen, T.A.; Wood, B.; Heckenberg, N.R.; Rubinsztein-Dunlop, H. Synthesis and surface modification of birefringent vaterite microspheres. *Langmuir* **2009**, *25*, 11672–11679. [CrossRef]
17. Parkin, S.J.; Vogel, R.; Persson, M.; Funk, M.; Loke, V.L.Y.; Nieminen, T.A.; Heckenberg, N.R.; Rubinsztein-Dunlop, H. Highly birefringent vaterite microspheres: production, characterization and applicaion for optical micromanipulation. *Opt. Express* **2009**, *17*, 721–727. [CrossRef] [PubMed]
18. Mori, Y.; Enomae, T.; Isogai, A. Application of Vaterite-Type Calcium Carbonate Prepared by Ultrasound for Ink Jet Paper. *J. Imaging Sci. Technol.* **2010**, *54*, 020504-1–020504-6. [CrossRef]
19. Udrea, I.; Capat, C.; Olaru, E.A.; Isopescu, R.; Mihai, M.; Mateescu, C.D.; Bradu, C. Vaterite synthesis via gas-liquid route under controlled pH conditions. *Ind. Eng. Chem. Res.* **2012**, *51*, 8185–8193. [CrossRef]
20. Nehrke, G.; Van Cappellen, P. Framboidal vaterite aggregates and their transformation into calcite: A morphological study. *J. Cryst. Growth* **2006**, *287*, 528–530. [CrossRef]
21. Tai, C.Y.; Chen, F.B. Polymorphism of CaCO$_3$, precipitated in a constant-composition environment. *AIChE J.* **1998**, *44*, 1790–1798. [CrossRef]
22. Zhan, J.; Lin, H.-P.; Mou, C.-Y. Biomimetic Formation of Porous Single-Crystalline CaCO$_3$ via Nanocrystal Aggregation. *Adv. Mater.* **2003**, *15*, 621–623. [CrossRef]
23. Li, Q.; Ding, Y.; Li, F.; Xie, B.; Qian, Y. Solvothermal growth of vaterite in the presence of ethylene glycol, 1,2-propanediol and glycerin. *J. Cryst. Growth* **2002**, *236*, 357–362. [CrossRef]
24. Chen, J.; Xiang, L. Controllable synthesis of calcium carbonate polymorphs at different temperatures. *Powder Technol.* **2009**, *189*, 64–69. [CrossRef]
25. Kawano, J.; Shimobayashi, N.; Miyake, A.; Kitamura, M. Precipitation diagram of calcium carbonate polymorphs: Its construction and significance. *J. Phys. Condens. Matter* **2009**, *41*, 425102. [CrossRef]
26. Tas, A.C. Monodisperse calcium carbonate microtablets forming at 70°C in prerefrigerated CaCl$_2$-Gelatin-Urea solutions. *Int. J. Appl. Ceram. Technol.* **2009**, *6*, 53–59. [CrossRef]
27. Christy, A.G. A Review of the Structures of Vaterite: The Impossible, the Possible, and the Likely. *Cryst. Growth Des.* **2017**, *17*, 3567–3578. [CrossRef]
28. Burgess, K.M.N.; Bryce, D.L. On the crystal structure of the vaterite polymorph of CaCO$_3$: A calcium-43 solid-state NMR and computational assessment. *Solid State Nucl. Magn. Reson.* **2015**, *65*, 75–83. [CrossRef]
29. Plummer, N.L.; Busenberg, E. The solubilities of calcite, aragonite and vaterite in CO$_2$-H$_2$O solutions between 0 and 90°C, and an evaluation of the aqueous model for the system CaCO$_3$-CO$_2$-H$_2$O. *Geochim. Cosmochim. Acta* **1982**, *46*, 1011–1040. [CrossRef]
30. Ševčík, R.; Šašek, P.; Viani, A. Physical and nanomechanical properties of the synthetic anhydrous crystalline CaCO$_3$ polymorphs: vaterite, aragonite and calcite. *J. Mater. Sci.* **2018**, *53*, 4022–4033. [CrossRef]
31. Manoli, F.; Dalas, E. Spontaneous precipitation of calcium carbonate in the presence of ethanol, isopropanol and diethylene glycol. *J. Cryst. Growth* **2000**, *218*, 359–364. [CrossRef]
32. Barhoum, A.; Ibrahim, H.M.; Hassanein, T.F.; Hill, G.; Reniers, F.; Dufour, T.; Delplancke, M.P.; Van Assche, G.; Rahier, H. Preparation and characterization of ultra-hydrophobic calcium carbonate nanoparticles. *IOP Conf. Ser. Mater. Sci. Eng.* **2014**, *64*, 012037. [CrossRef]
33. Njegić-Džakula, B.; Falini, G.; Brečević, L.; Skoko, Ž.; Kralj, D. Effects of initial supersaturation on spontaneous precipitation of calcium carbonate in the presence of charged poly-l-amino acids. *J. Colloid Interface Sci.* **2010**, *343*, 553–563. [CrossRef] [PubMed]
34. Vdović, N.; Kralj, D. Electrokinetic properties of spontaneously precipitated calcium carbonate polymorphs: The influence of organic substances. *Colloids Surfaces A Physicochem. Eng. Asp.* **2000**, *161*, 499–505. [CrossRef]
35. Perić, J.; Vučak, M.; Krstulović, R.; Brečević, L.; Kralj, D. Phase transformation of calcium carbonate polymorphs. *Thermochim. Acta* **1996**, *277*, 175–186. [CrossRef]
36. Nassrallah-Aboukaïs, N.; Jacquemin, J.; Decarne, C.; Abi-Aad, E.; Lamonier, J. F.; Aboukaïs, A. Transformation of vaterite into calcite in the absence and the presence of copper(II) species. Thermal analysis, IR and EPR study. *J. Therm. Anal. Calorim.* **2003**, *74*, 21–27. [CrossRef]
37. Wolf, G.; Günther, C. Thermophysical investigations of the polymorphous phases of calcium carbonate. *J. Therm. Anal. Calorim.* **2001**, *65*, 687–698. [CrossRef]

38. Maruyama, K.; Kagi, H.; Komatsu, K.; Yoshino, T.; Nakano, S. Pressure-induced phase transitions of vaterite, a metastable phase of CaCO$_3$. *J. Raman Spectrosc.* **2017**, *48*, 1449–1453. [CrossRef]
39. Vagenas, N.V.; Gatsouli, A.; Kontoyannis, C.G. Quantitative analysis of synthetic calcium carbonate polymorphs using FT-IR spectroscopy. *Talanta* **2003**, *59*, 831–836. [CrossRef]
40. Kontoyannis, C.G.; Vagenas, N.V. Calcium carbonate phase analysis using XRD and FT-Raman spectroscopy. *Analyst* **2000**, *125*, 251–255. [CrossRef]
41. Bryce, D.L.; Bultz, E.B.; Aebi, D. Calcium-43 chemical shift tensors as probes of calcium binding environments. Insight into the structure of the vaterite CaCO$_3$ polymorph by ^{43}Ca solid-state NMR spectroscopy. *J. Am. Chem. Soc.* **2008**, *130*, 9282–9292. [CrossRef]
42. Rodriguez-Blanco, J.D.; Shaw, S.; Benning, L.G. The kinetics and mechanisms of amorphous calcium carbonate (ACC) crystallization to calcite, viavaterite. *Nanoscale* **2011**. [CrossRef]
43. Gebauer, D.; Gunawidjaja, P.N.; Ko, J.Y.P.; Bacsik, Z.; Aziz, B.; Liu, L.; Hu, Y.; Bergström, L.; Tai, C.W.; Sham, T.K.; Edén, M.; Hedin, N. Proto-calcite and proto-vaterite in amorphous calcium carbonates. *Angew. Chem. Int. Ed.* **2010**, *49*, 8889–8891. [CrossRef] [PubMed]
44. Andreassen, J.-P.; Lewis, A.E. Classical and Nonclassical Theories of Crystal Growth. In *New Perspectives on Mineral Nucleation and Growth*; van Driessche, A., Kellermeier, M., Benning, L.G., Gebauer, D., Eds.; Springer International Publishing Switzerland: Cham, Switzerland, 2017; pp. 137–154.
45. Yang, M.; Jin, X.; Huang, Q. Facile synthesis of vaterite core-shell microspheres. *Colloids Surfaces A Physicochem. Eng. Asp.* **2011**, *374*, 102–107. [CrossRef]
46. Zhao, D.; Jiang, J.; Xu, J.; Yang, L.; Song, T.; Zhang, P. Synthesis of template-free hollow vaterite CaCO$_3$ microspheres in the H$_2$O/EG system. *Mater. Lett.* **2013**, *104*, 28–30. [CrossRef]
47. Andreassen, J.P.; Flaten, E.M.; Beck, R.; Lewis, A.E. Investigations of spherulitic growth in industrial crystallization. *Chem. Eng. Res. Des.* **2010**, *88*, 1163–1168. [CrossRef]
48. Andreassen, J.-P. Formation mechanism and morphology in precipitation of vaterite-nano-aggregation or crystal growth? *J. Cryst. Growth* **2005**, *274*, 256–264. [CrossRef]
49. Mori, Y.; Enomae, T.; Isogai, A. Preparation of pure vaterite by simple mechanical mixing of two aqueous salt solutions. *Mater. Sci. Eng. C* **2009**, *29*, 1409–1414. [CrossRef]
50. Mahtout, L.; Sánchez-Soto, P.J.; Carrasco-Hurtado, B.; Pérez-Villarejo, L.; Takabait, F.; Eliche-Quesada, D. Synthesis of vaterite CaCO$_3$ as submicron and nanosized particles using inorganic precursors and sucrose in aqueous medium. *Ceram. Int.* **2018**, *44*, 5291–5296. [CrossRef]
51. Jiang, J.; Zhao, H.; Wang, X.; Xiao, B.; Chen, C.; Wu, Y.; Yang, C.; Xu, S. A novel route to prepare the metastable vaterite phase of CaCO$_3$ from CaCl$_2$ ethanol solution and Na$_2$CO$_3$ aqueous solution. *Adv. Powder Technol.* **2018**, *29*, 2416–2422. [CrossRef]
52. Hou, W.; Feng, Q. Morphology and formation mechanism of vaterite particles grown in glycine-containing aqueous solutions. *Mater. Sci. Eng. C* **2006**, *26*, 644–647. [CrossRef]
53. Chen, Y.; Ji, X.; Wang, X. Microwave-assisted synthesis of spheroidal vaterite CaCO$_3$ in ethylene glycol-water mixed solvents without surfactants. *J. Cryst. Growth* **2010**, *312*, 3191–3197. [CrossRef]
54. Svenskaya, Y.I.; Fattah, H.; Inozemtseva, O.A.; Ivanova, A.G.; Shtykov, S.N.; Gorin, D.A.; Parakhonskiy, B.V. Key Parameters for Size- and Shape-Controlled Synthesis of Vaterite Particles. *Cryst. Growth Des.* **2018**, *18*, 331–337. [CrossRef]
55. Han, Y.S.; Hadiko, G.; Fuji, M.; Takahashi, M. Crystallization and transformation of vaterite at controlled pH. *J. Cryst. Growth* **2006**, *289*, 269–274. [CrossRef]
56. Popescu, M.A.; Isopescu, R.; Matei, C.; Fagarasan, G.; Plesua, V. Thermal decomposition of calcium carbonate polymorphs precipitated in the presence of ammonia and alkylamines. *Adv. Powder Technol.* **2014**, *25*, 500–507. [CrossRef]
57. Konopacka-Łyskawa, D.; Kościelska, B.; Karczewski, J.; Gołąbiewska, A. The influence of ammonia and selected amines on the characteristics of calcium carbonate precipitated from calcium chloride solutions via carbonation. *Mater. Chem. Phys.* **2017**, *193*, 13–18. [CrossRef]
58. Prah, J.; Maček, J.; Dražič, G. Precipitation of calcium carbonate from a calcium acetate and ammonium carbamate batch system. *J. Cryst. Growth* **2011**, *324*, 229–234. [CrossRef]
59. Sun, J.; Wang, L.; Zhao, D. Polymorph and morphology of CaCO$_3$ in relation to precipitation conditions in a bubbling system. *Chin. J. Chem. Eng.* **2017**, *25*, 1335–1342. [CrossRef]

60. Donnelly, F.C.; Purcell-Milton, F.; Framont, V.; Cleary, O.; Dunne, P.W.; Gun'ko, Y.K. Synthesis of CaCO3 nano- and micro-particles by dry ice carbonation. *Chem. Commun.* **2017**, *53*, 6657–6660. [CrossRef] [PubMed]
61. Massi, M.; Ogden, M.I.; Jones, F. Investigating vaterite phase stabilisation by a tetrazole molecule during calcium carbonate crystallisation. *J. Cryst. Growth* **2012**, *351*, 107–114. [CrossRef]
62. Lyu, S.G.; Park, S.; Sur, G.S. The Synthesis of Vaterite and Physical Properties of PP/CaCO$_3$ Composites. *Korean J. Chem. Eng.* **1999**, *16*, 538–542. [CrossRef]
63. Hirai, T.; Hariguchi, S.; Komasawa, I.; Davey, R.J. Biomimetic Synthesis of Calcium Carbonate Particles in a Pseudovesicular Double Emulsion. *Langmuir* **2002**, *13*, 6650–6653. [CrossRef]
64. Ganguli, A.K.; Ahmad, T.; Vaidya, S.; Ahmed, J. Microemulsion route to the synthesis of nanoparticles. *Pure Appl. Chem.* **2008**, *80*, 2451–5477. [CrossRef]
65. Walsh, D.; Lebeau, B.; Mann, S. Morphosynthesis of calcium carbonate (vaterite) microsponges. *Adv. Mater.* **1999**, *11*, 324–328. [CrossRef]
66. Huang, J.H.; Mao, Z.F.; Luo, M.F. Effect of anionic surfactant on vaterite CaCO$_3$. *Mater. Res. Bull.* **2007**, *42*, 2184–2191. [CrossRef]
67. Wang, L.; Sondi, I.; Matijević, E. Preparation of Uniform Needle-Like Aragonite Particles by Homogeneous Precipitation. *J. Colloid Interface Sci.* **1999**, *218*, 545–553. [CrossRef]
68. Gopi, S.; Subramanian, V.K.; Palanisamy, K. Aragonite-calcite-vaterite: A temperature influenced sequential polymorphic transformation of CaCO3 in the presence of DTPA. *Mater. Res. Bull.* **2013**, *48*, 1906–1912. [CrossRef]
69. Kitamura, M. Strategy for control of crystallization of polymorphs. *CrystEngComm* **2009**, *11*, 949–964. [CrossRef]
70. Beck, R.; Andreassen, J.P. The onset of spherulitic growth in crystallization of calcium carbonate. *J. Cryst. Growth* **2010**, *312*, 2226–2238. [CrossRef]
71. Vučak, M.; Perić, J.; Krstulović, R. Precipitation of calcium carbonate in a calcium nitrate and monoethanolamine solution. *Powder Technol.* **1997**, *91*, 69–74. [CrossRef]
72. Ogino, T.; Suzuki, T.; Sawada, K. The formation and transformation mechanism of calcium carbonate in water. *Geochim. Cosmochim. Acta* **1987**, *51*, 2757–2767. [CrossRef]
73. Flaten, E.M.; Seiersten, M.; Andreassen, J.P. Polymorphism and morphology of calcium carbonate precipitated in mixed solvents of ethylene glycol and water. *J. Cryst. Growth* **2009**, *311*, 3533–3538. [CrossRef]
74. Oral, Ç.M.; Ercan, B. Influence of pH on morphology, size and polymorph of room temperature synthesized calcium carbonate particles. *Powder Technol.* **2018**, *339*, 781–788. [CrossRef]
75. Zhao, Y.; Du, W.; Sun, L.; Yu, L.; Jiao, J.; Wang, R. Facile synthesis of calcium carbonate with an absolutely pure crystal form using 1-butyl-3-methylimidazolium dodecyl sulfate as the modifier. *Colloid Polym. Sci.* **2013**, *291*, 2129–2202. [CrossRef]
76. Bots, P.; Benning, L.G.; Rodriguez-Blanco, J.-D.; Roncal-Herrero, T.; Shaw, S. Mechanistic Insights into the Crystallization of Amorphous Calcium Carbonate (ACC). *Cryst. Growth Des.* **2012**, *12*, 3806–3814. [CrossRef]
77. Kralj, D.; Brecević, L.; Nielsen, A.E. Vaterite growth and dissolution in aqueous solution II. Kinetics of dissolution. *J. Cryst. Growth* **1994**, *143*, 269–276. [CrossRef]
78. Isopescu, R.; Mihai, M.; Capat, C.; Olaru, A.; Mateescu, C.; Dumitrescu, O.; Udrea, I. Modelling of Calcium Carbonate Synthesis by Gas-Liquid Reaction Using CO_2 from Flue Gases. *Chem. Eng. Trans.* **2011**, *25*, 713–718. [CrossRef]
79. Konopacka-Łyskawa, D.; Kościelska, B.; Łapiński, M. Precipitation of Spherical Vaterite Particles via Carbonation Route in the Bubble Column and the Gas-Lift Reactor. *JOM* **2019**, *71*, 1041–1048. [CrossRef]
80. Flaten, E.M.; Seiersten, M.; Andreassen, J.P. Growth of the calcium carbonate polymorph vaterite in mixtures of water and ethylene glycol at conditions of gas processing. *J. Cryst. Growth* **2010**, *312*, 953–960. [CrossRef]
81. Zhang, L.; Yue, L.H.; Wang, F.; Wang, Q. Divisive effect of alcohol-water mixed solvents on growth morphology of calcium carbonate crystals. *J. Phys. Chem. B* **2008**, *112*, 10668–10674. [CrossRef] [PubMed]
82. Chuajiw, W.; Takatori, K.; Igarashi, T.; Hara, H.; Fukushima, Y. The influence of aliphatic amines, diamines, and amino acids on the polymorph of calcium carbonate precipitated by the introduction of carbon dioxide gas into calcium hydroxide aqueous suspensions. *J. Cryst. Growth* **2014**, *386*, 119–127. [CrossRef]
83. Wang, C.; Piao, C.; Zhai, X.; Hickman, F.N.; Li, J. Synthesis and character of super-hydrophobic CaCO3 powder in situ. *Powder Technol.* **2010**, *200*, 84–86. [CrossRef]

84. Vucak, M.; Peric, J.; Pons, M.-N. The Influence of Various Admixtures on the Calcium Carbonate Precipitation from a Calcium Nitrate and Monoethanolamine Solution. *Chem. Eng. Technol.* **1998**, *21*, 71–75. [CrossRef]
85. Konopacka-Łyskawa, D.; Czaplicka, N.; Kościelska, B.; Łapiński, M.; Gębicki, J. Influence of Selected Saccharides on the Precipitation of Calcium-Vaterite Mixtures by the CO_2 Bubbling Method. *Crystals* **2019**, *9*, 117. [CrossRef]
86. Štajner, L.; Kontrec, J.; NjegićDžakula, B.; Maltar-Strmečki, N.; Plodinec, M.; Lyons, D.M.; Kralj, D. The effect of different amino acids on spontaneous precipitation of calcium carbonate polymorphs. *J. Cryst. Growth* **2018**, *486*, 71–81. [CrossRef]
87. Liu, Y.; Chen, Y.; Huang, X.; Wu, G. Biomimetic synthesis of calcium carbonate with different morphologies and polymorphs in the presence of bovine serum albumin and soluble starch. *Mater. Sci. Eng. C* **2017**, *79*, 457–464. [CrossRef] [PubMed]
88. Płaza, G.; Legawiec, K.; Bastrzyk, A.; Fiedot-Toboła, M.; Polowczyk, I. Effect of a lipopeptide biosurfactant on the precipitation of calcium carbonate. *Colloids Surfaces B Biointerfaces* **2018**. [CrossRef]

© 2019 by the author. Licensee MDPI, Basel, Switzerland. This article is an open access article distributed under the terms and conditions of the Creative Commons Attribution (CC BY) license (http://creativecommons.org/licenses/by/4.0/).

Article

A Complex Assemblage of Crystal Habits of Pyrite in the Volcanic Hot Springs from Kamchatka, Russia: Implications for the Mineral Signature of Life on Mars

Min Tang [1] and Yi-Liang Li [2],*

1 Department of Geology, School of Earth Sciences, Yunnan University, Kunming 650500, China; mtang@ynu.edu.cn
2 Department of Earth Sciences, The University of Hong Kong, Hong Kong 999077, China
* Correspondence: yiliang@hku.hk; Tel.: +852-25176912

Received: 30 April 2020; Accepted: 21 June 2020; Published: 23 June 2020

Abstract: In this study, the crystal habits of pyrite in the volcanic hot springs from Kamchatka, Russia were surveyed using scanning electron microscopy. Pyrite crystals occur either as single euhedral crystals or aggregates with a wide range of crystal sizes and morphological features. Single euhedral crystals, with their sizes ranging from ~200 nm to ~40 μm, exhibit combinations of cubic {100}, octahedral {111}, and pyritohedral {210} and {310} forms. Heterogeneous geochemical microenvironments and the bacterial activities in the long-lived hot springs have mediated the development and good preservation of the complex pyrite crystal habits: irregular, spherulitic, cubic, or octahedral crystals congregating with clay minerals, and nanocrystals attaching to the surface of larger pyrite crystals and other minerals. Spherulitic pyrite crystals are commonly covered by organic matter-rich thin films. The coexistence of various sizes and morphological features of those pyrite crystals indicates the results of secular interactions between the continuous supply of energy and nutritional elements by the hot springs and the microbial communities. We suggest that, instead of a single mineral with unique crystal habits, the continuous deposition of the same mineral with a complex set of crystal habits results from the ever-changing physicochemical conditions with contributions from microbial mediation.

Keywords: Kamchatka; hot springs; pyrite; complexity of crystal habits; Mars

1. Introduction

Numerous morphological, molecular, and geochemical biosignatures have been proposed over recent decades in order to identify records of past life in the 'sedimentary archives' of the ancient Earth or Mars [1–5]. Among these signatures, a variety of mineralogical biosignatures formed directly or indirectly by bacterial activity provide records of biogenesis in certain environments [6–10]. Microbially-mediated mineral precipitation happens in hot springs because microbial activities may change the concentration of ions in the micro-environments and provide nucleation sites for mineralization. However, the characterization of a biosignature based on the morphology of a single mineral often needs to be used cautiously because of the possibly confusing abiogenic imitators of biosignature [11]. As it is insufficient to take only one single mineral as a biosignature, a suite of parameters that may consistently indicate a biological origin must be considered [1]. It was recently proposed that the synthetic features of a mineral assemblage, including size, crystallinity, and morphology, could be a reference for a specific environment with a certain microbial community. In a study on the diversity of the crystal habits of gypsum, Tang et al. [12] described various morphologies and sizes of gypsum that uniquely coexisted in a square-meter sized volcanic hot spring on the Kamchatka Peninsula of Russia, and suggested that it was mainly due to the secular interactions between microbial metabolism and geochemical environments. Pyrite is also one of the most common biogenic minerals that are observed as euhedral or framboidal crystals in sediments or sedimentary rocks [8,13]. Laboratory studies of the crystallization of pyrite take physicochemical parameters, such as temperature, pressure, and ion concentrations into consideration and have established the relationship between morphologies of pyrite crystals and their depositional environments [14–21]. Though the chemical pathway of pyrite formation is still in a debate [14,22,23], the direct and indirect effects of biological processes on pyrite crystallization is commonly accepted [20,24]. Microorganisms obtain energy from the geochemical environment and release metabolic products that may affect the chemical composition of their aqueous environments and initiate subsequent mineralization [25–28]. For instance, coupling to the oxidation of organic matter, sulfate-reducing bacteria enzymatically reduce sulfate to hydrogen sulfide, which further reacts with iron in euxinic environments and leads to the precipitation of iron sulfides [29]. Thiel et al. [24] identified a novel type of microbial metabolism that favors energy conservation by oxidizing S^{2-} in FeS to S^- in FeS_2 as a syntrophy coupling to the hydrogenotrophic methanogenesis ($4FeS + 4H_2S + CO_2 \rightarrow 4FeS_2 + CH_4 + 2H_2O$). Microorganisms not only reduce the activation energy barrier for mineral nucleation, but also offer their cell walls as substrates to facilitate the nucleation of crystals [25,30]. Thus, microorganisms are important agents that may induce pyrite mineralization. However, in most cases, it is known that microorganisms have little control over the specific crystal habit of pyrite [26,31]; nevertheless, the biological processes that induce the precipitation of pyrite crystals and/or their assemblages should still carry information about past ecophysiological environments.

Framboid and euhedron are the two dominant morphologies of pyrite crystals in low-temperature sedimentary environments [32,33]. Pyrite is also one of the dominant iron sulfides in natural high-temperature sedimentary environments, such as deep-sea hydrothermal vents and terrestrial hot springs, which are analogs to the early environments for life on Earth [34–36], or possibly early Mars [6]. Pyrite nanoparticles of irregular sizes and shapes in deep-sea hydrothermal vents are considered important sources of iron for the deep ocean biosphere [37,38]. Framboidal and euhedral pyrite crystals were assumed to be generated as a consequence of microbial sulfate-reduction in active shallow submarine vents [39]. Microorganisms also thrive in hot springs that are usually characterized by extreme conditions such as high temperature and low pH through deep geological time [40–43]. Combined with the active iron, the production rate of biogenic hydrogen sulfide influences the amount of iron monosulfide in the system, which may eventually transfer to pyrite [19,24]. Many studies on pyrite in hot spring sediments have focused on the sulfur isotopic signatures (biogenic or abiogenic) [44]. Few data are available on the textures and crystal morphologies of pyrite and the relation to their microbially-mediated environments [45].

Looking for mineral biosignatures in the sedimentary rocks of Earth and Mars has long been an effort because biogenic minerals have a much higher chance of surviving the changing planetary environments during their multi-billion-year of evolution [6,7,46]. For instance, the single domain magnetite of 35–120 nm produced by magnetotactic bacteria has clear protein-modulated mineralization mechanisms and has very well-defined ecophysiological significance [47]. However, a completely inorganic process can also produce single domain magnetite with the same morphology [11]. Textural structures [2,9,48] or the complexity represented by a set of crystal habits, including morphology and size (e.g., gypsum, [12]), were suggested as a new type of biosignature of possible past Martian life. In this paper, we provide a detailed description of the high diversity of mineralogical features of pyrite identified from the Kamchatka volcanic hot springs, which we suggest to be a signature of microbiologically-mediated mineral deposition.

2. Materials and Methods

The Kamchatka Peninsula is located in the transition zone between the Eurasian, North American, and Pacific plates, and is one of the most tectonically active regions on Earth, featuring volcanoes and earthquakes [49]. There are 31 active volcanoes and hundreds of craters in Kamchatka, but hydrothermal activity is mostly located in the central and eastern volcanic zones [50,51]. As the largest living sulfide ore-forming hydrothermal system in Kamchatka, the Uzon caldera (54°26′–54°31′N, 159°55′–160°07′E) is located in the center of the eastern volcanic zone with thick Paleogene-Neogene sedimentary rocks. It formed after the collapse of the volcanic crater about 40,000 years ago and is underlain by Pliocene volcanogenic sediments [52–54].

Of the hundreds of hot springs in Kamchatka, five were chosen for this study (Figure 1, [55–57]): Burlyashii, Zavarzin, Thermophile, Jen's Vent, and Oil Pool (Oil Pool lacks data on location and chemistry but is in the same area). Based on geochemical data listed in Table 1, Jen's Vent and Burlyashii hot springs have the highest temperatures, while Thermophile has the lowest among the hot springs studied. All hot springs are predominantly in reduced geochemical conditions and with pH values varying at large scales (pH = 4.4–7.5). Concentrations of soluble Fe and S^{2-} species of the Burlyashii hot spring are much higher than the other hot springs. The general geochemistry of these volcanic hot springs can be found in Table 1 and Taran [55].

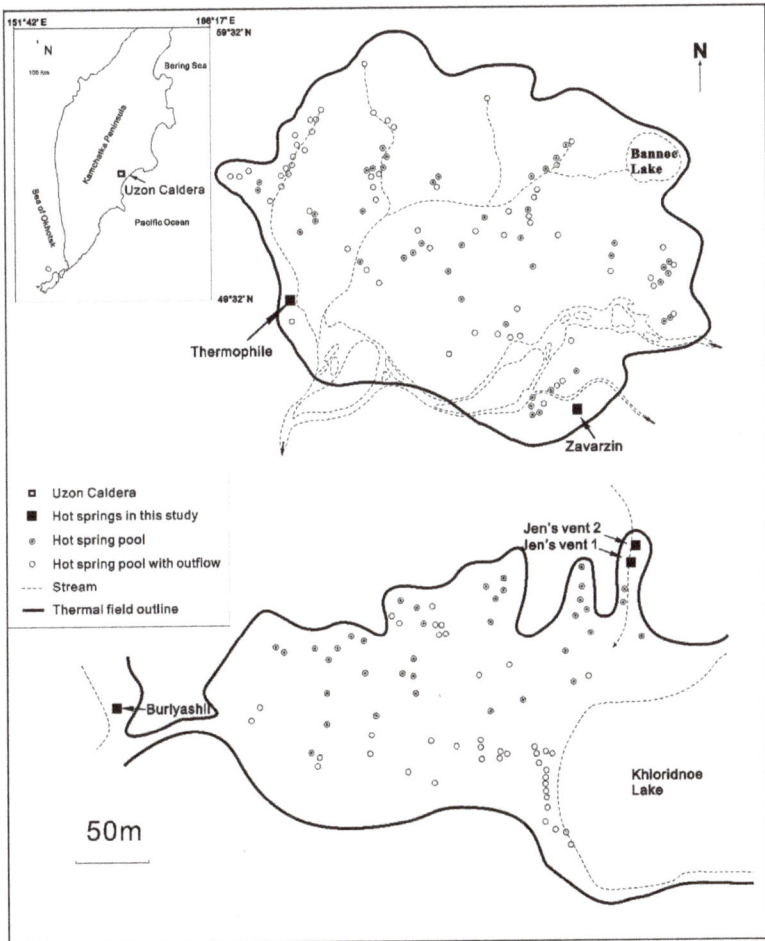

Figure 1. Locations of hot springs in Uzon Caldera, Kamchatka Peninsula (after 56,57).

Table 1. The geochemistry of Kamchatka hot springs. All concentration values are in mmol/l unless otherwise noted.

Parameters	Burlyashii	Zavarzin	Thermophile	Jen's Vent 1	Jen's Vent 2
Temperature (°C)	51–87	54–74	42–70	83	85
Eh (mV)	−90	−96		−240	−240
pH	6–6.5	5.5–7.5	4.4–7	5.3–5.9	5.3–5.9
Alkalinity	1.18–1.23			2.2	0.16–0.18
Soluble Fe	3.75×10^{-3}			0.18×10^{-3}	0.54×10^{-3}
SO_4^{2-}	0.23–2.3	0.335–0.557	0.1–0.3	1.35–1.96	1.29–3.125
S^{2-}	$(6.3–43.8) \times 10^{-3}$		$(0.6–43.1) \times 10^{-3}$		
NO_3^-		0.5		0.063	0.011
NO_2^-	$(0.1–0.3) \times 10^{-3}$		$(0.2–0.6) \times 10^{-3}$	0.41×10^{-3}	0.54×10^{-3}
NH_4^+	1.1–1.5	0.84	0.2–4		
References	[58–60]	[57–59,61–63]	[57,58,60]	[64,65]	[64,65]

Samples were collected with sterilized bottles by researchers from the University of Georgia [56], transported to the University of Hong Kong with dry ice, and stored at −21 °C. For scanning electron microscope (SEM) measurements, samples were dehydrated with anhydrous ethanol several times and spread onto silicon chips. The silicon chips were sputtered with gold/palladium for 20 seconds for electron microscopic observation. A Hitachi S4800 SEM in the Electron Microscope Center of the University of Hong Kong was used for morphological and structural characterizations using the secondary electron mode at low voltage (5 kV). Equipped energy-dispersive X-ray spectroscopy (EDS) was used to measure the in-situ chemical composition of each sample to identify minerals based on the primary results of SEM observations.

For Mössbauer spectroscopic measurements, samples were ground to a 200-mesh powder using agate mortar after being freeze-dried. Each sample was mounted onto an acrylic holder (10 mm^2) with a 5 mg Fe/cm^2 thickness. The ^{57}Fe Mössbauer spectra were collected at room temperature (293 K) in transmission mode with a 25mCi ^{57}Co/Pb source at the University of Hong Kong. The Mössbauer spectroscopic hyperfine parameters were calibrated by the fitted hyperfine parameters of the spectrum of a 25-μm α-Fe film measured after every a few samples.

3. Results and Discussion

The Mössbauer spectra of Oil Pool and Jen's Vent 1 hot spring sediments are shown in Figure 2 and the fitting results are listed in Table 2. Samples from Jen's Vent 1 showed two fitted doublets with chemical isomer shifts (IS) of 0.30 mm/s and 1.06 mm/s, and quadrupole splitting (QS) of 0.60 mm/s and 1.99 mm/s, respectively. The doublet with small IS and QS values was assigned to low-spin Fe^{2+} in pyrite [66], while that with large IS and QS values was assigned to high-spin Fe^{2+} on the lattice of clay minerals. Although Fe^{2+} in pyrite has similar hyperfine parameters to those of Fe^{3+} on the lattice of ferric iron oxides, detailed SEM observations and previous geochemical measurements (Table 1; [41,55]) confirmed the existence of pyrite rather than ferric iron oxides. The parameters in agreement with those of pyrite and Fe^{2+} in silicates [67,68] suggest that iron in Jen's Vent1 mainly existed as Fe^{2+} in pyrite and Fe^{2+}-bearing silicates. For the Oil Pool, two doublets with similar parameters were also observed, but showed the high level of Fe^{2+} in pyrite (96.65%).

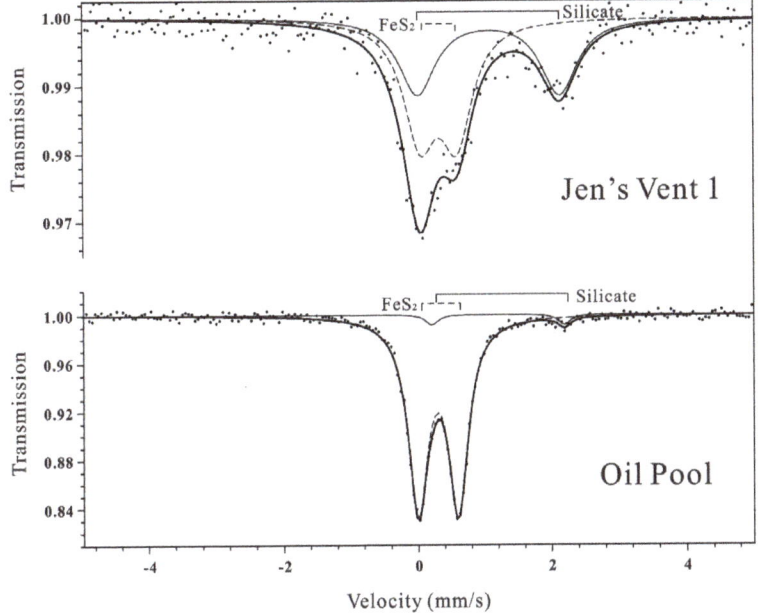

Figure 2. Room temperature ^{57}Fe Mössbauer spectra of low-spin Fe^{2+} in pyrite and high-spin Fe^{2+} on the lattice of silicates (Jen's Vent 1 and Oil Pool).

Table 2. Mössbauer spectroscopic parameters of hot spring sediments.

Sample	Mineral	IS^a (mm/s)	QS^b (mm/s)	Area (%)
Jen's Vent 1	FeS$_2$	0.31	0.55	56.66
	Silicate	1.06	2.13	43.34
Oil Pool	FeS$_2$	0.30	0.60	96.65
	Silicate	1.22	1.99	3.35

a. IS = isomer shift. b. QS = quadrupole splitting.

Eh-pH diagrams were plotted based on geochemical data of Jen's Vent 1 hot springs (83 °C, activity $[Fe^{2+}] = 10^{-6.7}$, $[SO_4^{2-}] = 10^{-2.8}$, according to [65]) which show the thermodynamic stabilities of sulfur species with the current geochemical conditions of these hot springs (Figure 3). It can be seen that the influences of the temperature and ion concentrations on thermodynamic equilibrium in the hot springs studied were insignificant. The electrochemical potentials of Kamchatka hot springs (Table 1) favor the stability of Fe(II) in silicates and pyrite, which is consistent with the Mössbauer spectroscopic results. If only ideal thermodynamic equilibrium is taken into account (Figure 3b), elemental sulfur seems unable to exist in the current springs. However, elemental sulfur was commonly observed in the sediments. The SEM observations showed that some of the elemental sulfur crystals were irregular (Figure 4a) and needed confirmation by EDS analysis (Figure 4b), while others could only be detected based on EDS microanalysis (Figure 4c,d). We also observed a small number of monoclinic sulfur crystals with well-developed crystal faces (Figure 4e) that were chemically confirmed by EDS analysis (Figure 4f). They imply that conditions in favor of sulfur deposition have existed previously.

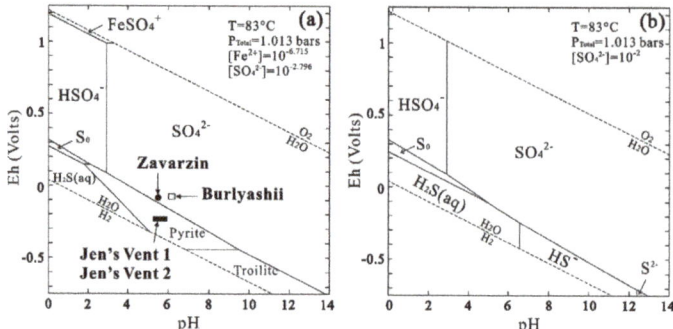

Figure 3. Eh-pH diagrams calculated using Geochemist's Workbench illustrating [69]: (**a**) Eh, pH, temperature and estimated ion concentrations in Kamchatka hot springs Jen's Vent 1, Vent 2, Zavarzin and Burlyashii (83 °C, Activities: $Fe^{2+} = 10^{-6.715}$, $SO_4^{2-} = 10^{-2.796}$); (**b**) The thermodynamic stability of elemental sulfur under the current geochemical conditions (83 °C, Activity: $SO_4^{2-} = 10^{-2}$).

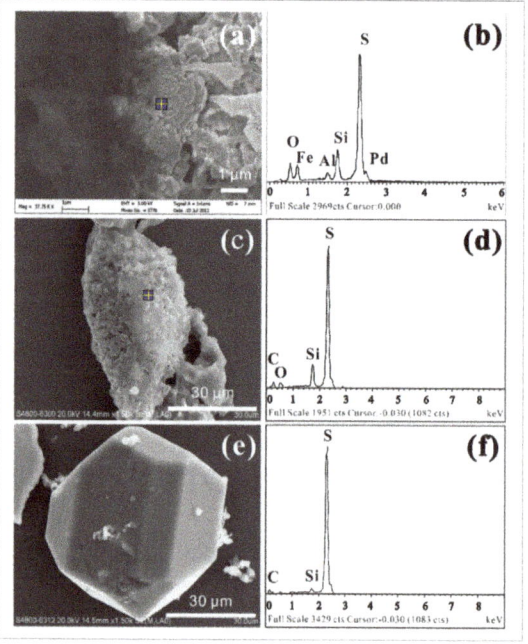

Figure 4. Scanning electron microscope (SEM) images (**a,c,e**) and energy-dispersive X-ray spectroscopy (EDS)-measured chemical composition of element sulfur (**b,d,f**) in different morphologies observed in Kamchatka hot springs. The signal of Pd in the spectrum of (**b**) should be ignored because it was from the instrumental background.

Anaerobic chemoorganoheterotrophic and chemolithoautotrophic bacteria and archaea have been identified and isolated from the Kamchatka hot springs [43,56,70]. Thermophilic sulfate-reducing bacteria (e.g., *Thermoanaerobacterium aciditolerans*), sulfur-reducing bacteria (e.g., *Thermanaerovibrio velox*), sulfur-reducing archaea (e.g., *Thermoproteus uzoniensis, Thermoplasmatales*), along with the other thermophilic microorganisms, build up a biological system that interplays with the geochemical system in those hot springs [43,61,71]. Besides those bacterial sulfur redox processes, *Thermoanaerobacter ethanolicus* and *Carboxydocella manganica* sp. nov. isolated from the hot springs in Kamchatka can reduce Fe(III) for respiration [70,72].

The coexistence of pyrite crystals with a variety of sizes and morphologies were observed and characterized in all the studied hot spring sediments. The size of single euhedral crystals is in a range from ~100 nm to ~40 µm, with various crystal habits including cubic {100}, pyritohedral {210}, octahedral {111}, pyritohedral {310} forms, and their combinations. Irregular or spherical aggregates of pyrite crystals appearing loosely or tightly were also common in the sediments. Most pyrite crystals showed smooth surfaces, while rough surfaces were observed on pyrite covered by clay minerals, organic matter, or the even finer pyrite nanocrystals. However, the framboidal structure of pyrite that is very common in sedimentary rocks and modern marine or lake sediments [19,29,73] was absent in the hot spring sediments. Below are some detailed descriptions of crystal habits.

3.1. Single Crystals

The combination of a{100} and o{111} forms making cubo-octahedron crystals is the most common combination form of pyrite crystals (Figures 5 and 6a,b). The dominant forms show a trend of transformation between a{100} and o{111} forms (Figure 5). Such crystals are sometimes elongated cubes because one pair of faces developed to a greater extent (e.g., Figure 5c). The size of single crystals is in a range from ~200 nm (Figure 5h) to >40 µm (Figure 5d). Some crystals have smooth surfaces (e.g., Figure 5a), some are covered by clay minerals (e.g., Figure 5d), and others show etched structures on the surface, especially on faces of o{111} (e.g., Figure 5c). Sometimes incomplete transformations between two forms result in face going missing during crystal development (Figure 6a,b).

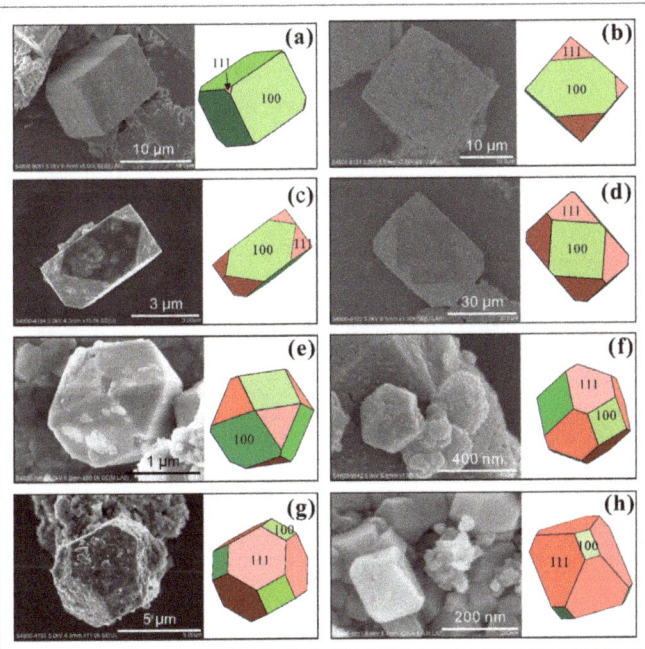

Figure 5. SEM images of single pyrite with combinations of a{100} and o{111} faces and their corresponding crystal shapes (color drawings on the right), indicating preferential orientation growth in the (100) and (111) directions from (**a**–**h**). (**a**) Pyrite crystals have smooth surfaces. (**b**,**d**,**g**) Pyrite crystals covered by clay minerals. (**c**,**e**) Etching pits on the pyrite crystal surface, especially on faces of o{111}.

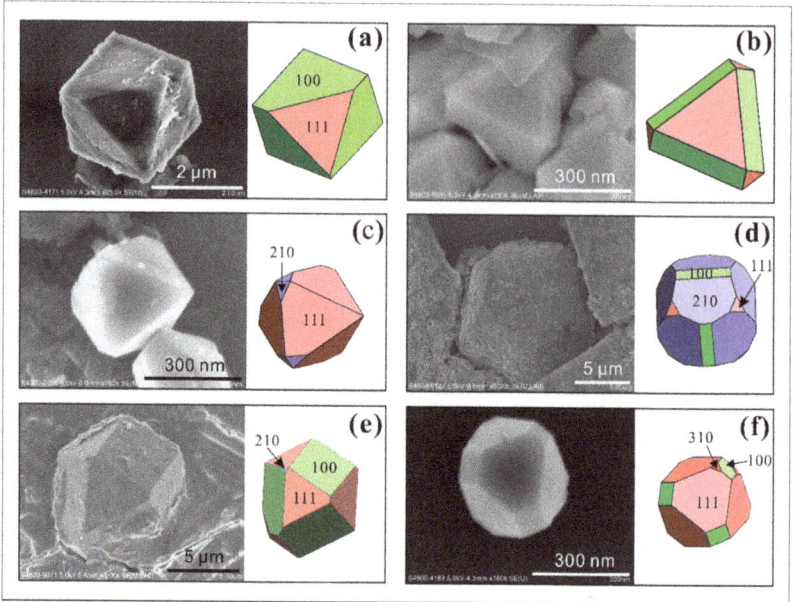

Figure 6. SEM images of single pyrite with combinations of a{100}, o{111}, e{210}, and f{310} faces and their corresponding crystal shapes (color drawings on the right) which show incomplete transformations between two forms result in face going missing during crystal development (**a**–**f**).

Combination forms of a{100}, o{111}, e{210}, and f{310} were also observed (Figure 6c–f). The octahedral form is dominant in these crystal habits and shows combinations with the e{210} form (Figure 6c) or with both a{100} and e{210} forms (Figure 6e). The pyritohedral form is dominant in only a few crystals (Figure 6d). A combination of a{100}, o{111}, and f{310} forms in one crystal of pyrite was observed, though the edges were ambiguous owing to its ultrafine size (300 nm, Figure 6f).

3.2. Pyrite Crystal Aggregates

Except those crystals that appeared clearly as single crystals in the sediments, most of them are in the form aggregates. Five types of pyrite crystal aggregates were observed in the spring sediments:

I   yrite crystals (single crystals with their sizes ranging from 5 to 10 μm) forming aggregates of ~20 μm together with clay minerals (Figure 7a,b). The pyrite crystals in these aggregates have combinations of {100}, o{111}, and e{210} habits. This type of aggregate, with sizes ranging from 10 to 100 μm, was commonly found in the samples studied.

II   Parallel intergrowths of pyrite nanocrystals (<300 nm) were observed, which attach to, or nucleate on the o{111} surface of larger pyrite crystals (>10 μm) (Figure 7c,d). The habits of these pyrite nanocrystals are mostly dominated by their o{111} form, which is sometimes slightly modified by e{210}.

III   Pyrite crystal aggregates attaching to the other minerals (Figure 7e,f). They are tiled on the surfaces of the other larger crystals and commonly appear in irregular crystal morphologies. These larger minerals offer surfaces for small pyrite crystals to stick onto.

IV   Massive pyrite nanocrystals (<100 nm) were observed to attach to, or crystallize on the surface of large pyrite crystals (>5 μm) (Figure 8a–g). The habits of these nanocrystals are octahedral (Figure 8d), cubic (Figure 8g), and irregular (Figure 8b). Pyrite nanocrystals do not just attach to or overgrow some surfaces of larger crystals like type II, they also tile the surface. Different stages of pyrite nanocrystal development (irregular crystals with or without obscure edges) are shown in Figure 8b. Octahedral nanocrystals prefer to attach to o{111} faces while cubic ones prefer a{100} faces (Figure 8e–g). Some of them grow in a certain direction (Figure 8g).

V Irregular pyrite nanocrystals aggregate as spherulites (Figure 8e,f,h). Some of the small single aggregates (~500 nm) attach to the surfaces of large pyrite crystals (white arrows in Figure 8e,f). Some large aggregates (1–5 μm) attaching to other mineral surfaces are covered by thin films containing organic carbon and sulfur, as measured by EDS (Figure 8h).

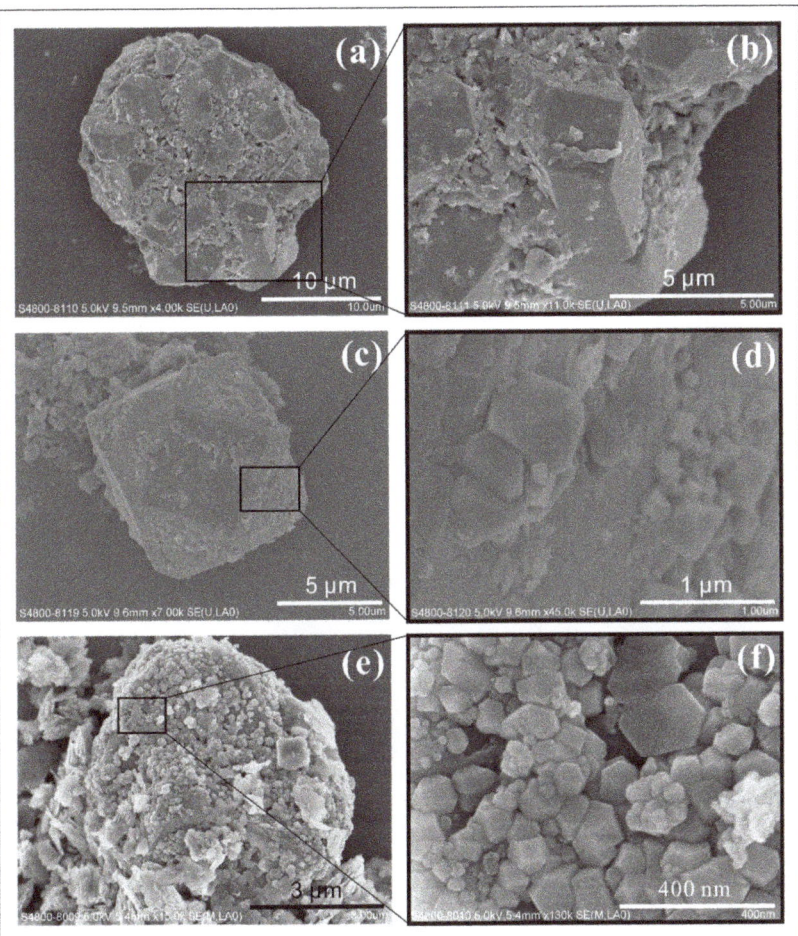

Figure 7. SEM images of pyrite aggregates. (**b,d,f**) are amplifications of the highlighted areas in images (a,c,e), respectively.

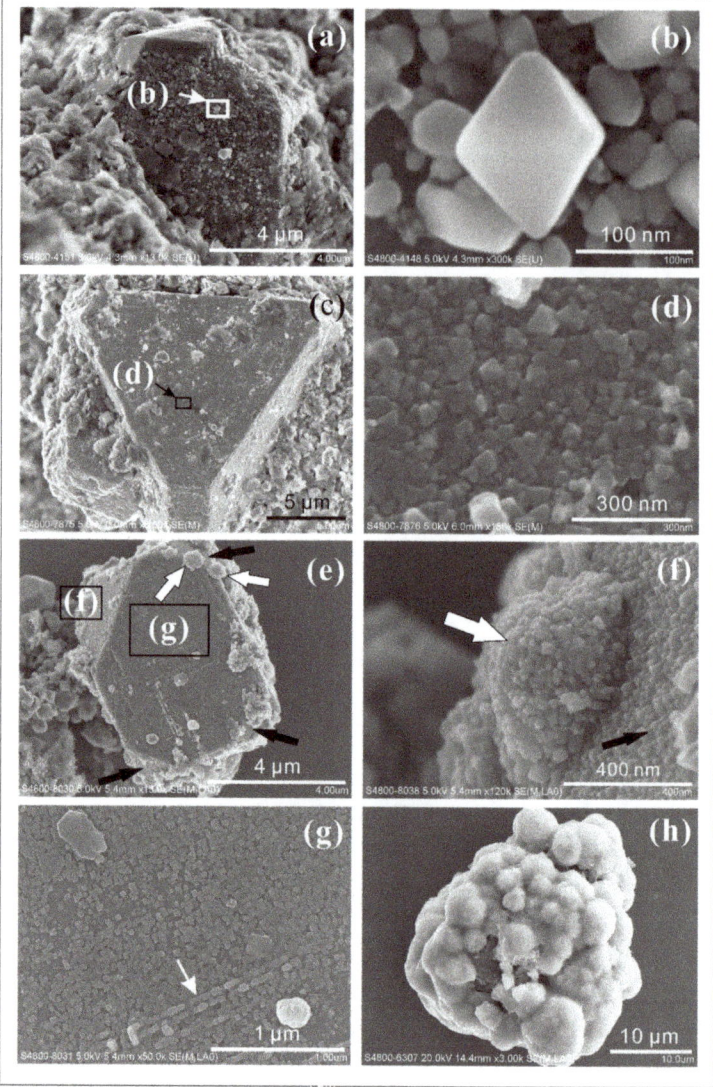

Figure 8. SEM images of pyrite aggregates. (**b**) showed octahedral pyrite crystals highlighted in (**a**). (**d**) An aggregate of octahedral pyrite crystals on the surface of one bigger crystal in (**c**). (**f,h**) Irregular pyrite nanocrystals aggregate as spherulites. (**g**) The linear arrangement of pyrite nanocrystals on the surface of a big pyrite crystal (**e**).

3.3. Intergrowth Texture

Crystal Intergrowths Appear in Four Types:

I   Intergrowth of single crystals. The cubical pyrite intergrowth texture was very common in the hot spring sediments. Cubical pyrite crystals with a size range of 5 to 10 μm show intergrowth with each other, which are sometimes coated by clay minerals (Figure 9a). The octahedral crystals ranging from 300 nm to 1 μm were observed to have intergrowth with each other (Figure 9b) and were covered by thin biofilms, as indicated by EDS measurements.

II Intergrowth of crystal combinations. Combination crystals of o{111} and a{100} show intergrowth with each other (Figure 9c). The individual crystals are around 3 µm in size. All faces of a{100} form are striated in a specific direction.

III Twin crystals appear as mirror images across the boundary where each crystal is combined with octahedron and cube habits (Figure 9e). The size of a whole crystal is about 3 µm. There are also other small pyrite crystals attached to the edges of twin crystals.

IV Parallel growth with relative smooth a{100} forms and rough o{111} forms, which can be covered by a thin layer of clay minerals (Figure 9f). The dimension of a whole crystal is about 700 nm.

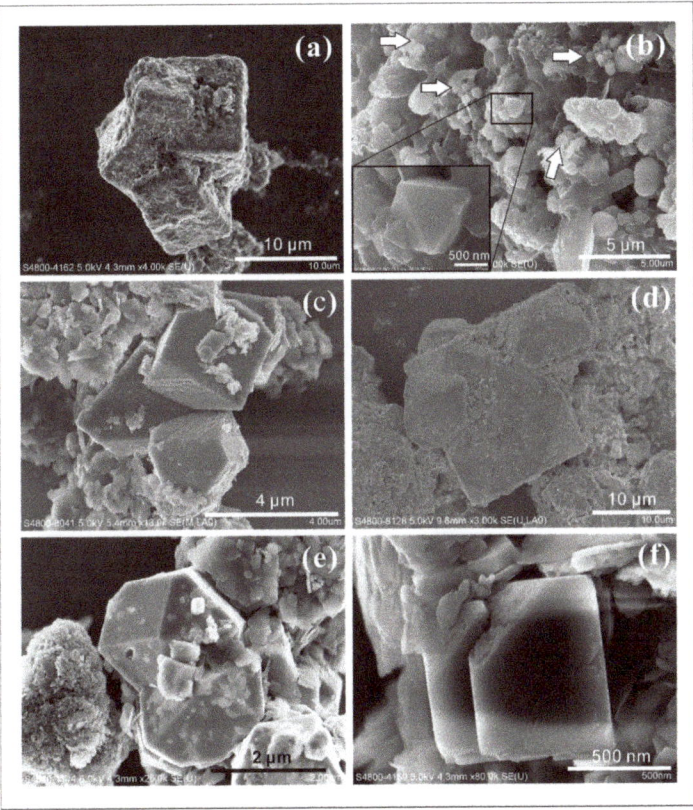

Figure 9. SEM images of different types of intergrowths of pyrite crystals (**a**–**f**). The inset at the bottom-left of (**b**) was the amplification of the highlighted area. (**c**) Pyrite crystals intergrowth. (**e**) Pyrite twin crystals. (**f**) Pyrite crystal parallel growth.

The single forms commonly observed that make pyrite polyhedrons include: cube a{100}, pyritohedron e{210}, and octahedron o{111} [74]. There are other single forms, such as pyritohedron f{310}, {210}, {211}, {321}, and a small quantity of {221}. Combination forms of pyrite crystals were found to be common in these hot spring sediments. As reported in previous studies, cubical pyrite crystals are the most common, while octahedrons are rare among all single forms [75], yet the octahedron form is not rare in the hot spring sediments studied. Normally, the habit of a crystal is confined by the crystallographic structure and defects, and its crystallization environments (mainly temperature and the degree of supersaturation, e.g., [16]). In terrestrial hot springs, the fluctuation of temperature in microenvironments is uncommon, and therefore it has negligible impacts on crystal habits [14].

3.4. Spherulite Pyrite Crystals

Pyrite spherules were observed in sediments of all hot springs (Figure 10). The size of pyrite spherules is usually around 500 nm to 1 μm, aggregated either loosely (Figure 10a) or tightly (Figure 10b). Their assemblages are different from framboidal pyrite, which is made of more euhedral pyrite nanocrystals arranged in a particular order [20]. Framboidal pyrites dominate the crystal forms in black shales (a hydrothermal environment without biologically induced pyrite mineralization, e.g., [76–78]). However, no pyrite framboid was observed in any of the studied hot springs. All of these pyrite spherules are covered by a relatively thick crust of clay rich in organic matter, as indicated by EDS measurements (Figure 10c–f).

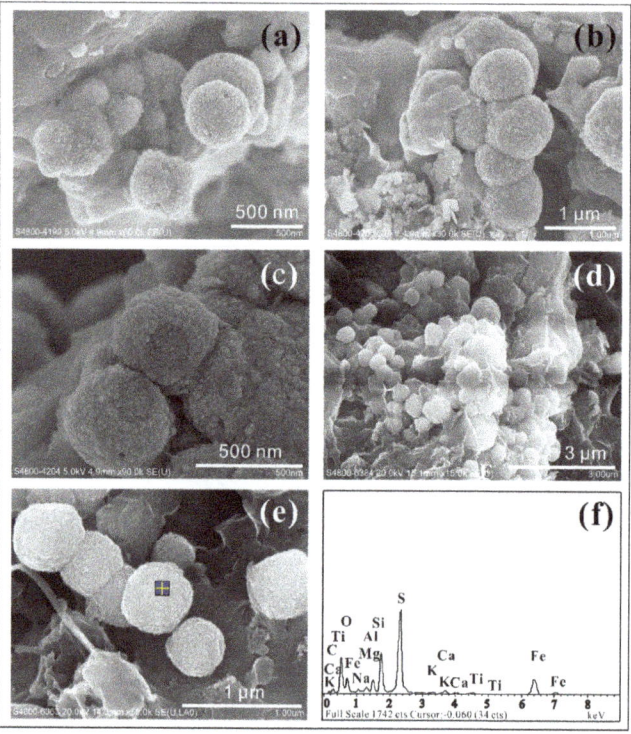

Figure 10. SEM images and the EDS result of spherulitic pyrite crystals that are characterized by biofilm covering materials (**a–e**), aggregated either loosely (**a**) or tightly (**b**). (**f**) is the EDS profile of the pyrite crystal in (**e**).

In summary, the diversity of pyrite crystal habits described in this study was found to be much higher than that of authigenic pyrites in deep-sea hydrothermal vents [41,79]. This diversity has been preserved in those small hot springs for >40,000 years [52–54]. The record of the complex crystallographic features over such a long time reflects secular interactions between the continuous supply of energy and nutritional elements by the active hot springs and the metabolisms of the microbial communities.

We can consider a complex assemblage of pyrite forms as the reflection of the interactions between the microbial communities with their geochemical environments, even though there is no direct record of a biologically mediated mineralization. The complex pyrite crystal habits is coincident with the thriving of microbial communities in the Kamchatka volcanic hot springs through time.

Further evaluation of the complexity may develop links between the microbial physiology (e.g., energy metabolism, metal respiration, and nutritional cycles) and the evolution of iron sulfide mineralogy.

This complex set of pyrite crystals may also be a useful reference for the identification of biogenic iron sulfides on Mars. Potential Martian biosignatures using crystal morphology and traits include the deposition of the digenetic apatite 'flowers' as the result of the biological cycle of phosphorus [48], single-domain magnetite formed at low temperatures [80–82], etched pits on the surface of crystals that have cell characteristics [83], abnormally tiny crystal sizes (~2–10 nm) [6,81], unusual crystal lengths in one or more dimensions [84,85], and mineral casts or encrustations preserving biological characteristics [6,85,86]. The formation of a jarosite-goethite-gypsum assemblage was considered to be the result of oxidation of pyrite [87]. The Chemistry and Mineralogy (CheMin) x-ray diffraction (XRD) analyses confirmed the existence of pyrite in mudstone at John Klein by Curiosity rover [88]. However, detailed observations of pyrite crystal on Mars have been limited by in situ measurement methods so far. Some of researchers have tried to look for clues in Marian meteorites. Euhedral pyrite crystals have not often been observed in many meteorites [89,90]. Pyrites in Northwest Africa (NWA) 7533 are cubic or octahedral crystals with average grains sizes of 30–40 μm [91]. Euhedral octahedral pyrite crystals (~50 μm) were observed in Marian meteorite NWA 7475 [92]. These assemblages of pyrite are not as complex as those we observed in the Kamchatka hot springs and their possible hydrothermal genesis [91] may have made the difference. Putative microbial activity in the hot spring environments on Mars in the distant past [93,94] might also have formed similar complex pyrite deposits. Due to the lack of dynamic geological activity and the freezing temperatures on Mars, the microstructures of any complex pyrite deposits in the near-subsurface sediments could be well-preserved for multiple billions of years. However, to fully understand the validity of a well-preserved pyrite complex on Mars, further experiments will be required. Such experiments will need to concentrate on the stability of complex pyrite deposits under various environmental conditions, such as burial, desiccation, heating, and other processes that are likely to have occurred throughout Martian history. Therefore, this presents a new avenue in the search for signs of ancient biota on Mars, and these pyrite complexes should be added to the list of potential signatures of Martian life.

4. Conclusions

The active volcanic hot springs on the Kamchatka Peninsula are extreme environments for microbial ecosystems characterized by the persistent supply of nutrients and geothermal energy. The electron microscopic observations presented in this paper show diverse crystal habits and a wide range of pyrite crystal sizes in the hot springs studied. We propose that it is the continuous interplay between the geochemical environments of the volcanic hot springs and the microbial ecophysiological activities that sustain the continuous precipitation of pyrite and preserve the diverse crystal sizes and habits. We suggest that the complexity of crystal habits of pyrite in those hot springs represents a combined biological and geochemical contribution to the kinetics of pyrite mineralization and its preservation, and thus implies a biologically mediated process.

Author Contributions: Conceptualization, Y.-L.L. and M.T.; methodology, M.T.; validation, Y.-L.L. and M.T.; formal analysis, Y.-L.L. and M.T.; writing—original draft preparation, M.T.; writing—review and editing, Y.-L.L. and M.T.; supervision, Y.-L.L. All authors have read and agreed to the published version of the manuscript.

Funding: This research was funded by RGC, grant number 17312016, NSFC, grant number 31970122 and Joint Foundation Project between Yunnan Science Technology Department and Yunnan University, grant number C176240210019.

Acknowledgments: We thank Juergen Wiegel of the University of Georgia for his generosity in sharing those precious samples.

Conflicts of Interest: The authors declare no conflict of interest.

References

1. Boston, P.; Spilde, M.; Northup, D.; Melim, L.; Soroka, D.; Kleina, L.; Lavoie, K.; Hose, L.; Mallory, L.; Dahm, C. Cave biosignature suites: Microbes, minerals, and Mars. *Astrobiology* **2001**, *1*, 25–55. [CrossRef] [PubMed]
2. Douglas, S.; Yang, H. Mineral biosignatures in evaporites: Presence of rosickyite in an endoevaporitic microbial community from Death Valley, California. *Geology* **2002**, *30*, 1075–1078. [CrossRef]
3. Seager, S.; Turner, E.; Schafer, J.; Ford, E. Vegetation's red edge: A possible spectroscopic biosignature of extraterrestrial plants. *Astrobiology* **2005**, *5*, 372–390. [CrossRef] [PubMed]
4. Chan, C.S.; Fakra, S.; Emerson, D.; Fleming, E.J.; Edwards, K.J. Lithotrophic iron-oxidizing bacteria produce organic stalks to control mineral growth: Implications for biosignature formation. *ISME J.* **2011**, *5*, 717–727. [CrossRef]
5. Summons, R.E.; Amend, J.P.; Bish, D.; Buick, R.; Cody, G.D.; Des Marais, D.J.; Dromart, G.; Eigenbrode, J.L.; Knoll, A.H.; Sumner, D.Y. Preservation of Martian organic and environmental records: Final report of the Mars Biosignature Working Group. *Astrobiology* **2011**, *11*, 157–181. [CrossRef]
6. Banfield, J.F.; Moreau, J.W.; Chan, C.S.; Welch, S.A.; Little, B. Mineralogical biosignatures and the search for life on Mars. *Astrobiology* **2001**, *1*, 447–465. [CrossRef] [PubMed]
7. Cady, S.L.; Farmer, J.D.; Grotzinger, J.P.; Schopf, J.W.; Steele, A. Morphological biosignatures and the search for life on Mars. *Astrobiology* **2003**, *3*, 351–368. [CrossRef]
8. Dove, P.M.; De Yoreo, J.J.; Weiner, S. *Biomineralization*; Mineralogical Society of America and the Geochemical Society: Washington, DC, USA, 2003; Volume 54, pp. 10–17.
9. Douglas, S. Microbial biosignatures in evaporite deposits: Evidence from Death Valley, California. *Planet. Space Sci.* **2004**, *52*, 223–227. [CrossRef]
10. Fortin, D.; Langley, S. Formation and occurrence of biogenic iron-rich minerals. *Earth Sci. Rev.* **2005**, *72*, 1–19. [CrossRef]
11. Golden, D.C.; Ming, D.W.; Morris, R.V.; Brearley, A.J.; Lauer, H.V.; Treiman, A.H.; Zolensky, M.E.; Schwandt, C.S.; Lofgren, G.E.; McKay, G.A. Evidence for exclusively inorganic formation of magnetite in Martian meteorite ALH84001. *Am. Mineral.* **2004**, *89*, 681–695. [CrossRef]
12. Tang, M.; Ehreiser, A.; Li, Y.-L. Gypsum in modern Kamchatka volcanic hot springs and the Lower Cambrian black shale: Applied to the microbial-mediated precipitation of sulfates on Mars. *Am. Mineral.* **2014**, *99*, 2126–2137. [CrossRef]
13. Wilkin, R.T.; Barnes, H.L.; Brantley, S.L. The size distribution of framboidal pyrite in modern sediments: An indicator of redox conditions. *Geochim. Cosmochim. Acta* **1996**, *60*, 3897–3912. [CrossRef]
14. Xian, H.; Zhu, J.; Liang, X.; He, H. Morphology controllable syntheses of micro- and nano-iron pyrite mono- and poly-crystals: A review. *RSC Adv.* **2016**, *6*, 31988–31999. [CrossRef]
15. Luther, G.W., III. Pyrite synthesis via polysulfide compounds. *Geochim. Cosmochim. Acta* **1991**, *55*, 2839–2849. [CrossRef]
16. Morse, J.W.; Wang, Q. Pyrite formation under conditions approximating those in anoxic sediments: II. Influence of precursor iron minerals and organic matter. *Mar. Chem.* **1997**, *57*, 187–193. [CrossRef]
17. Rickard, D. Kinetics of pyrite formation by the H_2S oxidation of iron (II) monosulfide in aqueous solutions between 25 and 125 °C: The rate equation. *Geochim. Cosmochim. Acta* **1997**, *61*, 115–134. [CrossRef]
18. Rickard, D.; Luther, G.W., III. Kinetics of pyrite formation by the H_2S oxidation of iron (II) monosulfide in aqueous solutions between 25 and 125 °C: The mechanism. *Geochim. Cosmochim. Acta* **1997**, *61*, 135–147. [CrossRef]
19. Schoonen, M.A. Mechanisms of sedimentary pyrite formation. In *Sulfur Biogeochemistry—Past and Present*; Amend, J.P., Edwards, K.J., Lyons, T.W., Eds.; The Geological Society of America, Inc.: Boulder, CO, USA, 2004; pp. 117–134.
20. Ohfuji, H.; Rickard, D. Experimental syntheses of framboids—A review. *Earth Sci. Rev.* **2005**, *71*, 147–170. [CrossRef]
21. Wang, D.; Wang, Q.; Wang, T. Shape controlled growth of pyrite FeS_2 crystallites via a polymer-assisted hydrothermal route. *CrystEngComm* **2010**, *12*, 3797–3805. [CrossRef]
22. Goldhaber, M.B. Sulfur-rich sediments. In *Treatise on Geochemistry*; Holland, H.D., Turekian, K.K., Eds.; Pergamon: Oxford, UK, 2003; Volume 7, pp. 257–288.

23. Rickard, D.; Luther, G.W., III. Chemistry of iron sulfides. *Chem. Rev.* **2007**, *107*, 514–562. [CrossRef] [PubMed]
24. Thiel, J.; Byrne, J.M.; Kappler, A.; Schink, B.; Pester, M. Pyrite formation from FeS and H_2S is mediated by a novel type of microbial energy metabolism. *bioRxiv* **2018**, 396978. [CrossRef]
25. Fortin, D.; Beveridge, T.J. Microbial sulfate reduction within sulfidic mine tailings: Formation of diagenetic Fe sulfides. *Geomicrobiol. J.* **1997**, *14*, 1–21. [CrossRef]
26. Konhauser, K.O. Diversity of bacterial iron mineralization. *Earth Sci. Rev.* **1998**, *43*, 91–121. [CrossRef]
27. Baeuerlein, E. *Biomineralization: From Biology to Biotechnology and Medical Application*; Wiley-VCH: Weinheim, Germany, 2004; pp. 159–176.
28. Panda, A.K.; Bisht, S.S.; De Mandal, S.; Kumar, N.S. Bacterial and archeal community composition in hot springs from Indo-Burma region, North-east India. *AMB Express* **2016**, 6. [CrossRef] [PubMed]
29. Berner, R.A. Sedimentary pyrite formation: An update. *Geochim. Cosmochim. Acta* **1984**, *48*, 605–615. [CrossRef]
30. Severmann, S.; Mills, R.A.; Palmer, M.R.; Telling, J.P.; Cragg, B.; John Parkes, R. The role of prokaryotes in subsurface weathering of hydrothermal sediments: A combined geochemical and microbiological investigation. *Geochim. Cosmochim. Acta* **2006**, *70*, 1677–1694. [CrossRef]
31. Weiner, S.; Dove, P.M. An overview of biomineralization processes and the problem of the vital effect. In *Biomineralization*; Dove, P.M., De Yoreo, J.J., Weiner, S., Eds.; Mineralogical Society of America and Geochemical Society: Washington, DC, USA, 2003; Volume 54, pp. 1–29.
32. Sweeney, R.E.; Kaplan, I.R. Pyrite framboid formation: Laboratory synthesis and marine sediments. *Econ. Geol.* **1973**, *68*, 618–634. [CrossRef]
33. Maclean, L.C.W.; Tyliszczak, T.; Gilbert, P.U.P.A.; Zhou, D.; Pray, T.J.; Onstott, T.C.; Southam, G. A high-resolution chemical and structural study of framboidal pyrite formed within a low-temperature bacterial biofilm. *Geobiology* **2008**, *6*, 471–480. [CrossRef] [PubMed]
34. Haymon, R.M.; Kastner, M. Hot spring deposits on the east pacific rise at 21° N: Preliminary description of mineralogy and genesis. *Earth Planet. Sci. Lett.* **1981**, *53*, 363–381. [CrossRef]
35. Vargas, M.; Kashefi, K.; Blunt-Harris, E.L.; Lovley, D.R. Microbiological evidence for Fe(III) reduction on early Earth. *Nature* **1998**, *395*, 65–67. [CrossRef] [PubMed]
36. Westall, F. Early life on earth and analogies to mars. In *Water on Mars and Life*; Tokano, T., Ed.; Springer: Berlin/Heidelberg, Germany, 2005; Volume 4, pp. 45–64.
37. Yücel, M.; Gartman, A.; Chan, C.S.; Luther, G.W., III. Hydrothermal vents as a kinetically stable source of iron-sulphide-bearing nanoparticles to the ocean. *Nat. Geosci.* **2011**, *4*, 367–371. [CrossRef]
38. Gartman, A.; Luther, G.W., III. Comparison of pyrite (FeS_2) synthesis mechanisms to reproduce natural FeS_2 nanoparticles found at hydrothermal vents. *Geochim. Cosmochim. Acta* **2013**, *120*, 447–458. [CrossRef]
39. Prol-Ledesma, R.M.; Canet, C.; Villanueva-Estrada, R.E.; Ortega-Osorio, A. Morphology of pyrite in particulate matter from shallow submarine hydrothermal vents. *Am. Mineral.* **2010**, *95*, 1500–1507. [CrossRef]
40. Barghoorn, E.S.; Nichols, R.L. Sulfate-reducing bacteria and pyritic sediments in Antarctica. *Science* **1961**, *134*, 190. [CrossRef] [PubMed]
41. Schieber, J. Sedimentary pyrite: A window into the microbial past. *Geology* **2002**, *30*, 531–534. [CrossRef]
42. Elshahed, M.S.; Senko, J.M.; Najar, F.Z.; Kenton, S.M.; Roe, B.A.; Dewers, T.A.; Spear, J.R.; Krumholz, L.R. Bacterial diversity and sulfur cycling in a mesophilic sulfide-rich spring. *Appl. Environ. Microbiol.* **2003**, *69*, 5609–5621. [CrossRef] [PubMed]
43. Wagner, I.D.; Wiegel, J. Diversity of thermophilic anaerobes. *Ann. N. Y. Acad. Sci.* **2008**, *1125*, 1–43. [CrossRef]
44. Rice, C.; Ashcroft, W.; Batten, D.; Boyce, A.; Caulfield, J.; Fallick, A.; Hole, M.; Jones, E.; Pearson, M.; Rogers, G. A Devonian auriferous hot spring system, Rhynie, Scotland. *J. Geol. Soc. Lond.* **1995**, *152*, 229–250. [CrossRef]
45. Westall, F.; de Wit, M.J.; Dann, J.; van der Gaast, S.; de Ronde, C.E.J.; Gerneke, D. Early Archean fossil bacteria and biofilms in hydrothermally-influenced sediments from the Barberton greenstone belt, South Africa. *Precambrian Res.* **2001**, *106*, 93–116. [CrossRef]
46. Westall, F. Morphological biosignatures in early terrestrial and extraterrestrial materials. In *Strategies of Life Detection*; Botta, O., Bada, J.L., Gomez-Elvira, J., Javaux, E., Selsis, F., Summons, R., Eds.; Springer: Boston, MA, USA, 2008; pp. 95–114.
47. Bazylinski, D.; Frankel, R. Magnetosome formation in prokaryotes. *Nat. Rev. Microbiol.* **2004**, *2*, 217–230. [CrossRef]

48. Sun, S.; Chan, L.; Li, Y.-L. What lurks in the Martian rocks and soil? Investigations of sulfates, phosphates, and perchlorates. Flower-like apatite recording microbial processes through deep geological time and its implication to the search for mineral records of life on Mars. *Am. Mineral.* **2014**, *99*, 2116–2125. [CrossRef]
49. Gorbatov, A.; Kostoglodov, V.; Suarez, G.; Gordeev, E. Seismicity and structure of the Kamchatka subduction zone. *J. Geophys. Res. Solid Earth* **1997**, *102*, 17883–17898. [CrossRef]
50. Okrugin, V.M.; Tazaki, K.; Bel'kova, N.L. Hydrothermal mineral formation systems of Kamchatka and the biomineralization. In *Proceedings: International Symposium of the Kanazawa University 21st-Century COE Program*; Kamta, N., Ed.; Kanazawa University: Kanazawa, Japan, 2003; Volume 1, pp. 235–238.
51. Kozhurin, A.; Acocella, V.; Kyle, P.R.; Lagmay, F.M.; Melekestsev, I.V.; Ponomareva, V.; Rust, D.; Tibaldi, A.; Tunesi, A.; Corazzato, C.; et al. Trenching studies of active faults in Kamchatka, eastern Russia: Palaeoseismic, tectonic and hazard implications. *Tectonophysics* **2006**, *417*, 285–304. [CrossRef]
52. Karpov, G.A.; Naboko, S.I. Metal contents of recent thermal waters, mineral precipitates and hydrothermal alteration in active geothermal fields, Kamchatka. *J. Geochem. Explor.* **1990**, *36*, 57–71. [CrossRef]
53. Waltham, T. A guide to the volcanoes of southern Kamchatka, Russia. *Proc. Geol. Assoc.* **2001**, *112*, 67–78. [CrossRef]
54. Kontorovich, A.E.; Bortnikova, S.B.; Karpov, G.A.; Kashirtsev, V.A.; Kostyreva, E.A.; Fomin, A.N. Uzon volcano caldera (Kamchatka): A unique natural laboratory of the present-day naphthide genesis. *Russ. Geol. Geophys.* **2011**, *52*, 768–772. [CrossRef]
55. Taran, Y. Geochemistry of volcanic and hydrothermal fluids and volatile budget of the Kamchatka–Kuril subduction zone. *Geochim. Cosmochim. Acta* **2009**, *73*, 1067–1094. [CrossRef]
56. Zhao, W. Diversity and Potential Geochemical Functions of Prokaryotes in Hot Springs of the Uzon Caldera, Kamchatka. Ph.D. Thesis, The University of Georgia, Athens, GA, USA, 2008.
57. Hollingsworth, E.R. Elemental and Isotopic Chemistry of the Uzon Caldera: The Evolution of Thermal Waters, Gas, and Mineral Precipitation. Master's Thesis, The University of Georgia, Athens, GA, USA, 2006.
58. Goin, J.C.; Cady, S.L. Biosedimentological processes that produce hot spring sinter biofabrics: Examples from the Uzon Caldera, Kamchatka Russia. In *From Fossils to Astrobiology*; Seckbach, J., Walsh, M., Eds.; Springer: Dordrecht, The Netherlands, 2009; Volume 12, pp. 159–179.
59. Kochetkova, T.; Rusanov, I.; Pimenov, N.; Kolganova, T.; Lebedinsky, A.; Bonch-Osmolovskaya, E.; Sokolova, T. Anaerobic transformation of carbon monoxide by microbial communities of Kamchatka hot springs. *Extremophiles* **2011**, *15*, 319–325. [CrossRef]
60. Zhao, W.; Song, Z.; Jiang, H.; Li, W.; Mou, X.; Romanek, C.S.; Wiegel, J.; Dong, H.; Zhang, C.L. Ammonia-oxidizing Archaea in Kamchatka Hot Springs. *Geomicrobiol. J.* **2011**, *28*, 149–159. [CrossRef]
61. Burgess, E.A. Geomicrobiological Description of Two Contemporary Hydrothermal Pools in Uzon Caldera, Kamchatka, Russia, as Models for Sulfur Biogeochemistry. Ph.D. Thesis, The University of Georgia, Athens, GA, USA, 2009.
62. Burgess, E.; Unrine, J.; Mills, G.; Romanek, C.; Wiegel, J. Comparative geochemical and microbiological characterization of two thermal pools in the Uzon Caldera, Kamchatka, Russia. *Microb. Ecol.* **2012**, *63*, 471–489. [CrossRef] [PubMed]
63. Gumerov, V.M.; Mardanov, A.V.; Beletsky, A.V.; Prokofeva, M.I.; Bonch-Osmolovskaya, E.A.; Ravin, N.V.; Skryabin, K.G. Complete genome sequence of "*Vulcanisaeta moutnovskia*" strain 768-28, a novel member of the hyperthermophilic crenarchaeal genus *Vulcanisaeta*. *J. Bacteriol.* **2011**, *193*, 2355–2356. [CrossRef] [PubMed]
64. Kyle, J.E. Mineral-Microbe Interactions and Biomineralization of Siliceous Sinters and Underlying Rock from Jenn's Pools in the Uzon Caldera, Kamchatka, Russia. Master's Thesis, The University of Georgia, Athens, GA, USA, 2005.
65. Kyle, J.E.; Schroeder, P.A.; Wiegel, J. Microbial silicification in sinters from two terrestrial hot springs in the Uzon Caldera, Kamchatka, Russia. *Geomicrobiol. J.* **2007**, *24*, 627–641. [CrossRef]
66. McCammon, C. Mössbauer spectroscopy of minerals. In *Mineral Physics and Crystallography: A Handbook of Physical Constants*; Ahrens, T.J., Ed.; American Geophysical Union: Washington, DC, USA, 1995; Volume 2, pp. 332–347.
67. Evans, B.J.; Johnson, R.G.; Senftle, F.E.; Cecil, C.B.; Dulong, F. The ^{57}Fe Mössbauer parameters of pyrite and marcasite with different provenances. *Geochim. Cosmochim. Acta* **1982**, *46*, 761–775. [CrossRef]
68. Dyar, M.D.; Agresti, D.G.; Schaefer, M.W.; Grant, C.A.; Sklute, E.C. Mössbauer spectroscopy of earth and planetary materials. *Annu. Rev. Earth Planet. Sci.* **2006**, *34*, 83–125. [CrossRef]

69. Bethke, C.M.; Yeakel, S. *The Geochemist's Workbench, Release 7.0: GWB Essentials Guide*; Hydrogeology Program; University of Illinois: Urbana, IL, USA, 2007; pp. 1–98.
70. Kublanov, I.V.; Perevalova, A.A.; Slobodkina, G.B.; Lebedinsky, A.V.; Bidzhieva, S.K.; Kolganova, T.V.; Kaliberda, E.N.; Rumsh, L.D.; Haertlé, T.; Bonch-Osmolovskaya, E.A. Biodiversity of thermophilic prokaryotes with hydrolytic activities in hot springs of Uzon Caldera, Kamchatka (Russia). *Appl. Environ. Microbiol.* **2009**, *75*, 286–291. [CrossRef]
71. Mardanov, A.V.; Gumerov, V.M.; Beletsky, A.V.; Perevalova, A.A.; Karpov, G.A.; Bonch-Osmolovskaya, E.A.; Ravin, N.V. Uncultured archaea dominate in the thermal groundwater of Uzon Caldera, Kamchatka. *Extremophiles* **2011**, *15*, 365–372. [CrossRef]
72. Slobodkina, G.B.; Panteleeva, A.N.; Sokolova, T.G.; Bonch-Osmolovskaya, E.A.; Slobodkin, A.I. Carboxydocella manganica sp nov., a thermophilic, dissimilatory Mn(IV)- and Fe(III)-reducing bacterium from a Kamchatka hot spring. *Int. J. Syst. Evol. Microbiol.* **2012**, *62*, 890–894. [CrossRef]
73. Wilkin, R.T.; Barnes, H.L. Pyrite formation in an anoxic estuarine basin. *Am. J. Sci.* **1997**, *297*, 620–650. [CrossRef]
74. Anthony, J.W.; Bideaux, R.A.; Bladh, K.W.; Nichols, M.C. *Handbook of Mineralogy*; Mineralogical Society of America: Chantilly, CA, USA, 2005; p. 4129.
75. Blackburn, W.H.; Dennen, W.H. *Principles of Mineralogy*; William C Brown Pub: Dubuque, IA, USA, 1994; p. 413.
76. Steiner, M.; Wallis, E.; Erdtmann, B.-D.; Zhao, Y.; Yang, R. Submarine-hydrothermal exhalative ore layers in black shales from South China and associated fossils—Insights into a lower Cambrian facies and bio-evolution. *Palaeogeogr. Palaeoclim. Palaeoecol.* **2001**, *169*, 165–191. [CrossRef]
77. Jiang, S.-Y.; Chen, Y.-Q.; Ling, H.-F.; Yang, J.-H.; Feng, H.-Z.; Ni, P. Trace- and rare-earth element geochemistry and Pb–Pb dating of black shales and intercalated Ni–Mo–PGE–Au sulfide ores in Lower Cambrian strata, Yangtze Platform, South China. *Min. Depos.* **2006**, *41*, 453–467. [CrossRef]
78. Xu, J.; Li, Y.-L. An SEM study of microfossils in the black shale of the lower Cambrian niutitang formation, southwest China: Implications for the polymetallic sulfide mineralization. *Ore Geol. Rev.* **2015**, *65*, 811–820. [CrossRef]
79. Juniper, S.K.; Thompson, J.A.J.; Calvert, S.E. Accumulation of minerals and trace elements in biogenic mucus at hydrothermal vents. *Deep Sea Res. Part A Oceanogr. Res. Pap.* **1986**, *33*, 339–347. [CrossRef]
80. Thomas-Keprta, K.L.; Bazylinski, D.A.; Kirschvink, J.L.; Clemett, S.J.; McKay, D.S.; Wentworth, S.J.; Vali, H.; Gibson, E.K., Jr.; Romanek, C.S. Elongated prismatic magnetite crystals in ALH84001 carbonate globules: Potential Martian magnetofossils. *Geochim. Cosmochim. Acta* **2000**, *64*, 4049–4081. [CrossRef]
81. Thomas-Keprta, K.L.; Clemett, S.J.; Bazylinski, D.A.; Kirschvink, J.L.; McKay, D.S.; Wentworth, S.J.; Vali, H.; Gibson, E.K.; McKay, M.F.; Romanek, C.S. Truncated hexa-octahedral magnetite crystals in ALH84001: Presumptive biosignatures. *Proc. Natl. Acad. Sci. USA* **2001**, *98*, 2164–2169. [CrossRef] [PubMed]
82. Thomas-Keprta, K.L.; Clemett, S.J.; Bazylinski, D.A.; Kirschvink, J.L.; McKay, D.S.; Wentworth, S.J.; Vali, H.; Gibson, E.K., Jr.; Romanek, C.S. Magnetofossils from ancient Mars: A robust biosignature in the Martian meteorite ALH84001. *Appl. Environ. Microbiol.* **2002**, *68*, 3663–3672. [CrossRef]
83. Fisk, M.R.; Giovannoni, S.J.; Thorseth, I.H. Alteration of oceanic volcanic glass: Textural evidence of microbial activity. *Science* **1998**, *281*, 978–980. [CrossRef]
84. MacLachlan, M.; Manners, I.; Ozin, G.A. New (inter)faces: Polymers and inorganic materials. *Adv. Mater.* **2000**, *12*, 675–681. [CrossRef]
85. Barbieri, R.; Stivaletta, N.; Marinangeli, L.; Ori, G.-G. Microbial signatures in sabkha evaporite deposits of Chott el Gharsa (Tunisia) and their astrobiological implications. *Planet. Space Sci.* **2006**, *54*, 726–736. [CrossRef]
86. Hofmann, B.; Farmer, J. Filamentous fabrics in low-temperature mineral assemblages: Are they fossil biomarkers? Implications for the search for a subsurface fossil record on the early Earth and Mars. *Planet. Space Sci.* **2000**, *48*, 1077–1086. [CrossRef]
87. Zolotov, M.Y.; Shock, E.L. Formation of jarosite-bearing deposits through aqueous oxidation of pyrite at meridiani planum, mars. *Geophys. Res. Lett.* **2005**, *32*, L21203. [CrossRef]
88. Vaniman, D.T.; Bish, D.L.; Ming, D.W.; Bristow, T.F.; Morris, R.V.; Blake, D.F.; Chipera, S.J.; Morrison, S.M.; Treiman, A.H.; Rampe, E.B.; et al. Mineralogy of a mudstone at Yellowknife Bay, Gale crater, Mars. *Science* **2014**, *343*, 1243480. [CrossRef] [PubMed]

89. Greenwood, J.P.; Mojzsis, S.J.; Coath, C.D. Sulfur isotopic compositions of individual sulfides in Martian meteorites ALH84001 and Nakhla: Implications for crust–regolith exchange on Mars. *Earth Planet. Sci. Lett.* **2000**, *184*, 23–35. [CrossRef]
90. Lorand, J.-P.; Pont, S.; Chevrier, V.; Luguet, A.; Zanda, B.; Hewins, R. Petrogenesis of Martian sulfides in the chassigny meteorite. *Am. Mineral.* **2018**, *103*, 872–885. [CrossRef]
91. Lorand, J.-P.; Hewins, R.; Remusat, L.; Zanda, B.; Pont, S.; Leroux, H.; Marinova, M.; Jacob, D.; Humayun, M.; Nemchin, A.; et al. Nickeliferous pyrite tracks pervasive hydrothermal alteration in Martian regolith breccia: A study in nwa 7533. *Meteorit. Planet. Sci.* **2015**, *50*, 2099–2120. [CrossRef]
92. Wittmann, A.; Korotev, R.L.; Jolliff, B.L.; Irving, A.J.; Moser, D.E.; Barker, I.; Rumble, D., III. Petrography and composition of Martian regolith breccia meteorite northwest Africa 7475. *Meteorit. Planet. Sci.* **2015**, *50*, 326–352. [CrossRef]
93. Boston, P.; Ivanov, M.; McKay, C. On the possibility of chemosynthetic ecosystems in subsurface habitats on Mars. *Icarus* **1992**, *95*, 300–308. [CrossRef]
94. Allen, C.C.; Oehler, D.Z. A case for ancient springs in Arabia Terra, Mars. *Astrobiology* **2008**, *8*, 1093–1112. [CrossRef]

© 2020 by the authors. Licensee MDPI, Basel, Switzerland. This article is an open access article distributed under the terms and conditions of the Creative Commons Attribution (CC BY) license (http://creativecommons.org/licenses/by/4.0/).

MDPI
St. Alban-Anlage 66
4052 Basel
Switzerland
Tel. +41 61 683 77 34
Fax +41 61 302 89 18
www.mdpi.com

Crystals Editorial Office
E-mail: crystals@mdpi.com
www.mdpi.com/journal/crystals

www.ingramcontent.com/pod-product-compliance
Lightning Source LLC
LaVergne TN
LVHW070153120526
838202LV00013BA/1049